THE BIBLE AS LITERATURE

THE BIBLE
AS LITERATURE

An Introduction

THIRD EDITION

JOHN B. GABEL
CHARLES B. WHEELER
ANTHONY D. YORK

New York Oxford   OXFORD UNIVERSITY PRESS   1996

Oxford University Press

Oxford   New York
Athens   Auckland   Bangkok   Bombay
Calcutta   Capetown   Dar es Salaam   Delhi
Florence   Hong Kong   Istanbul   Karachi
Kuala Lumpur   Madras   Madrid   Melbourne
Mexico City   Nairobi   Paris   Singapore
Taipei   Tokyo   Toronto

and associated companies in
Berlin   Ibadan

Library of Congress Cataloging-in-Publication Data
Gabel John B.
The Bible as literature : an introduction /
John B. Gabel, Charles B. Wheeler,
and Anthony D. York. –3rd ed.
p.   cm. Includes bibliographical references and index.
ISBN 0–19–509285–6
1. Bible as literature.   I. Wheeler, Charles B.
II. York, Anthony Delano.   III. Title.
BS535.G25   1995   809'.93522–dc20   94–39201

9 8 7 6 5 4
Printed in the United States of America
on acid-free paper

Collegis optimis qui nos totiens de
his rebus audiverunt

Contents

To the Reader, *ix*

I The Bible as Literature, *3*
II Literary Forms and Strategies in the Bible, *17*
III Ancient Near Eastern Literature and the Bible, *43*
IV The Bible and History, *63*
V The Physical Setting of the the Bible, *75*
VI The Formation of the Canon, *93*
VII The Composition of the Pentateuch, *108*
VIII The Prophetic Writings, *124*
IX The Wisdom Literature, *137*
X The Apocalyptic Literature, *155*
XI The Intertestamental Period, *171*
XII Apocrypha and Pseudepigrapha: The Outside Books, *192*
XIII The Gospels, *209*
XIV Acts and the Letters, *230*
XV The Text of the Bible, *253*
XVI Translating the Bible, *267*
XVII The Religious Use and Interpretation of the Bible, *290*

Appendix I: The Name of Israel's God, *311*
Appendix II: Writing in Biblical Times, *315*
Index, *325*

To the Reader

This book is a systematic general introduction to the study of the Bible as literature. It is intended to support such study by providing essential background information of the sort that few students have either the time or ability to piece together out of the enormous mass of published material on the Bible. We have tried throughout to make our work accessible to literate adults, feeling that the subject already has difficulties enough without our imposing others upon it. The particular audience we have in mind—one that we have addressed for many years as teachers—is college undergraduates enrolled in a Bible course offered by a department of literature. We believe that the book can also be used in introductory courses offered in seminaries and theological schools as well as extramurally by persons studying the Bible on their own.

Before going further, we should make clear what this book is not. It is not a commentary on the Bible, either generally or on a book-by-book, chapter-by-chapter basis. Nor is it an attempt to impose an interpretive scheme or point of view on the Bible, for that would usurp the function of religion. Nor, finally, is it an advocate for the value of the Bible as a vehicle of moral instruction or a provider of religious insights or a source of inspiration for the conduct of daily life. We do not deny the legitimacy of these uses, but we shall not take them into account either. It is sufficient for our purpose that the Bible be—as it is—a fascinating human document of enormous importance to the culture and history of the modern world, a document that can speak volumes to humans about their own humanity. It would be inappropriate for us to go beyond this view, for everything beyond it is in the area of personal belief and is subject to sectarian controversy. Whatever else it may be, for the purposes of the present book the Bible is a human product. Its contents evolved and came together as a result of activity by real people living in actual places over a period of more than a thousand years of human history, people subject to all the influences on culture that then existed. Hence the Bible is timebound. This much is certainly known and should be acknowledged. To what extent the Bible may also be timeless—suprahistorical and supernatural—is up to the individual reader or the religious interpreter, for whom we cannot speak and do not attempt to speak.

In the last half-century the world of biblical scholarship has been shaken by a number of major revisions in heretofore generally accepted views about the Bible and its background, thanks in part to the great increase in archaeological and epigraphical materials from the ancient Near East. We have taken into account these new developments insofar as they affect the reading of the Bible as literature. In general we have drawn upon the latest and most respected impartial scholarship on the Bible. The views set forth here thus represent a consensus. In some cases we have suppressed our own inclinations in order to represent this consensus.

We begin the book with some background chapters on literary aspects of the Bible, the influence of nonbiblical literature upon it, the relation between the Bible and history, the physical setting in which biblical events occurred, and the process by which the Bible took shape. We then conduct a literary-historical examination of the Bible, proceeding in approximate sequence through the two testaments, with individual chapters on the Pentateuch and on the major types of biblical literature: prophecies, wisdom writing, apocalypses, gospels, and "acts" and letters. Between the Old and New Testament blocks are chapters on the intertestamental period and the Bible-like writings not found in the Bible; following the New Testament block are chapters on biblical translation, on the text of the Bible, and on the way the Bible has characteristically been made use of in religion.

Although the book has a logical overall order, we hasten to say that it need not be read in that order. Each of the chapters is sufficiently self-contained that it can be read independently. The instructor of a college course on the Bible may well decide at the outset to assign the chapter on the canonization of the Bible or the one on translation. Whatever the order followed, we can assure those readers who finish this book that they will have gained an understanding of how the Bible came into being, why it took the shape it did, and what has been made of it during the centuries of its existence.

The source-text for our quotations from the Bible is not the King James Version (KJV, also known as the Authorized Version). Though honored by time and the affection of countless readers, it is unfortunately not satisfactory for our purposes. As a translation it is too often inaccurate, its archaic language frequently obscures the meaning for modern readers, and its New Testament is based upon inferior originals. There are now a number of superior modern translations, the choice of any one of which would be easily defensible. Of these we have elected to quote from the New Jerusalem Bible (NJB), principally because it alone represents the personal name of the deity of Israel as "Yahweh," the generally accepted rendering in English of the Hebrew original. (For a full discussion of this matter, see appendix 1, "The Name of Israel's God.")

A brief preliminary remark needs to be made about certain of our terms of reference. For the sake of convenience, throughout this book we shall speak of "the Bible," although—as we explain at length in various places—there is not just one Bible but at least four. By the term we shall generally mean simply what most people mean when they speak of the Bible: the volume that is read from in religious services or that one swears on in court or that lies in the bureau drawer in one's motel room. When it is important to make distinctions, we shall refer to the Jewish or Catholic or Eastern Orthodox or Protestant Bible. For the sake of convenience, likewise, we shall in many places refer to the "Old Testament" and "New Testament," though of course the Jewish Bible has no "New" Testament and Jews do not think of their sacred text as being the "Old" Testament—it is simply the Bible. But the terms "Old Testament" and "New Testament" are established and familiar and serve as a kind of shorthand that will save us from bogging down in descriptive terminology.

It should be noted that we make a distinction in this book between the Hebrew Bible and the Jewish Bible. The latter is simply the Bible recognized by Jews, the scriptures of Judaism, and can be found in any of the world's languages. The Hebrew Bible, on the other hand, is *in* the Hebrew language and is the *source* of the Jewish Bible (as well as the Christian Old Testament). Only for those who read the Jewish Bible in Hebrew are the Jewish and the Hebrew Bibles the same thing. We have carefully considered every occasion in the following pages where "Hebrew Bible," "Jewish Bible" (or "Jewish scriptures"), or "Old Testament" might be appropriate and have made choices among those terms with what we hope are consistent results.

In contexts where language and biblical translation are under discussion, we refer to the "Hebrew Bible." In contexts where Jewish religion is under discussion—particularly Jewish religion of the period before the New Testament existed—we use the terms "Jewish Bible" and "Jewish scriptures." In contexts where Christian activity with respect to the Jewish canon is under discussion, we use the term "Old Testament."

In historical references we employ the abbreviations "B.C.E." and "C.E." These stand for, respectively, "before the common era" and "common era." They replace the traditional and familiar "B.C." and "A.D." Our representation of dates will thus conform to the system routinely used by most biblical scholars. It is true that the whole world employs a chronology that takes the supposed birth date of Christ as its starting point; but to the religious sensibilities of non-Christians, the neutral references to the "common" era are more acceptable.

At the end of most of our chapters there is a brief list of books and articles entitled "Suggested Further Reading." The lists are by no means bibliographies; true bibliographies for the very large subjects we pursue in this book would each run to many hundreds of items. The lists merely provide references for readers who are particularly interested in the subject of a given chapter and would like to learn more about it. Although these lists are certainly not complete or definitive, the works included have been chosen with some care. We have excluded items primarily religious in nature or containing scholarship too technical for the lay reader; and we have passed over a good many valuable older studies in favor of others that represent the state of the art in biblical scholarship. Such scholarship is now abundantly represented in the *Anchor Bible Dictionary*, and the reader will see that we have cited it after nearly every chapter. But we have retained references to the earlier *Interpreters' Dictionary of the Bible*, a useful work that has not been entirely superseded by the later one. Finally, many of the works we cite provide their own reading lists and thus can serve as the means for directing readers continually deeper into the subjects with which we have dealt.

This third edition includes new chapters on the influence of Near Eastern literatures on the Bible and on the textual sources of our printed Bibles. A number of minor changes have been made in the text of the remainder of the book to accommodate the new material. Beyond that, the older chapters have been altered in places to sharpen their focus and to expand the treatment of their topics.

THE BIBLE AS LITERATURE

I

The Bible as Literature

What does it mean to read the Bible "as literature"? Primarily that for the time being one looks at the Bible in the same way that one would look at any other book: as a product of the human mind. In this view the Bible is a collection of writings produced by real people who lived in actual historical times. Like all other authors, these persons used the languages native to them and the literary forms then available for self-expression, creating, in the process, material that can be read and appreciated under the same conditions that apply to literature in general, wherever it is found. This view is not necessarily in conflict with the traditional religious one, namely, that the Bible was written under the direct inspiration of God and given to humans to serve as a guide to their faith and conduct. But it is clearly different, with its own requirements and its own aims.

Reading the Bible as literature should not be uncomfortable for persons who hold the religious view (though it may seem a little strange at first), and it places no demands upon the many persons who, for reasons of their own, take a skeptical or noncommittal view of the Bible. The Bible is the common heritage of us all, whatever our religious beliefs, and we should be able to study it, up to a point, without getting into religious controversy. Later—and separately—

anyone who chooses should be able to return to viewing the Bible as a repository of religious truth. The important thing is to know what one is doing, to make one's choice explicit, and to follow it consistently. We are here going to look at a group of literary texts *as* literary texts.

Our position is that the Bible in some fundamental respects is not different from the works of, let us say, Shakespeare or Emily Dickinson or Henry Fielding or Ernest Hemingway. If we were actually studying the works of these authors, such a chapter as this would not be necessary—for who can imagine needing to read something called "Shakespeare as Literature" or "Emily Dickinson as Literature"? We assume that their work is literature; it needs no demonstration. But different assumptions have historically been applied to the Bible, and in many circles they are still in force. For millions of persons the Bible was and still is *the* book. In many households it was the only book to be found, and it was displayed as a proud possession—its mere physical presence was assumed to have some beneficent power. Such a household might also have owned the works of Shakespeare, similarly displayed, but the crucial difference is that no one in that household or any other would have thought to ask of Shakespeare's works, "Will they save us?" Even persons with no religious commitment, who do not believe the Bible at all, tend to assume that this work demands to be treated in a special way, a way peculiar to itself. Hence merely to say "the Bible is literature," as though this answered all questions, is not enough. As a prerequisite to further study, we must attempt to make it clear why and how the Bible, as literature, belongs in the same category with all these other pieces of writing.*

We are using the term "literature" in its broadest sense. There is a narrower sense of the term that encompasses only what is known as belles lettres: poetry, short stories, novels, plays, essays. Although the Bible does contain this kind of material, it also contains genealogies, laws, letters, royal decrees, instructions for building, prayers, proverbial wisdom, prophetic messages, historical narratives, tribal lists, archival data, ritual regulations, and other kinds of material more difficult to classify. We must acknowledge this remarkable diversity and be careful not to exclude any of it from the scope of our study. Otherwise we could not honestly claim to be considering the Bible as a whole.

---

*The term "writing" as used in this chapter refers to the substance of the communication in language, not to the technology used in communicating it. What went into making the Bible as a physical object is discussed in appendix 2.

WRITING AS THE
EXPRESSION OF A
SUBJECT

But if the literature of the Bible (or the Bible as literature) is, indeed, so diverse, can anything at all be said about it that will apply across the board? Fortunately, one thing of fundamental importance can be said: Every piece of writing in the Bible expresses a *subject*, not an object. The difference between the two is crucial. As ordinarily understood, objects are things that exist externally to ourselves and independently of us. They do not have to be material—objects can be ideas, events, even possibilities—but they are "out there." In respect to a piece of writing, the object would be whatever portion of this external existence the author captured and put on paper. Such writing normally comes to us with at least some kind of implied truth claim: "This thing that I am telling you is so; it really happened." We judge it, if we can, by its closeness to truth. This approach works poorly with most belles lettres, however, because such writing makes no claim to truth that can be taken seriously; in fact, we usually recognize such writing for what it is and do not even attempt the judgment. But the Bible does not so easily escape, for its writing seems to make constant and serious truth claims that are taken at face value by a great many readers who look to the Bible as a true record of God's dealings with humankind—that is, as accurate, objective reporting. Nevertheless, this approach works no better with the Bible than with most belles lettres because in so many cases we have no knowledge of the objects represented in the Bible apart from what the authors have written about them; thus we have nothing by which to judge the writing. In the absence of any objective means of determining the apparent truth claims of biblical writing, it might seem that we are left with the prospect of nothing beyond the usual futile arguments between believers and nonbelievers that never change anyone's mind and that monopolize attention to the exclusion of everything else. The only way to escape this dead end is to rethink our conception of the literary situation, and it is for this reason that the alternative term "subject" has been introduced.

A subject is not something "out there" but something "in here." It exists in the author's consciousness; it is a conception of what the author wishes to express. This subject may be a private whim or fancy with no reference to objective reality, or it may have reference to something as solid, tangible, and generally shared as Solomon's Temple. This matters not at all. Any communication about the Temple requires that this object first enter the author's mind as a group of perceptions. These perceptions are modified by the author's individual point of view and past experience and are further transformed when they emerge because now they are in the form of words, not stones and mortar. What do these words tell us? They do not necessarily report what the Temple was really like, although that may be their apparent purpose, but rather they tell us what the author thought about the Temple and wished readers to think about it. The appropriate questions now have nothing to do with whether or not the words correspond to an objective reality but with their purpose and effect as literary devices. What was the author trying to accomplish? How was this done? Were the means adequate to this end? What can we learn from watching this author at work?

Consider, for example, the first creation story in Genesis. The object here is entirely unknowable, for it is a series of cosmic events witnessed by no human being who ever lived. The *subject*, however, is a conception of how the universe was created. There is no point in asking, "Was the universe really created in six twenty-four-hour days?" or "Could light have been created before there was any light-emitting heavenly body?" It is unimportant whether the answer to such questions be positive or negative because neither answer gets one anywhere. The process comes to a dead halt.

On the other hand, if one asks questions about the subject, one can learn a great deal. What is the point, in the first place, of carefully separating the acts of creation from one another and presenting them in a cumulative series? Why is God shown creating things merely by saying, "Let there be . . ."? Why does creation of the sun and moon follow creation of the earth rather than precede it? Why is each act of creation finished off by a divine appraisal and judgment? And so on. The answers to these and other such questions get us into the mental world of the author, which is the immediate cause of all that we see on the page. In this case we begin to understand the concepts of deity and creation held by the so-called Priestly authors, who were supposedly responsible for this story along with much else in the Pentateuch. For them God was awesome and remote, sharing power with no one and no thing, bringing the universe into being simply because he wished to do so. His creative acts were supremely orderly, both in their form and their effect. When he finished, there was absolutely nothing left to do: The work was perfected. This conception was so important to the Priestly authors that they prefixed their account to an already existing account of Adam and Eve in the Garden of Eden, thus forming Genesis 1. So in the opening chapters of Genesis we now have two very different accounts of what was presumably the same event: the Creation. There is no reason to suppose that the

Priestly authors believed that there had actually been two creations. No, there had been only one set of divine acts in the beginning—one object—but there is more than one perspective from which to view it, that is, more than one subject.

An example from the New Testament will further illustrate the issue. In chapter 7 of Acts, Stephen, a member of the Church in Jerusalem, delivers a speech before the Jewish Council that is trying him for blasphemy. The speech, reported at length by the author of Acts, is a summary of the history of Israel from Abraham to Solomon; it is capped with an accusation that the Jews of Stephen's time have betrayed their own faith by murdering Jesus of Nazareth, just as their ancestors had done by persecuting and murdering the prophets sent by God. This accusation, climaxed by Stephen's vision of Jesus standing at God's right hand, leads to his death by stoning, and he becomes the first Christian martyr.

The event is a crucial one, and the speech is obviously the key element in it. Considered as an object, though, the speech is most strange. It has nothing to do with the charge of blasphemy—indeed, it is not a speech of defense at all but rather a bitter indictment of Stephen's accusers and everything they stand for. In it Stephen presumes to review Jewish history for the Council, a matter that its members were certainly thoroughly versed in; and though it charges them with being bad Jews, the speech itself gives a very eccentric view of Jewish tradition. It looks as though Stephen is deliberately courting martyrdom by being as offensive to his audience as possible. Indeed, how can he expect that his message will be understood or accepted by men whom he himself believes to be deaf to the truth? We might also wonder how Luke (who composed Acts about half a century after this event) obtained the verbatim text of Stephen's speech. Was there, perhaps, a sympathizer in this audience who memorized it?

We have these problems as long as we look at the speech as something that was once "out there" and that was transmitted bodily to the written page without filtering through the mind of an author. As long as our only concern is whether the speech is exact and authentic—a real object—we are limited to either admiring Stephen's courage or deploring the fabrication of history. But if we look at the speech as a subject, we are liberated. Now we can recognize that it is, in fact, a carefully structured literary composition, even though the story presents it as a burst of spontaneous oratory. Now we can acknowledge that its real author was Luke, who wrote the speech that he believed Stephen would have (or should have) given on that occasion—as all ancient historians were accustomed to doing—perhaps following traditional accounts but supplying the exact wording himself. In the process he made use of one of the oldest literary forms in Judaism, the historical recital. Judaism does not have a creed in the usual sense of that term: a list of beliefs to which members of the faith subscribe. Instead, that function is served by the historical recital: a summary of the covenant relationship of the Jews with God, concentrating on the high points of the Jewish past. It may be long and detailed, as in this instance, or very brief, as in Deuteronomy 26:5–9 ("My father was a wandering Aramaean . . ."). In any case it indirectly answers the question, "Who are we and what do we stand for?" by answering another question, "Where did we come from and how did we get here?" In Acts 7 this familiar form is put to unorthodox use, for the content of Stephen's recital is hostile to Judaism and is designed to support the claims of the Christians. Why? Because Luke is writing to his own Christian audience, not to Stephen's Jewish one. It is Luke's conception of the subject that we see and Luke's purpose that is served by it. The speech is not meant to offend the Council but rather to convince Luke's audience. The trial of Stephen, who is long dead and gone by this time, can serve as an example for Christian readers of Acts, which Luke is in the process of composing with the literary skill that he had already put to good use in his gospel.

We can now reasonably compare chapter 7 with the other historical recitals in Acts—those by Peter in chapter 3 and by Paul in chapter 13—on the basis of their language, rhetorical structure, content, and function without being encumbered by the question of whether they are to be believed as factual records. With this new perspective we can see what Luke was "really" up to here: He was drawing the closest of parallels between Stephen and Christ as victims of religious bigotry. Like Jesus, Stephen works miracles and produces "signs" during his ministry. Stephen makes converts but arouses the hatred of the orthodox, as did Jesus. Like Jesus, Stephen is an opponent of the Temple cult. He is tried by the Jewish Council, as was Jesus. Stephen's charge that the Jews have always persecuted their prophets echoes that of Jesus in Luke 11:49–51. Stephen calls Jesus the "Son of Man," Jesus' own term—the only time this usage occurs in the New Testament outside the four gospels. As Jesus did in Luke 23:34, Stephen asks forgiveness for his murderers (though the authenticity of the earlier passage is in some dispute), and Stephen's prayer, "Lord Jesus, receive my spirit," parallels Jesus' own words in Luke 23:46. It is clear that for Luke the martyrdom of Stephen was a decisive event second in importance only to the Crucifixion, for it marked the beginning of the Church's destined mission toward the Gentiles and away from the Jews. Hence he took pains to emphasize the resemblance of Stephen's sacrifice to that of the Christ. This, in general, is Luke's subject. What Stephen's sacrifice "really" was and "really" meant are questions, respectively, for the historian and the theologian. We are not concerned with these questions because our attention is focused on the means through which all of this comes to us: a composition in which the normal literary processes of selection, emphasis, wording, and organization operate, and which must be studied as a literary composition. This vehicle is not the event—it never is. It is its own kind of thing.

BIBLICAL AUTHORSHIP

Though the biblical author, like any other author, is a person who gives expression to a subject through the medium of language, biblical literature itself cannot be accounted for simply by saying that so-and-so wrote such-and-such. Most biblical works have quite a complex history of authorship behind them. The prophetic books of the Jewish Bible are the easiest ones to account for because a large part of their contents is believed to be the words of the persons whose names are attached to them, persons who lived and prophesied in historic times. But the form of these books—the selection and arrangement of their contents—seems to have been the responsibility of others. In the process, material from different prophets and from different times occasionally got lumped together in the same book. This is conspicuously so in Isaiah, chapters 40–66 of which were not written by the eighth-century prophet who wrote most of the first thirty-nine chapters. Although some of the books in the division of the Jewish scriptures called the Writings—Ruth, Esther, Job, Ecclesiastes, and Jonah—seem to be by single authors, in no case do we know anything about the author. Moreover, Job and Ecclesiastes both contain additions to the text by other persons who were unsympathetic to the original authors' aims. The rest of the Jewish Bible is almost entirely the product of collaboration—collaboration of a special kind because the various authors were widely separated in space and time, had no knowledge of one another, and certainly had no conception of the form that their work would finally take. The best example of this is the Pentateuch, which contains material from at least four different sources. When the sources themselves are the product of an oral, collaborative tradition, as the so-called "J" and "E" documents are supposed to have been, this complexity is enormously increased. At first the New Testament appears to be different inasmuch as all but one of its books have traditionally borne the name of an author. Most of the attributions of authorship in the New Testament, however, are merely traditional and have no historical basis. Each of the four gospels (as chapter 13 of this book will discuss in more detail) was written by a particular man, but we have no idea who he really was. We do have seven or so genuine letters of Paul and, perhaps, the Revelation of John; the rest of the New Testament is either anonymous or pseudonymous.

THE BIBLE AS AN
ANTHOLOGY

All of our discussion points to a cardinal fact: The Bible is not a book at all, in the usual sense of the term, but an anthology—a set of selections from a library of religious and nationalistic writings produced over a period of some one thousand years.* The Bible cannot have the kind of unity that we normally expect in a book from our own period. There is no such thing as a biblical style or a biblical point of view or a biblical message: There are styles, points of view, messages. "It says in the Bible" is a common and handy way of making biblical references and does no harm if not taken seriously, but we should remember that there is no "it" or "he" or "she" to say things "in the Bible," because the *whole* Bible has no author. If one insists on looking at the Bible as a unified and homogeneous work, planned from the beginning, then one is forced not only to ignore what is known about its origins and composition but also to explain away a host of textual problems—duplications of material, omissions, interpolations, contradictions—that are most sensibly accounted for as the result of multiple authorship over a long period of time. Far from simplifying the problems, the dogma that the Bible is a unity multiplies and magnifies them. It is true that an appearance of unity results at times when the Bible quotes or derives something from itself, such as when the New Testament brings in prophecies from the Old Testament. But retrospective agreement alone is not sufficient: Agreement should work both ways. That is, it is not enough that Paul quotes Genesis 15:6 on the selection of Abraham in his letter to the Romans. One would have to show also that the author of Genesis 15:6 knew that his statement about Abraham would be used later by Paul and that he deliberately phrased it with that use in mind. (Whether he knew that Paul would be quoting him not in the Hebrew original but in the Greek translation is still another question.) Paul himself obviously believed that the history of Israel as displayed

---

*Indeed, its name comes from a plural Greek noun, *ta biblia* ("the books").

in the Jewish scriptures had been arranged for the benefit of his own generation, so that not only the biblical text but the events behind that text were part of a grand design; but this is the view of faith, not the view of literary criticism, and we cannot consider it here.

The notion that the Bible speaks with one voice has been powerfully, though accidentally, encouraged in our own culture by the long preeminence of the King James Version. If there were stylistic variations from one section to the next when this version was written, such variations have been pretty much obscured during the intervening four centuries: It all sounds alike to us now. What we notice—and what many persons particularly value—is the flavor of antiquity, the solemn and measured tone, the sense of something *behind* the words of this version. Champions of the King James Version feel that if God spoke English, this is how he would sound. Modern vernacular translations of the Bible are unpopular with such persons because, in attempting to reflect the individual characteristics of the original Hebrew and Greek texts, the translators violate this assumption. (This matter will be treated further in chapter 16.)

THE REDACTORS

The literary history of biblical texts is complicated not only by the circumstances of authorship that we have already discussed but also by the activities of a set of persons nearly as important as the original authors: the redactors. Unfortunately, the redactors are just as mysterious as most of the authors are. Not one of them is identified in any manner in the Bible, and the evidences of redaction are all inferential. Still, redaction did take place, for there is no other way to explain the condition of the texts as we have them. Redactors were persons who made up finished versions of texts out of sources available to them—sources that may have consisted of complete alternative versions or several partial versions or, perhaps, one substantially complete version that required only minor changes. The redactors added transitions or links where needed, attempting to produce what they viewed as final documents that could be preserved and copied. In these respects we imagine redactors as having had essentially conservational and editorial functions. But we must admit that this picture may be oversimplified: The redactors may well not have been limited in their sources to the manuscripts on the desks before them. It is equally possible that at times they contributed material of their own to the texts, behaving in effect like authors.

This then is what redactors did, but what did redactors think they were doing? At many stages in the study of the Bible as literature we have to ask this question, but the answer to it can be tantalizingly elusive. It is, after all, the same question that we ask of the author of any work, modern or ancient, for the meaning of any work is intimately related to its author's intention. It is not necessarily true that authors achieve what they intend; but even if they fail (or, perhaps, succeed in spite of themselves), we cannot interpret the result without hypothesizing something about authorial intention. The issue is particularly acute when there are duplications or apparent contradictions in the biblical text, for in such instances, it seems, it would have been quite easy for redactors simply to silently eliminate the problem. If they were willing to adapt and modify their materials as far as they did, why did they stop short of making a "finished" text? Were they so constrained by reverence for their documents that they dared not leave anything out?* Such a feeling may account for the three complete versions of the judgment on the Gibeonites in Joshua 9 or for the flat disagreement in Genesis 37 as to whether the Ishmaelites or the Midianites brought Joseph to Egypt. On the other hand, as biblical scholars know, the sacredness of a text is no guarantee that it will not be tampered with; paradoxically, its very importance may cause well-meaning persons to attempt to correct its presumed faults and bring it closer to the ideal.

Of course, it is possible that the redactors in cases like the two cited did not see any duplication or contradiction, just as modern conservative interpreters may refuse to see differences between the two creation accounts in Genesis and so deny the need for two sources to account for their existence. It is also possible—and this possibility was suggested earlier—that the redactors had a more sophisticated understanding of their role than we think and that they excused such apparent defects according to some overriding principle of rightness that we have difficulty in seeing. About all we can say for sure is that no one ever touched a biblical text with the conscious aim of making it worse.

The history of redactional activity begins with the J and E sources in the Pentateuch (as chapter 7 will detail), and these five books remain the most thoroughly studied group of literary texts in the Old

---

*It is possible, too, that what operated was not so much reverence for the documents per se as reverence toward a great historical figure whose name had encouraged the creation of fabulous stories that later on acquired the status of biographical fact. These stories differ in no respect except date from what is called "haggadah" in rabbinic literature; and their earlier date, of course, is what put them inside the scriptures rather than outside. Such, for example, would be the story of David and Goliath.

Testament from this point of view. But the Pentateuch is by no means the only or even the best place to look for evidence of the redactor's hand. The Book of Judges offers some particularly clear examples of redactional activity because the basic materials of the book—folktales about heroes from the early years of Israel in Canaan—had to be assembled and presented in some kind of framework before there could be any book at all. What seems to have happened is that a redactor in the eighth century B.C.E., someone with a strong theological bent, put together the original collection. This redactor's theme was the consequences of faithfulness or unfaithfulness to Yahweh, the deity of Israel. When the Israelites disobey, Yahweh delivers them to their enemies. They plead for help and Yahweh sends a rescuer, a charismatic leader. This leader (the "judge") succeeds for a time, but after the leader's departure, the people lapse again and suffer for it. The stories per se have little in common and are held together only by the redactor's efforts. At one point an ancient cultic poem, the Song of Deborah, was included. Later a redactor of the so-called Deuteronomic school added the moralizing introduction of chapters 2:6–3:6 and made other changes, perhaps adding chapters 9 and 16, which look as if they did not originally belong there; this redactor also brought the number of judges to the significant total of twelve by listing six minor figures. The final redaction of Judges occurred during the Priestly period and accounts for chapters 17–21, which illustrate what happened as a result of Israel's not having a king. Not only are there no judges in this pro-monarchical narrative but their absence is not even remarked on. At this time the historical prologue of chapters 1–2:5 was added. Its opening words, "Now after Joshua's death," were the redactor's attempt to connect the opening of Judges with the end of Joshua (in Judges 2:6 Joshua is still alive). The history in 1–2:5 itself, however, comes from a very old document and is not the redactor's own composition.

In the New Testament, evidence of redactional activity is especially prominent in the synoptic gospels—Matthew, Mark, and Luke—although in these instances the authors were their own redactors. According to the most accepted view, Matthew and Luke each adapted the earliest gospel, Mark, to his own purposes, working independently. But each author also drew on other sources not so easily identifiable by us (see chapter 13). How and why the synoptic authors put these sources together is the province of a technical discipline, redaction criticism. There is perhaps only a fine line between genuinely original composition—such as the gospel of Mark may have been—and authorship by redaction, since Mark no less than the others was drawing on traditions of the life of Jesus that he had inherited. On their part, Matthew and Luke, though following various sources, had wholly personal conceptions of their subject and went about presenting it in ways that are strikingly different.

The importance of redactors should be obvious by now: Without them we would have no Bible. Documents and sources do not come together of themselves to form literary units spontaneously. The intention of the redactors in bringing these documents and sources together as they did is as much a part of the total meaning as were the intentions of the original authors. We cannot understand the final product properly without taking both into account. Along the way we have to do a great deal of poking, prodding, decomposition, and looking behind the final product to discover earlier stages of its existence, much as a geologist or an archaeologist has to dig down through strata of the earth to learn the history of a site. This kind of activity is much resented and opposed by theological conservatives, who see it as reducing the Bible to the level of an ordinary human work, filled with human imperfections, thus denying the Bible's divine authority. Most of these people are comfortable enough with the work of "lower criticism"—the study of the text as such and its transmission—but they object to the "higher criticism," which this chapter presents, because it apparently destroys the integrity of the text. They would prefer to believe that each book of the Bible was written in its entirety by one inspired author and that the manuscript, at the time the writer finished it, was whole and perfect. We have already pointed out that this approach creates more problems than it solves.

THE LITERATURE-OF-THE-BIBLE APPROACH

We said earlier that we use the term "literature" in its broadest sense and do not limit its coverage to belles lettres. Readers may be aware of other literary treatments of the Bible that employ (or imply) a narrower definition of "literature" and thus give attention only to the Bible's belletristic portions. Such treatments, we might say, exemplify a literature-of-the-Bible rather than a Bible-as-literature approach. These treatments make much—and rightly so—of such narratives as those of Adam and Eve, Cain and Abel, Isaac and Rebecca, Joseph and his brothers, and Saul and David; of such poems as the Song of Moses, the Song of Deborah, David's lament over Jonathan, and Psalms 1 and 23; and of such moral essays (or collections of moral statements) as portions of Ecclesiastes and the Sermon on the Mount. But the literature-of-the-Bible approach, even though it appropriately celebrates the Bible's glories, is in the end a limited approach. We shall conclude this chapter by pointing out why that is so.

There is no doubt, as we have already acknowledged, that the Bible contains a great amount of rather dull and routine stuff, of interest only to specialists or to certain kinds of believers. One can sympathize with the wish to avoid it and to spend one's time on material that is intrinsically more appealing—especially material that has a recognizable literary shape. The Old Testament is particularly suited to this selective treatment because it contains some of the most engaging as well as some of the dullest material in the Bible. There is a natural temptation to extract from it those famous narratives that seem to stand on their own as human documents, that present characters with psychological realism, and that are plotted with such subtlety and skill that they yield impressive results to literary analysis. It is no wonder that the story of Abraham and the three angels or of Jacob's tricking Esau or of Judah and Tamar as well as many other stories like them have so often been taken out of context and studied as if they were independent compositions—and, what is even more objectionable, as if doing so were fair to the Bible as a whole and fully adequate to the needs of students of the Bible as literature.

This mode of treatment is unfair to the Bible not only because it leaves out much of the content of the Bible but also because it misrepresents the status of the pieces it selects. These literary compositions came to be included in the Bible not for their literary qualities but for their usefulness to religion as perceived by redactors and by religious communities. Though the Bible is an anthology—as we shall continue to insist—it is not an anthology assembled according to the principles that govern anthology makers nowadays. Without in the least yielding in our appreciation of the humor, drama, and psychological interest of such choice portions of the text as the account of Jacob and Esau— qualities that were probably appreciated in biblical times even as now—we have to insist that the value of such portions to those responsible for their inclusion had nothing to do with literary merit. For us to look on the account of Jacob and Esau as only a good story is to misunderstand it quite seriously.

A further limitation of the literature-of-the-Bible approach is that, although recognizing the literary status and characteristics of individual portions of the Bible, it denies this recognition to the Bible as a whole. The Bible as a collection or anthology has its own existence as literature: It was composed, compiled, shaped, added to, edited, copied, translated, and interpreted in ways quite recognizable to literary scholars. And all of these activities and processes deserve attention from anyone who wishes to have more than merely an aesthetic appreciation of the Bible. A considerable proportion of what follows will concern these activities and processes, for it is our intention that readers know not only what is in the Bible, but what shape it takes, how it got that way, and how it has come down to us in the form in which we read it.

II

Literary Forms and Strategies in the Bible

The Bible is an anthology, to the making of which many hands contributed over centuries of human history. Some of the contributors were original authors, their identities for the most part lost in the mists of the past, and some—even more completely removed from our view—were redactors, who patched and revised and combined literary materials to form the whole documents that ultimately became the biblical books we now have in our canons. Some aspects of this process are strange to us now, especially the process of redaction, because Western literary traditions have evolved along different lines. Yet there is a great deal in the process of biblical writing that we can recognize, feel at home with, and treat in the terms that modern literary critics apply to literature of their own time.

LITERARY FORMS IN
THE BIBLE

Every piece of writing is a kind of something. It takes its place within a particular formal tradition and in itself exemplifies that tradition. This was no less true during biblical times than it is now. There are of course innovators, pioneers, who seek to do things in a way that no one has ever done before—particularly in the twentieth century, a period in which so much value is placed on originality. But even in the twentieth century the innovations have succeeded only in stretching the boundaries of the traditional forms, not in doing away with them altogether. James Joyce's *Ulysses* is still recognizably a novel and cannot be approached without a strong sense of what the traditional novel has already led us to expect as appropriate to that form. Before the modern period, and certainly in biblical times, writers who had some conception of a subject they wished to give expression to would turn naturally and as a matter of course to a traditional literary form as a vehicle for doing so. (Not that the two—conception and vehicle—were necessarily separate in writers' minds; it is more likely that even as writers thought of a subject, these thoughts themselves took form in a traditional way.) It follows that the modern reader of the Bible cannot hope to make sense of it as literature without knowing something about these forms. Though the Bible in a general sense is literature, just as the products of modern writers are literature, its literary forms are different enough from ours to require particular study.

All literary forms quickly become public property. Even in our own age—the age of innovation—successful forms are quickly absorbed into the general culture and become available to anyone who wants to use them. This was the case in biblical times—but with one important difference: Biblical writers seem, for the most part, to have wanted to submerge their individuality in the chosen form and made no effort to give the result a personal stamp. Surely there were writers who spoke out of deep personal feelings, but they tend to disappear as persons. For example, the author of Psalm 22 ("My God, my God, why have you forsaken me?") may well have written out of a personal crisis, perhaps a serious illness. But the author's feelings of estrangement and despair are nevertheless traditionally expressed, utilizing a form that modern scholars call the "lament," of which there are nearly forty other instances in the book of Psalms, thus making it by far the most common type of psalm. As a comparison of lament psalms will show, they tend to follow a stereotyped pattern: The speakers invoke God, describe their trouble (which often includes persecution by enemies), assert faith in God, petition for help (sometimes offering a vow), and thank God for the rescue that they foresee. Psalm 13, a less famous lament, offers in miniature a particularly clear specimen of the form.

If the forms of biblical writing were the property of the whole culture, we must go on to ask why they were popular—in other words, what general significance, what role or function in the life of the nation did they have? For example, why would anyone have composed a psalm? If anyone did so, how would it have been used? The prevalent theory now is that most (if not all) of the poems in the psalmbook were used in ceremonies at the Second Temple—sung or chanted with musical accompaniment at various points in the ritual. Not all the poems were used all the time. But as with our modern hymnbooks, there were no doubt old favorites that did extra duty. Also, as with our modern hymnbooks, the collection included poems from different periods of time, written under varying circumstances but all in the same tradition. Indeed, the analogy of the book of Psalms with our own hymnbooks—its lineal descendants—is quite exact. Although most modern hymns bear the names of authors, we do not normally pay much attention to who wrote what, and few worshippers would be able to name the persons responsible for the words of even their most beloved hymns. Much as we honor the genius of an Isaac Watts or a Martin Luther or a Thomas of Celano, the hymn form itself transcends them. The proper place to begin studying the hymn is with that form, not with the individual authors who used it.

Liturgical forms, dictated by the needs of public worship ceremonies, are by no means limited to the psalms. Once we learn how to look for such forms, we find them to be abundant in the Jewish scriptures. Numbers 6:24–26 is a blessing that essentially consists of the threefold repetition of the name of Yahweh. It is inserted into the narrative of the law-giving at Sinai, but its actual source is much more likely to have been in the services at Solomon's Temple before the Exile. The language of the covenant-renewal ceremony in Joshua 24:14–24 seems to have come from an annual ceremony of that sort held at Shechem. This is preceded in 24:2–13 by a historical recital, itself a traditional form, as we can see by looking at Deuteronomy 6:20–25 or 1 Samuel 12:6–15. The historical recital (to which we have already referred) is an expanded version of the prologue that traditionally began a pact or covenant, known technically as a "suzerainty treaty," between a ruler and his people. The Decalogue is a condensed version of such a treaty, with many analogues in other Near Eastern cultures older than that of the Bible (see chapter 3). To have the treaty affirmed or reaffirmed in public would be a natural practice not only because the people collectively were signatories but also because in the ancient world no agreement was valid without witnesses. The more witnesses there were, the more secure the treaty would be (see, for example, Exodus 24:3–8).

Although the "literary" prophets in the Old Testament operated independently of the national cult and frequently in direct opposition to it, their speech was public and it took traditional forms. They did not see themselves as individuals but as vehicles for the word of God. Already in Amos, the first of such vehicles, prophetic oracles are a highly stereotyped literary composition. Many of these oracles can be identified in the works of Amos and those who followed him. They are marked by having certain standard functions (denouncing the people for their sins, promising punishment from Yahweh), utilizing a more-or-less coherent set of central images (for example, Israel as a disobedient child and Yahweh as the parent), and including linguistic formulas ("These are the very words of Yahweh"). We like to believe that the best of these oracles bear distinctive marks of the individual prophet's thinking and literary style; but many of them could be moved from one prophetic book to another without creating any problems—some were, in fact, moved during the redaction of these books (for example, Isaiah 2:2–4 is almost identical with Micah 4:1–3). It takes no credit away from the original prophetic authors to point out that their forms can easily be imitated or parodied by a modern reader. The result would at best be a kind of curiosity with no real relevance, but it would still demonstrate something important about the usefulness of traditional forms.

The Old Testament is also the repository for several specimens of ancient patriotic poetry: the victory songs of Moses in Exodus 15, Moses' blessing on Israel in Deuteronomy 28, Jacob's blessing of his sons in Genesis 49, the Song of Deborah in Judges 5. We know nothing about the setting of these compositions—how they were used or what part they played in the life of the people—but of their character as public speech there is no doubt. Even the magnificent lament of David for Saul and Jonathan (2 Sam. 1:19–27), which shows all the signs of being an original composition full of personal grief, speaks as much of the loss to Israel as it does of the loss to David himself. And, though the meaning of the Hebrew text of 2 Samuel 1:18 is not clear, the indication is that the lament for Saul and Jonathan was to be taught to and recited by the people on certain occasions.

Two other significant Old Testament literary forms, the wisdom saying and the apocalypse, will receive chapters of their own. It remains to add a few words about the commonest of all Old Testament forms, the narrative. There is no single literary form that can be called "Old Testament narrative," for the narratives in the Old Testament are quite diverse in nature—as well they might be, coming from so many different authors writing in such different times. About the only thing they have in common is that none of them was ever composed in the first place merely to preserve knowledge that certain things happened. All the Old Testament stories are tendentious, that is, they serve to uphold a theological point or to illustrate a significant theme in the unfolding drama of the covenanted people. This is obvious enough in the Deuteronomic History (roughly, Joshua through Kings), which is continuously and openly biased, but it is also true of the ancient stories in Genesis, which we are in the habit of reading today for their local color and narrative skill, forgetting that it took more than qualities like these to ensure their preservation.

Unfortunately, it is often difficult for modern readers to see the meaning that may have been obvious to the ancient authors and their audience. For example, the three "wife-sister" tales in Genesis (12:10–20, 20:1–18, and 26:1–11) each tell of a patriarch living temporarily in a foreign country who passes his wife off as his sister, thereby fooling the foreign king and, more important, saving his own neck. Twice the hero of the story is Abraham and once it is Isaac. The first of these, in chapter 12, is probably the earliest and the source or inspiration for the other two. But what does the story mean? Are we supposed to deplore the willingness of Abram (as he is then called) to buy security for himself by lying to the pharaoh about his wife and allowing her to be taken into the pharaoh's harem? Are we supposed to admire Abram's cleverness at outwitting the foreign king and gaining material prosperity for himself through the king's favor? Are we supposed to sympathize with Abram's dilemma in a situation where there is no obvious right way to act? Or, perhaps, is the whole point of the story to emphasize the providential intervention of Yahweh, who stepped in and rescued the Covenant with the descendants of Abram and Sarai from imminent danger of collapse? Behind the written story is doubtless an oral tradition of considerable age, a tradition that might have been mainly directed to the celebration of the beauty of Sarai, which would have made her desirable to others (notice that it is kings who desire her) and of course put her marriage to Abram in jeopardy. In any case the author of the second version, in chapter 20, retold the story with some basic changes, presumably to substitute the author's own emphases; the author of the third version did the same thing in chapter 26. The final redactor who wove all three into the texture of the Genesis narrative may well have understood their meaning no better than we do, but since he could have regarded them all as history, he did not have to feel obliged to figure them out.

Among the varieties of narrative form in the Old Testament are etiologies (stories explaining the origin of something, especially names), birth narratives (which typically speak of a barren wife, a divine guest, an annunciation, and a "sign"), miracle stories (such as those associated with Elisha), accounts of theophanies (the appearance of Yahweh in the burning bush to Moses or to Abraham before the destruction of Sodom), and hero stories (the exploits of a Samson, a Jacob, a Daniel). Often we understand these stories better when we view them along with other examples of the same form rather than as they come to us imbedded in a context provided separately and later by a redactor. Jacob has much more in common with Samson than he does with Isaac, his own father. The themes of the strong man and the trickster, which these stories share, create a formal resemblance between the stories that transcends their many differences.

The New Testament is also rich in traditional literary forms, but their setting is the life of a comparatively small group, the Church, rather than the life of a nation. The four gospels were created for this Church, using written and oral sources that by then had been in existence for a full generation and more. Unquestionably, the most famous literary form in the gospels is the parable, the use of which especially characterized Jesus' teaching ("He would not speak to them except in parables," says Mark 4:34). Yet Jesus did not invent the parable, however much he may have stamped it with his own individuality, for there was an extensive tradition, going back to Old Testament days, of teaching by such indirect means (the *mashal* in Hebrew).

Other traditional forms in the gospels are the pronouncement story, the story of healing, the "saying," the birth narrative (in Matthew and Luke only), the beatitude, the "woe," the legal commentary ("You have heard . . . but I say this to you . . ."), the allegory, the commissioning of apostles, the Transfiguration scene. These and other common elements are intricately modified as they are used by one or another of the gospel writers, following their individual tastes and needs. In every case the element itself originated within the Church as an oral tradition, the common property of a group of believers. But though we may be sure of the context, we cannot easily reconstruct the use of that element within it. The Last Supper narrative is a notable exception, because a communal meal among the believers that featured a recitation of the words and an imitation of the actions of Jesus on that occasion must very soon have become standard within the Church. The earliest version that we have is from Paul, who claims to have received it by a private revelation "from the Lord" (1 Cor. 11:23–26). Whatever one may think of Paul's claim, it is clear that the Last Supper narrative existed and functioned independently of any literary context such as a gospel and that its appearance in the gospels (Matt. 26, Mark 14, Luke 22) reflects its importance to the Church.

Literary forms are large-scale structures: They reflect authors' primary choices of means for embodying a subject, but they do not as a rule determine in any detail the strategies for doing so. What kind of language should an author choose? What rhetorical devices would be helpful? Should the author proceed directly or indirectly? If the latter, by which of the possible ways? Biblical authors had to answer these and other such questions when they wrote, just as modern authors do, and here—as was not quite the case with literary forms—we can meet biblical authors on grounds familiar to us, for the same strategies are still in use today. To be sure, the languages are quite different, which means that certain effects cannot be duplicated from one language to another, but this is just as true of English vis-à-vis German, for example, as it is of English vis-à-vis Hebrew or Greek. The means for getting the effects, however, are the means that authors have used ever since the dawn of literary culture, and we can approach the literature of the Bible with the full confidence that biblical authors drew their weapons from the same armory that supplies us today.

HYPERBOLE

We have chosen to begin with hyperbole, that is, deliberate exaggeration for effect, because it is so common in biblical writings and yet so frequently overlooked. A very simple device, it clearly shows that the biblical authors assumed an audience and that the Bible text cannot properly be understood without taking this fact into account. For example, in 1 Kings 1:40 readers are told that the newly anointed King Solomon was brought home in procession, "with pipes playing and loud rejoicing and shouts to split the earth." Of course the earth did not split! There was not, nor could there have been, a geological convulsion from such a cause. Taken at its face value, the statement is untrue. But not even the most literal-minded reader of the Bible is going to defend the truth of this statement because he or she, like all of us, knows the author to be saying that the rejoicing and noise were very great; the reader understands that to communicate this effect, the author had to exaggerate in this fashion.

Once aware of hyperbole in the biblical text, anyone can find abundant examples. Perhaps the most famous hyperbole, certainly one of the most deliberately repeated, is in the Covenant between Yahweh and the descendants of Abraham, who will be "like the dust on the ground" (Gen. 13:16) or "as numerous as the stars of heaven and the grains of sand on the seashore" (Gen. 22:17). That such phrases were conventional devices is shown by their appearance in other contexts, for example, in Genesis 41:49, Joshua 11:4, 1 Samuel 13:5, and 2 Samuel 17:11. The adjective "all" is frequently used to express a similar conventional hyperbole. "That day, all the people and all Israel understood that the king had had no part in the murder of Abner son of Ner" (2 Sam. 3:37)—meaning that the fact was very widely known. In 1 Kings 18:19 "all Israel" is summoned to Mount Carmel to watch the contest between Elijah and the prophets of Baal. There was no intent to deceive in this, for the original author and his audience both assumed that at least a few persons were left back home on this occasion to mind the livestock. The point is, rather, that it was an immense crowd and one representative of the whole nation. Similarly, when we are told that Absalom had intercourse with his father's concubines "with all Israel watching" (2 Sam. 16:22), the meaning is that the treasonous act was done with much fanfare and publicity, not that everyone in the nation crowded around the site as spectators.

As we might anticipate, stories having to do with military activity are especially likely to show hyperbole. For example, of the Midianites, who began raiding Israel during the days of Gideon, it is said, "they came up as thick as locusts with their cattle and their tents; they and their camels were innumerable" (Judg. 6:5). In the battle between David and the rebel forces led by Absalom in the highlands of Gilead, 20,000 men are said to have been killed in one day—yet "that day the forest claimed more victims than the sword" (2 Sam. 18:8). For the punishment of the Benjaminites, 400,000 trained infantry from the other tribes of Israel are said to have been assembled at Mizpah (Judg. 20:1–2). These and other statistics should not be taken too seriously: behind them is the imagination of a storyteller, not the figures of a census taker. Whether the writer of these stories was always aware of the exaggeration is sometimes in doubt. It is possible that the author of 1 Kings 8:63 really believed that 22,000 oxen and 120,000 sheep were sacrificed by Solomon at the consecration of the Temple, but the numbers are inherently incredible. It would be idle, however, to dispute them on factual grounds. The same thing is true of the 600,000 adult male Israelites said to have begun the exodus from Egypt and—the most famous case of all—the ages of the patriarchs who lived before the Flood. We should recognize that whatever else lies behind these figures, they are intended to impress the reader with the magnitude of the subject: It is this that comes across to the reader, and this is the only important consideration.

Hyperbole reaches its climax in the narrative portions of Daniel and the entire book of Esther. In both of these, hyperbole is so constant that it can no longer be regarded as merely a device: It must be seen as an intrinsic part of the author's conception of his subject. The furnace heated up to seven times its usual temperature in Daniel 3:19 or the gallows seventy-five feet high in Esther 5:14 are not momentary and local exaggerations but are typical of the wholes within which they exist.

Hyperbole is not confined to narrative writings nor to the Old Testament. Jesus well understood the value of hyperbole, and we can do no better than close this section with three famous examples of it from his reported sayings: "Alas for you, scribes and Pharisees, you hypocrites! . . . You blind guides, straining out gnats and swallowing camels!" (Matt. 23:23, 24); ". . . it is easier for a camel to pass through the eye of a needle than for someone rich to enter the kingdom of Heaven" (Matt. 19:24); "If your hand or your foot should be your downfall, cut it off and throw it away: it is better for you to enter into life crippled or lame, than to have two hands or two feet and be thrown into eternal fire. And if your eye should be your downfall, tear it out and throw it away . . ." (Matt. 18:8–9).

METAPHOR

When the author of Psalm 1, pursuing the distinction between the righteous and the wicked, came to the point of characterizing their lives in general, he chose a method guaranteed to bring the two vividly and memorably before our eyes: The righteous, he wrote, are "like a tree planted near streams," whereas the wicked are "just like chaff blown around by the wind." This device, probably as old as language itself, is metaphor. Its mechanism is simple. A word that (as we say) is literal in the contexts within which it is usually found is taken out of those contexts and used in a context of some other kind. For "chaff" the usual context would be a conversation or piece of writing having to do with agriculture; the new context would be a didactic psalm having to do with human conduct. It is perfectly obvious that human beings are not really trees, not really chaff. What prevents such statements from being simply false is the fact that *in some sense* they are true. The two objects have something in common, an area of shared meaning, however much they also differ. (That the statements in Psalm 1 use an explicit term of comparison—"like"—and thus may technically be similes does not concern us here. For our purposes metaphor and simile are equivalent.)

The best way to visualize this process is with a simple diagram, like the one provided for the metaphor in Psalm 1:3 (fig. 1). Each circle in the diagram represents one of the terms (normally there are only two terms), and we can imagine that circle as containing all the connotations of its respective term: every possible meaning associated with the word or the idea. The two circles intersect because the terms share certain connotations. This area of intersection varies from metaphor to metaphor, depending on how close the terms were to begin with.

We don't, of course, think in intersecting circles when we read, only when we analyze. To the reader the metaphor is a unity, but a unity with a special resonance and richness because it extends meaning into areas where meaning would not normally be found. Overused, a metaphor tends to sink back into literality, where it is simply ignored or becomes a cliché. Whether a given metaphor, like the comparison of wicked men to chaff, has the effect of a cliché depends as much on the attitudes of the reader as it does on mere frequency of use: There is no objective standard one can apply. In any case our modern horror of the cliché was probably not shared by the ancients, for whom the familiarity of the metaphor may well have been the best thing about it.

The scope of biblical metaphor is well illustrated in the book of Psalms. According to the psalmists, Yahweh is a "stronghold for the oppressed" (9:9), a "rock" (28:1), a "shield" (3:3), a "fortress" (18:2), a "light" (27:1), a "shepherd" (23:1). He will give us refuge "in the shadow of [his] wings" (57:1). The honest "will ever see his face" (11:7), but his "face is set against those who do evil" (34:16). His "right hand is triumphant" (118:15). His word is a "lamp" to guide the feet (119:105); his promises are "natural silver which comes from the earth seven times refined" (12:6). Human life is "a mere puff of wind" (144:4); "dust," "wild flowers" (103: 14–15); and our days are "as fleeting as a shadow" (144:4). God's opponents are "smoke," "wax" (68:2); "thistledown," "chaff" (83:13). The righteous man does not "take a stand in the path that sinners tread" (1:1) and assures Yahweh that his "steps never stray from the paths you lay down" (17:5). The wicked man, on the other hand, lurks "unseen like a lion in his lair" (10:9); he is "smooth-tongued" and "speaks from an insincere heart" 12:2); though his words are "more soothing than oil," they are nevertheless "sharpened like swords" (55:21); "cursing has been the uniform he wore" (109:18); he sets "traps" and "snares" for others (141:9), but he himself will be "enveloped in a cloak of shame" (109:29).

One notices in such a list how again and again an abstract idea is replaced by something specific and concrete and how the everyday experience of the audience furnishes the source from which meaning

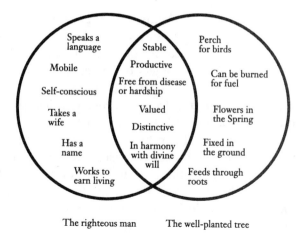

The righteous man   The well-planted tree

FIGURE 1. METAPHOR

is drawn. Who in Palestine had never seen a rock? Who would not have appreciated the force of this favorite metaphor of the psalmist, one that expresses the massive dependability of God in a world otherwise beset by uncertainty and danger? Every member of the contemporary audience would have felt the force of the warning that the Day of the Lord is coming "like a thief in the night" (1 Thess. 5:2), or of the prayer to "forgive us our debts, / as we have forgiven those / who are in debt to us" (Matt. 6:12), or of the reference to separating "sheep from goats" on Judgment Day (Matt. 25:32).

Our culture today is close enough to that of the biblical world—thanks in part to the influence of the Bible itself—for us to appreciate the force of such metaphors about as well as the original audience could. But there are still a few surprises in store for us, for example, in the traditional Old Testament metaphor in which lips, ears, and hearts are said to be "uncircumcised." In one book, The Song of Songs, metaphor seems to have gone all the way over into nonsense when the lover tells his beloved that her nose is like "the Tower of Lebanon" (7:5) and her hair "like a flock of goats / surging down the slopes of Gilead" (6:5). There is no literal respect in which a woman's nose could look like "the Tower of Lebanon"! But no literal comparison is intended. Instead, the reader's and the poet's attention is given solely to one aspect of the chosen object: the beauty and uniqueness of the tower (or mountain), disregarding all other characteristics. It is as though the two circles in figure 1 touched only at one point and did not intersect at all. In order not to misread such a metaphor, we have to lay aside our own notions of appropriateness and try to enter into the spirit of the original.

SYMBOLISM

Many of the metaphors we cited could as well be used to illustrate another device: symbolism. If a concrete object or an action is used to stand for something through metaphor, that object or action may be separated from the metaphorical statement and used independently of it and still bear this further significance wherever it goes. Although any of the human senses may be involved in symbolism, most often symbolism is visual, as we note again and again in apocalyptic writing—literally the work of "seers." The figure of Christ in Revelation 1:12–16 or the image from Nebuchadnezzar's dream in Daniel 2 are typical. Characteristically, these symbols do not yield their meaning until interpreted by the seer or his celestial guide even though they are clearly visible and are precisely described for us in words. Another form of nonverbal symbolism is found in so-called acted prophecy: for example, (1) when Ahijah in 1 Kings 11:30–31 tears his cloak into twelve pieces and gives ten of them to Jeroboam, symbolizing the coming division of the nation of Israel, or (2) when Hosea, on instructions from Yahweh, marries a prostitute in order to symbolize (or act out) the relationship then existing between Yahweh and the apostate Israel. Many more examples of this symbolism—and some stranger ones—can be found in stories about or in the writings of the prophets, especially in the books of Jeremiah and Ezekiel.

ALLEGORY

A device that depends on the bringing of two areas of meaning into relationship—as metaphor and symbolism do—always leaves open the opportunity for exploring this relationship systematically. Such a development produces allegory. The basic metaphor or symbol is analyzed into component parts, and these parts are brought together in a series of one-to-one relationships. For example, when the author of Ephesians 6:13 exhorted readers to "take up all God's armour" against the forces of evil, he might have stopped there with this metaphor; but he went on to turn it into allegory by drawing out its implications: The armor is broken down into belt, breastplate, shoes, shield, helmet, and sword, each of which is assigned a specific function in the anticipated struggle. The starting point of an allegory may be an artificial construct—like the figure in Nebuchadnezzar's dream, which is composed of five symbolic materials (gold, silver, bronze, iron, clay) and has no meaning independent of them—or it may be a more natural and inevitable image—like that of the human body, which in 1 Corinthians 12:14–31 symbolizes the structural harmony and interdependence of parts in the community of believers.

These three examples might be called vertical or static allegory because they have meaning simply by standing there, without any action ensuing. But allegory always has the potential for motion: If a stage is peopled with actors, it is almost inevitable that they will start doing things. Thus a story begins to unfold. To picture it, all we need do is take the two intersecting circles in the metaphor diagram (fig. 1) and unroll them to make two parallel horizontal lines, as in figure 2. On the first line are the connotations of one of the basic terms; on the second line are the connotations of the other term. The lines are connected to show their relationship. Normally, only one of these lines is expressed in words. Because this is what is presented in the text—what is immediately before us—the second line can be called the "ostensible" level. But the ostensible level is not the meaning the author is trying to get across, which lies on the other level, normally unexpressed. We may call this the "actual" level. The value of the allegory—indeed, the whole point of it—is found on the actual level. Finding this point, however, may require interpretive skill on the part of the reader because the ostensible level may serve as much to disguise meaning as to reveal it. Hence authors who want to make sure that their point is understood may add to the allegory a separate passage explaining it, as is done in Nathan's allegory for King David in 2 Samuel 12 or in the remarkable historical allegory of Ezekiel 17.

The problem of understanding is especially acute in the gospel accounts of the parables of Jesus. As the gospel authors present them, these stories often require interpretation because their hearers either do not realize that they are allegorical or cannot ravel out their component meanings. All nine of the major parables in Matthew, for example, are allegories; three of them have explicit interpretations added. Many scholars, seeking to explain the paradox of a popular teacher using a method that confused or mystified his listeners, argue that the parables of Jesus were originally just simple stories with a single point. However, they argue, after his death the stories were modified in the direction of allegory by a Church tradition that either did not grasp their original purpose or had reasons of its own for altering their thrust. (We shall have more to say about this problem in chapter 13.) The first parable recorded by Mark, the parable of the sower, rather plainly does not fit the allegorical interpretation added to it and so affords us a glimpse into what may have happened. In the added interpretation the focus is changed from the act of sowing the seed to

Two Allegories of the Vineyard:

Isaiah 5:1–7

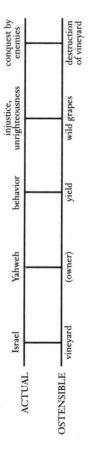

Matthew 21:33–43

The Allegory of the Wedding Feast:

Matthew 22:1–14

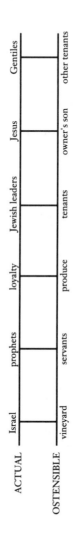

FIGURE 2. ALLEGORY

the fate of the seed after it has been sown. Now the sower disappears altogether, although the harvest should be his reward for having persevered in his work. Indeed, because the harvest can be of no concern to the soil it grew in, the concept of reward now becomes irrelevant. It is replaced by the concept of fitness or receptiveness to the growth of the seed (that is, the gospel), thus initiating a theme of paramount interest to the writer—as we can tell from everything else he wrote— but a theme that was surely not intended in the earliest form of the parable.

Although most biblical allegories are limited to one correct interpretation, the converse does not apply: There is no limit to the number of allegorical expressions that may be given to a single theme or idea. The more important it is, the more likely it is to show up in many different forms. Because no theme in the Old Testament surpasses in importance that of the relationship between God and humankind, we are not surprised to find it expressed in Old Testament writings through a whole battery of complementary but distinctive allegories. Each of these is developed consistently in terms dictated by the basic metaphor underlying it. If the basis is pastoral life, then God is a shepherd, the people are his sheep, and his role is to guide, nourish, and protect them through their lives. Their role, in return, is to give him gratitude and loyalty: thus Psalm 23. Not all of the implications of the pastoral metaphor are used, however, because God does not shear people for their wool and still less does he slaughter and roast them for dinner (not that many readers would ever think of such a possibility, for a good allegory will avoid calling attention to any aspects of the relationship that do not fit its purpose). If the basis is agricultural, then the allegory may present God as the owner of a vineyard, his chosen people as the vines, and their failure to produce the proper harvest (righteous deeds) as cause for abandoning them to their fate (Isa. 5:1–7). Or the same set of conditions may be expressed through the allegory of the law court, with God as the accuser, the people of Israel as the defendants, and (in Mic. 6:1–5) the earth itself as judge. The law court is especially prominent in the book of Job because divine justice is its central theme. The "defender" or "Vindicator" (*go'el* in Hebrew) for whom Job calls in 19:25–27 is the person who would step forward in court and clear him of the false charges that, as Job believes, have led to his punishment. In Hosea 2 the relationship between God and humans is most touchingly represented in the allegory of the family: Yahweh, the husband, speaking to Hosea as his son, accuses Israel, the wife, of infidelity because she has deserted him and lavished her care on the pagan gods, her lovers.

We must be careful not to confuse allegory as a method of writing, which is the subject here, with allegory as a method of reading, which is not. Allegory is written by writers who intend to write it, and normally the signs of their intention are quite plain. But some texts are ambiguous or suggestive, and some readers are more disposed than others to seek for hidden clues and artful disguises on the theory that the author must have had something more important in mind than what we see on the page. Thus The Song of Songs, which on the surface is a series of tender love poems celebrating a physical union, has traditionally been read allegorically: The lover and his beloved are not simply a man and a woman, but they represent Christ and his church (in the usual Christian interpretation) or Yahweh and Israel (in the usual Jewish interpretation), and subordinate details in the poem are allegorized according to a reader's taste. The fact that the Song can plausibly be read in such very different ways suggests the danger of allegorical interpretation and reminds us that, unless there is very clear evidence of the intention to write allegory, we should normally assume that it is absent. (For more on the matter of allegorical interpretation, see chapter 17.)

PERSONIFICATION

In personification an inanimate object or a group of persons, such as a tribe or nation, is spoken of as though it were a single person and is given human attributes. Thus in Hosea 11:1, "When Israel was a child I loved him, / and I called my son out of Egypt." The context makes it obvious that the nation of Israel is meant, not the individual of that name, and the reference of course is to the Exodus. In Isaiah 42:1 Israel is personified differently, as God's servant and the instrument of his will; and still differently in Jeremiah 3:6–13 and most notably Ezekiel 23, where the northern kingdom of Israel and the southern kingdom of Judah are personified as faithless wives who have turned to harlotry by worshipping other gods.

It is apparent that we are dealing with common metaphors of the relationship of God to humankind, just as we were in our discussion of allegory. The usual kind of allegory, which is narrative, has a dramatis personae, some of whom may be actual persons and some of whom may be abstractions or collective entities—but each of the latter will be transformed into a person for the sake of the allegory. These transformations may be routine: Much of the personification in the biblical text is traditional and contributes nothing important to its context. In the right hands, however, it is capable of enormous effect, as in the personification of Jerusalem in Lamentations 1:1–9 or—with even more eloquence—in Ezekiel 16, where the entire history of Jerusalem is laid out in allegorical terms from a bitterly critical point of view. Other striking effects of personification can be seen in the Psalms, for example, "Let the rivers clap their hands, / and the mountains shout for joy together . . ." (Ps. 98:8); or let "all the trees of the forest cry out for joy . . ." (Ps. 96:12); or "the mountains skipped like rams, / the hills like sheep" (Ps. 114:4). Obviously, it would be useless to judge these personifications for their credibility or literal appropriateness, for the whole point is that the event celebrated is not normal, and the poet's exuberant imagination makes certain that we share his feelings for it.

IRONY

Genesis 31 tells the story of Jacob's escape from his long servitude to Laban, his father-in-law. As Jacob prepares to depart in secret with his wives and children and all his movable possessions, Rachel steals her father's *teraphim* (household idols) and, unbeknown to her husband, hides them in the baggage. When Jacob's caravan is overtaken along the way by Laban ten days later, Laban reproaches Jacob for leaving unannounced and accuses him of having stolen the idols as well. This Jacob denies, not knowing of Rachel's trick. The tents are searched. Coming finally to Rachel's tent, the two men are, in fact, close to discovering the truth, but Rachel has cleverly concealed the idols in a camel bag and is sitting on it. She apologizes for not getting up, claiming that she is in her menstrual period; thus the camel bag is not searched, the idols are not found, and the two men depart, with Jacob indignant at having been falsely accused.

The point of this episode may have something to do with belittling the practice of idolatry, and it is certainly a chance to show that Jacob is not the only trickster in the family; but the ancient hearer of the story may well have enjoyed it mainly for the reason that we do, as a charming and effective example of dramatic irony. For we knew all along that Rachel had stolen the idols. So we were able to observe the proceedings—from above as it were—and appreciate both her quiet cleverness and the ignorance of the two men: one operating on suspicions that are more correct than he knows, the other displaying an injured innocence that he has no right to have. This is the situation diagramed in figure 3. Notice here, as in all such cases, that the audience (reader) has a superior point of view and sees the entire situation. The situation itself always involves two ideas or stages that are not just different but actually opposite and unreconcilable: Rachel cannot

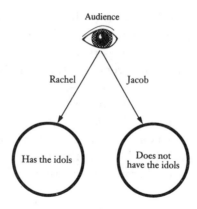

FIGURE 3. DRAMATIC IRONY

both have and not have the idols. The contrast between the audience's complete perception and the actors' partial one generates the irony, which is always tinged with satisfaction from the observer's superiority, the observer's omniscience. In this case Rachel, too, knows the full truth, but because she cannot watch herself, she cannot take the audience's point of view. She is involved in the action, not above it.

This kind of irony is traditionally called "dramatic" because it presents characters in action being observed by an audience, whether in a real play or in a narrative that readers can stage in their own imaginations. Because much of the most interesting Old Testament narrative was shaped with a listening audience in mind, it invites modern readers to do what the original listeners did many centuries ago: to participate in the narrative occasion, responding appropriately to the narrator's art. All that it requires is a little effort of the imagination. Thus if we place ourselves mentally in the world of Judges 3, we cannot help but feel the thrill of suspense as the hero, Ehud, arms himself in secret before his interview with the hated foreign king, Eglon. We delight in the way Eglon falls for Ehud's ruse and in the picture of the fat king, after the murder, lying on the floor of his roof-chamber in a pool of blood with the knife in his belly. As the crowning touch, we savor the dramatic irony of the following scene in which Eglon's servants, finding the doors locked, assume that the king is relieving himself (in the Hebrew text, "covering his feet") in the toilet and so wait respectfully below as the minutes pass by and Ehud makes good his escape.

If we move to the more sophisticated literary art that produced the book of Esther, there is a real feast of irony. Just to take one example: Having observed Haman plot to get rid of his enemy Mordecai the

Jew by having him hanged on a seventy-five-foot-high gallows, we are then taken into King Ahasuerus's bedchamber. There the king learns (by providential accident) that Mordecai once saved his life from some traitors and through oversight was never recognized for his service. At that moment Haman enters, planning to recommend the hanging of Mordecai. Before Haman can bring up this matter, the king asks, "What is the right way to treat a man whom the king wishes to honour?" (Esther 6:6). Supposing that the king must be referring to him (we, of course, know better), Haman recommends a public display of royal favor, whereupon the king tells him to "do everything you have just said to Mordecai the Jew" (Esther 6:10). With his vanity in ruins and his hatred of Mordecai even greater because he dares not mention it now to the king, Haman is forced to escort Mordecai in person through the ceremony of acclaim. From our superior position we have just watched a man destroy himself through ignorance combined with malice. That the king is unaware of the torment he has innocently caused his trusted adviser Haman is a further irony, making the principal one all the more delicious. Our feast is appropriately concluded when Haman is hanged on the very gallows he built for Mordecai.

The most challenging structure of irony in the entire Bible is the book of Job. One cannot easily, or perhaps ever, reduce it to a simple formula of the kind we have been using here. But at least readers should be made aware that the effect of the book depends fundamentally on its prose prologue, which establishes the dramatic irony by allowing readers to see the wager made in heaven between Yahweh and "the adversary" (the actual meaning of the Hebrew term usually translated "Satan"). If readers did not know this, they would be as much in the dark as Job is and might well assume, as Job does, that he is being unjustly punished rather than being tested. On the other hand, if Job himself knew that this was a test, the book could not even exist. His ignorance and the readers' knowledge alike are required.

Besides dramatic irony, which is a structural device, there is an irony purely of language, that is, an irony in which words are used in a double-edged way. Typically, the language is favorable or complimentary on the surface, but it is intended to have the opposite effect. When Elijah, in 1 Kings 18:27, suggests to the prophets of Baal that they "call louder" to get a response from their god, who has not manifested himself at the altar, Elijah's words superficially indicate sympathy for their embarrassment. In fact, he is mocking them, and they know it. The god will not appear because the god is impotent. When the cutting edge of irony is especially sharp and when the intention is exaggerated to the point of obviousness, we call it sarcasm.

Sarcasm advertises itself unmistakably. No one would make the mistake of supposing that Paul was attempting to flatter his audience when, in addressing the Corinthians (1 Cor. 4:10) he says, "Here we are, fools for Christ's sake, while you are the clever ones in Christ." When Yahweh finally speaks to Job out of the tempest, he brushes aside Job's case not only by demonstrating the comparative insignificance of this human creature but also by drenching his language with sarcasm: "Where were you when I laid the earth's foundations? / Tell me, since you are so well-informed! / Who decided its dimensions, do you know? / . . . If you do know, you must have been born when they were, / you must be very old by now!" (Job 38:4–5, 21). There is no subtlety here, nor should there be any. Yahweh's speech is meant to silence Job, and it does so.

WORDPLAY

If we were reading the Bible in its original languages, the section on puns, assonance, onomatopoeia, verbal patterning, and other such devices might well be the longest in this chapter. But there is no point in discoursing at length on a topic that is invisible to modern readers, except perhaps to ask them to take the evidence for it on faith. Ordinary translation is difficult enough; to carry over not only exact meanings but patterns of sound and verbal structures as well is usually impossible. If there is any choice, the pattern yields to the meaning, as it should.

Still, one hates to lose the effect of something like the Hebrew *adam la-hebel damah* in Psalm 144:4, where the sounds of *adam* ("man") are reversed and recombined in *damah* ("is like, resembles"). This gives the melancholy thought a dimension of meaning that is not communicated in the English of the NJB translation, "Human life, a mere puff of wind," since the Hebrew word-structure carries the suggestion of a circularity decreed by Providence. Again, the NJB rendering of Ecclesiastes 7:1, "Better a good name than costly oil," conveys the basic sense of the Hebrew *tov shem misshemen tov* (if one understands "name" to signify "reputation"); but it does not convey the pun on *shem-shemen* or represent the chiastic (crosswise) structure of the clause: literally, "Good / name // is-more-than // ointment / good." The resonance of the original proverb, with its implied criticism of those who place their values on material things, as well as its compactness and the ease with which it sticks in the memory, are all sacrificed in the transition to another language.

It is well known that the more ancient Old Testament narratives contain many puns (a good study edition of the Bible will call attention to them). Some puns are etiological; for example, in Genesis 11:9

"Babel" is derived from the Hebrew *balal* ("he mixed"). Many other puns connect the name of someone with an essential characteristic or prophetic circumstance; for example, Jacob *(ya'akov)* puns both on *akev* ("heel") and the verb *akav*, which means "to circumvent or over-reach" (Jacob, a twin, came into life grasping Esau's heel and eventually supplanted him in the elder brother's role). One of the most systematically pursued of these puns is that on the name of Isaac *(yitshaq* in Hebrew—the dot under the letter *h* indicates a rough pronunciation, as in the German *ach),* which is understood to have been derived from *tsahaq* ("to laugh, to play with"). Three times in the narrative, in Genesis 17:17, 18:12, and 21:6, laughter is connected with the birth of Isaac. Other puns seem to exist for the more general purpose of enriching meaning. Before their transgression, Adam and Eve are "the naked ones" *(arumin,* from *erom,* meaning "naked"), but they do not know that they are naked. The serpent, on the other hand, is knowing *(arum)*—but then he, too, in the view of the author and audience of this story, is naked, having neither fur nor feathers. And in Genesis 27:36, the tricked Esau cries that his brother is rightly named the supplanter, for he has stolen not only "my birthright" *(bekorati)* but also "my blessing" *(birekati).*

The most famous wordplay in the Old Testament—whether it is actually a pun is arguable—involves the very name of Yahweh, which in Exodus 3 is related in a complicated fashion to the verb *hayah* ("to be"). Another play on names, equally famous, occurs in the New Testament in Matthew 16:18, where Jesus is reported as saying to Peter, "You are Peter and on this rock I will build my community."* This is a double pun because it works in the two original languages: It plays on the similarity of Peter's Greek name, "Petros," to the Greek *petra* ("rock") as well as of Peter's Aramaic name, "Kephas," to the Aramaic *kepha* (also "rock"). Jesus' actual words would have been in Aramaic, the language he spoke, but they were translated into Greek before being included in Matthew's gospel.

POETRY

It has never been a secret that the Old Testament contains poetry, but until comparatively recent times no one suspected how much. (In fact, about a third of it is poetry, and few of its books contain no poetry at all.) The only portion that readers in the past would normally have called poetry is the book of Psalms—and this because the psalms were presented as songs to be sung, not be-

---

*The Greek word *ekklēsia,* here rendered "community," is translated as "church" in most English Bibles.

cause there were any identifying characteristics in their texts, which were (and sometimes still are) printed with lines run together as if they were prose. The problem came about because Hebrew poetry had no formal device like our rhyme to mark the ends of poetic lines and because its rhythm was too fluid to settle into patterns that unmistakably announced the presence of verse. Separated by centuries from its authors, readers and translators of this poetry had no way of recognizing it for what it was. The key to unlock this treasure store was finally provided by Bishop Robert Lowth in 1753 in his *Lectures on the Sacred Poetry of the Hebrews*. To a large extent we still depend on Lowth's original insights. The key to Hebrew poetry, he found, is that it is a structure of thought rather than of external form and that a Hebrew poem is composed by balancing a series of sense units against one another according to certain simple principles of relationship.

These sense units are formed into phrases or clauses, often complete sentences, with obvious grammatical coherence. In modern versions of the Old Testament, they are arranged into lines as the meaning of the original seems to require. Because we do not know how Hebrew poetry was arranged in the times when it was first written down, these reconstructions have no real authority. But they offer something with which to work, and they do immediately signal to the eye that one is looking at a poem, not at a piece of prose.

The general term for the relationship between these units is "parallelism." Of the several types of parallelism found in Hebrew poetry, the simplest consists in the repetition of the same thought in different words. From one unit to the next, the only change is that of language. Hence this type is called "synonymous." For an example of synonymous parallelism, we can look to Psalm 8, two verses of which are diagramed in figure 4. The sense of the first line, "I look up at your heavens, shaped by your fingers," is restated in the second line by substitution: "the moon and the stars" equals "your heavens," and "you set firm" equals "shaped by your fingers." In the second distich (the technical name for the two-line group) we find the same pattern, as figure 4 shows.

The modern reader, confronted with this sort of thing, has to make adjustments. Our own literary forms do not encourage repetition; still less are they built on it. But the Hebrew poet thought otherwise and worked within a different tradition. A modern poet, having said something, will be anxious to urge his or her composition forward to the next stage (perhaps with memories of high school or college papers handed back with "Rep" scrawled in red in the margins).

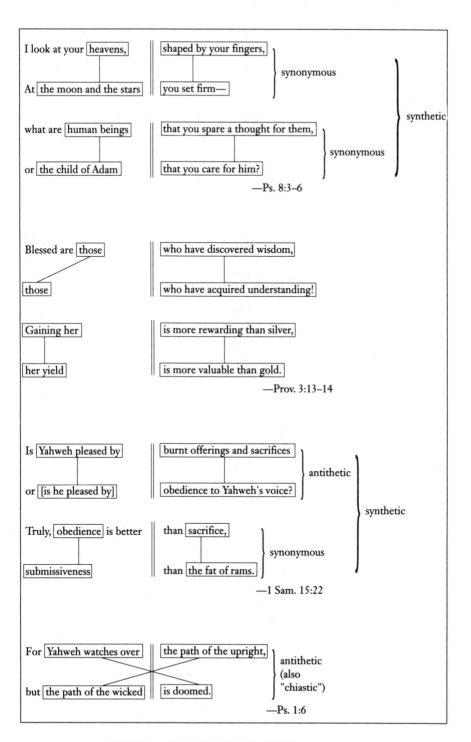

FIGURE 4. PARALLELISM IN HEBREW POETRY

The ancient Hebrew poet seems to have been in no hurry; if a thought was truly important, it could not be exhausted in one statement. Turning it in the hand and viewing it from different angles, as it were, the Hebrew poet could more fully demonstrate its latent significance.

Returning to the example from Psalm 8, we can see that the second distich also shows synonymous parallelism, inasmuch as "human beings" is equivalent to "the child of Adam," and "spare a thought for them" is equivalent to "care for him." The two distichs are connected by the pronouns "your" and "you," but we can also see that there is a logical relationship between them, the second one supplying the thought that comes to mind as a *result* of looking up at the heavens: they are related as cause and effect (that is, "This is what happens when I look up at the heavens"). Units related by logic or by the forward movement of the poet's thought, like this one, are obviously parallel in a different way. The term "synthetic" has been given to this type of parallelism. In the example from Proverbs 3, the pattern seen in Psalm 8 is repeated almost exactly. Two distichs are constructed by synonymous parallelism, the second one related to the first by synthetic parallelism (it tells why wisdom makes someone blessed). The only difference is that here the key term "blessed," which controls all that follows, stands outside the pattern as such.

The third major type of parallelism is "antithetic"; this occurs where a unit offers a thought that denies or provides an exception to the preceding one, as in the example from 1 Samuel 15:22 in figure 4. Here the overall opposition is between "sacrifices" and "obedience," and this is set up in antithetical fashion in the first distich. The second distich, itself constructed by synonymy, answers the first one by vigorously excluding sacrifice as a means of serving the deity. Note that "obedience" functions as a sort of hinge or pivot between the two distichs. The example from Psalm 1:6 in figure 4 also shows antithetic parallelism. Its four elements are arranged in a chiastic pattern, which gives a little additional emphasis to the distich because the thought is not completed until the last of the four elements falls into place. If the author had wanted to use synonymous parallelism here, he might have written for the second line something like, "Yahweh protects good men from evil." Had he wished to use synthetic parallelism, he might have written, "And causes their enemies all to perish."

Other types of parallelism are offshoots or variations of these basic ones. Two in particular are worth defining and illustrating. "Emblematic" parallelism is a variety of synonymous parallelism in which the thought is expressed half literally and half metaphorically:

> A golden ring in the snout of a pig
>> is a lovely woman who lacks discretion. (Prov. 11:22)

> Like apples of gold inlaid with silver
>> is a word that is aptly spoken.
> A golden ring, an ornament of finest gold,
>> is a wise rebuke to an attentive ear. (Prov. 25:11–12)

"Climactic" parallelism uses the method of synonymy to build up a thought by the repetition of short phrases toward some sort of climax, for example, in the prophet Zephaniah's description of the Day of the Lord:

> That Day is a day of retribution,
> a day of distress and tribulation,
> a day of ruin and devastation,
> a day of darkness and gloom,
> a day of cloud and thick fog,
> a day of trumpet blast and battle cry
> against fortified town
> and high corner-tower. (Zeph. 1:15–16)

Perhaps the most famous use of climactic parallelism in the whole Bible is in Deborah's victory song, celebrating the murder of Sisera by Jael:

> She reached her hand out to seize the peg,
> her right hand
>> to seize the workman's mallet.
> She hammered Sisera,
>> she crushed his head,
> she pierced his temple and shattered it.
> Between her feet, he crumpled,
>> he fell, he lay;
> at her feet, he crumpled, he fell.
> Where he crumpled,
>> there he fell, destroyed. (Judg. 5:26–27)

The Hebrew word for "destroyed" is not only stronger than any of the words preceding it but is also not anticipated by repetition and thus makes a very effective climax.

This passage from Deborah's song incidentally illustrates a problem that Hebrew poetic parallelism created for readers either unfamiliar with this device or inclined to take everything they read very literally, for the prose account of the murder in chapter 4 has Jael hammering the tent peg into Sisera's skull as he lay sleeping. The song more plausibly suggests that she simply hit him over the head with the heavy peg and killed him by fracturing his skull. But the author of the prose account, using this ancient poem as his source, assumed that the hammer and the tent peg were two separate instruments instead of poetic equivalents for the same one; thus he described their use in what seemed to him the logical way. The best-known example of this kind of misreading is Matthew's treatment (21:4–5) of Zechariah 9:9, a messianic prophecy that the gospel writer drew on in his account of Jesus' entry into Jerusalem. The prophet Zechariah describes the king entering Jerusalem as

> humble and riding on a donkey,
> on a colt, the foal of a donkey.

Ignoring the parallelism (the donkey and colt are the same beast), Matthew has the disciples bring two animals to Jesus, both of which Jesus proceeds to ride at the same time* (a detail not found in the other synoptic gospels).

It should be evident even from these few examples that parallelism is a device of great power. What is not so evident is the range of possibility that it offers, the marvelous variety that Hebrew poets were able to create within the apparently narrow limits that it set on poetic form. One begins to see this rather slowly, only after first mastering the basic principles we have outlined. Much of the best Hebrew poetry, such as that in Job, the psalms, and the prophetic oracles, nearly escapes the patterns altogether. Reducing it to a diagram is not easy and often impossible. Yet the sense of underlying pattern is always there, much as it is in the playing of traditional jazz, where the unvarying foursquare beat on which the music is built may never come to the surface as such, and where the art of the performers seems often to be gauged by the extent to which they can depart from this beat without ever forgetting their way back to it. This contest between rule and freedom is a polite one, which neither side wishes to win but both strive not to lose.

---

*"This was to fulfil what was spoken by the prophet: Say to the daughter of Zion: / Look your king is approaching, / humble and riding on a donkey / and on a colt, / the foal of a beast of burden. . . . So the disciples . . . brought the donkey and the colt, then they laid their cloaks on their backs and he took his seat on them" (Matt. 21:4–8).

III

Ancient Near Eastern Literature and the Bible

The preceding chapters of this book have made the case that the Bible is an anthology—a collection of works written by many hands over many centuries—and that evident in those works are the literary forms and strategies familiar to us from our reading outside the Bible. What remains to be considered along this line is the extent to which biblical authors were prompted to write as they did through the influence of older literature. It is no secret, of course, that *within* the Bible the later material reflects the earlier—indeed, that some of the later books and portions of books were written to update or extend or counter earlier material. But beyond this internal influence of biblical writers upon one another there was the influence of literature outside the Judaeo–Christian tradition, literature reflecting ideas and employing forms and strategies that were the stock in trade of all the literary cultures of the Near East.

The existence of this outside influence should be of no surprise—after all, individual literary cultures never spring into being in a geographical vacuum. But surprising to say, its existence was scarcely even guessed at until relatively late in the history of biblical scholarship. The problem was that, given the theological assumption that the Bible was unique, there was little awareness before the early nineteenth century that there might have *been* any literature outside the Bible before biblical times. That ignorance derived, in turn, from an ignorance of what form an ancient Near Eastern work of literature might have taken and, more basically, an ignorance of the languages in which it might have been written. That there were carvings on monuments and rock surfaces in the Near East had long been known, but it was far from certain that these were scripts representing languages. It was thought that Egyptian hieroglyphs ("sacred carvings") were only decorative devices or perhaps symbols of spiritual concepts but in any case not a form of writing. The language spoken by the Pharaohs and their subjects was completely unknown.

DECIPHERING ANCIENT
NEAR EASTERN
LANGUAGES

In 1799, during Napoleon's campaign to conquer Egypt, a French expeditionary unit discovered in the small village of Rosetta the key that would unlock the language and literature of ancient Egypt: a basalt slab now known as the Rosetta Stone. It was inscribed from top to bottom in three different scripts. The first was a passage in Egyptian hieroglyphics, the second in what is now called "demotic Egyptian" (a cursive form of hieroglyphics), and the third in Greek. Scholars of the time could of course read the Greek text, and they correctly assumed that the other two blocks of text said the same thing as the Greek. The demotic block was deciphered relatively easily, but the hieroglyphics resisted interpretation until a brilliant young French linguist named Jean-François Champollion undertook the task. After many false starts and fifteen years of labor, Champollion announced his findings in 1822. The Rosetta Stone, he said, carried a text honoring King Ptolemy V Epiphanes; it had been carved in 196 B.C.E., during the historical period between the composition of the two biblical Testaments. Champollion's decipherment of Egyptian writing, which met with skepticism at first, was confirmed in 1866 by the discovery of a second ancient inscription in two languages, the Decree of Canopus. During the succeeding century, building on the work of Champollion and his successors, Egyptologists produced massive grammars and dictionaries, even of ancient Egyptian dialects, thus opening a world of literature that would otherwise have been lost to us.

Recovering the dead languages of Mesopotamia proved to be a more formidable task, but again the key lay in trilingual inscriptions: the same text presented in three different languages. Scattered throughout what are now the countries of Iraq and Iran are numerous monuments and cliff faces carved with a cuneiform script. But unfortunately for the scholars attempting to decipher them, none of these bore any familiar Greek that they could go by: Both the characters and the languages they represented were mysteries. One of these inscriptions, a particularly impressive example, had been carved on a cliff face at Behistun, in northwest Iran, by Darius the Great, ruler of Persia in the seventh/sixth century B.C.E. Between 1835 and 1837, Henry Rawlinson, a British military adviser to the Shah, began the difficult task of copying it—difficult because Darius had destroyed all means of access to the cliff face in order to preserve the inscription from vandals. With the aid of ladders, rock-climbing equipment, and daredevil youths, Rawlinson finally got the text transcribed; but then the still more difficult task of decipherment began. By assuming that the record included proper nouns—kings have a fondness for repeating their own names as often as possible—and working from known data about Near Eastern history, Rawlinson was able to assign phonetic values to some and finally all of the Old Persian portion of the text. Eventually the three languages on the cliff face turned out to be Old Persian, Elamite, and Babylonian, the three of them quite dissimilar but written in what appeared to be the same script. Rawlinson's interpretations were confirmed by other scholars in 1857, after which one might say the floodgates opened and the discipline of modern Assyriology—so named from the great interest in "Assyrian" inscriptions in the early days—began.

Cuneiform script had been invented by the Sumerians of southern Mesopotamia sometime around 3000 B.C.E. Pressing the end of a small wedge into a soft clay tablet (see appendix 2 for a discussion of this procedure) was the simplest and potentially the most durable way of recording language at that time, and as a result the Sumerian invention became the vehicle for writing a number of languages besides the three already mentioned. These included Hurrian, from the northern part of the fertile crescent; Hittite, from Anatolia (modern Turkey); Urartean, from ancient Armenia; and Eblaite (discovered as recently as 1974), from ancient Syria. The writing on the Ugaritic tablets found in 1929 near the Phoenician coast and deciphered in 1930 proved to be an alphabetic use of the cuneiform script (which was otherwise employed to represent spoken syllables and not letters).

THE SEARCH FOR DOCUMENTARY MATERIALS   Once the ancient languages were yielding their secrets, archaeologists got to work to provide documentary materials for interpretation. Between 1842 and 1845 Richard Lepsius led an expedition to Egypt and the Sudan and eventually published twelve colossal volumes detailing the inscriptional and architectural material he found there. This great work (*Monuments from Egypt and Ethiopia* is its English title) was the first major study to combine a knowledge of the Egyptian language with on-site inspection of the monuments. In Mesopotamia the search for the capital of Assyria, Nineveh (a city mentioned in the biblical books Jonah and Nahum), began when Paul Emile Botta dug at Khorsabad in the 1840s. Actually, he was at the site of the royal city (Dur-Sharrukin) of Sargon II, who is mentioned in Isaiah 20:1. In 1849 Henry Layard, who was also looking for Nineveh, published the results of his digging at Nimrud, supposing it to be Nineveh, but the true site was found by Rawlinson three years later. In any case Layard's claim to fame is his discovery at Nimrud of the Black Obelisk of Shalmaneser III, monarch of Assyria from 858 to 824 B.C.E. This is a paneled bas-relief sculpture showing foreign princes bringing tribute to the great king, with inscriptions explaining the contents of the pictures. Near the top is a panel with a foreign ruler abjectly bowing before Shalmaneser as if to kiss his feet. With Rawlinson's translation of the inscription in 1850, the world learned the identity of the subject ruler: "Jehu, son of Omri." Here, for the first time, was a contemporaneous extra-biblical reference to a king of Israel, and not only a reference but a portrait as well!* Suddenly the Bible was thrust onto the stage of human history, and its position as a privileged documentary record was effectively ended.

One hundred and fifty years later, documents and artifacts are still being retrieved from the ancient Near East (though no other contemporary portraits like that of Jehu have been found), and the end is nowhere in sight. There is now an immense body of extra-biblical literature—more than half a million documents from the ancient Near East, many of which yet remain to be studied.

THE IMPACT OF THE RECOVERED NEAR EASTERN LITERATURE

The nineteenth century, the period in which the great work of recovering ancient languages and literatures began, was a time of immense intellectual ferment. The Mosaic authorship of the Pentateuch was very much in question; and scholars, particularly in Germany, were beginning to look critically at other aspects of the biblical text with a view to separating its mythical elements from whatever core of historical fact it might contain. Charles Lyell's *Principles of Geology* (1830–33) made it difficult to believe that the world had come into being during one week in 4004 B.C.E., as the Genesis

---

*In time scholars would learn the full significance of that inscription and why Jehu is identified in it as "the son of Omri" whereas the Bible knows him as "Jehu, son [or grandson] of Nimshi" (compare 2 Kings 9:2 and 9:20).

story was supposed to require. Not that the Bible text itself bore any date, but anyone working from genealogical tables, such as that in Luke 3, and adding together the ages of the patriarchs (see, for example, Genesis 5) as Archbishop James Ussher did in 1650, could make such a claim. (The 4004 figure became famous by being included in the notes to many editions of the King James Version.) If this date were an impossibility, could the Bible be believed when it told of other events and places? Under such circumstances it was inevitable that pious readers would seize upon the recent archaeological discoveries as a way of supporting what the Bible said, of reinforcing the reality of the world that was portrayed in its pages. Hence, at first the new evidence tended to be studied mainly for the light it cast upon the Bible, even if this meant using data selectively or distorting them; and the archaeologists themselves often tried to fulfil these expectations in their work, thus compromising its scientific value.

But the newly recovered documents that had been used initially to "prove" the truth of the Bible were soon being examined with more objective eyes. Consider the case of the nineteenth-century Englishman George Smith, with whom the modern comparative study of the Bible and ancient Near Eastern literatures can be said to have begun. Like so many of these pioneers, Smith had a remarkable linguistic aptitude (someone called him an "intellectual picklock" for his uncanny ability to read Assyrian inscriptions). He was an outsider, almost entirely self-taught, but such was his talent that the British Museum hired him to help in publishing the material now arriving from Mesopotamia. In the tablets that had been dug up at the site of Nineveh, Smith encountered a familiar story, which he presented to the newly formed Society of Biblical Archaeology late in 1872 in a lecture entitled "Chaldean Account of the Deluge." Though incomplete, the story was clearly that of a world-wide deluge like Noah's flood—but from a source outside the Bible! This sensational find inspired a hasty return to Nineveh to search for more documents, this time with Smith himself in the party; within half a year the missing fragments of the story were found.* Initially, the discovery was welcomed by traditionalists as confirming the reality of Noah's flood, an event that had increasingly come under suspicion in the new biblical criticism. Before long, however, it became clear that these newly

---

*Earlier Smith had also found the prototype of the story of a baby who is cast adrift on a river by his mother—"as Jochebed did the infant Moses on the Nile," Smith drily observed—but who nevertheless is rescued and grows up to become the leader of his people. In this version he is Sargon I, King of Akkad around 2000 B.C.E.

found flood stories were not independent journalistic accounts of a single, actual historical event but works of literature, complete with plot, character, dialogue, motivation, and point of view. Their historicity, if any, was irrelevant. We have since learned that the flood story goes back to a Sumerian original (dated about 2200 B.C.E.) and was told in many different Near Eastern cultures, with the substitution of local heroes and deities as characters. But even in Smith's day it was apparent that a wholly new category of ancient writing had been opened up. While much of the archaeological material could be tied to historical moments—like the monuments to Ptolemy V and Darius the Great or the thousands of records of business transactions preserved on cuneiform tablets—another considerable portion of it belonged to the timeless world of myth and legend.

Late-nineteenth-century scholars, led by the always industrious Germans, began publishing collections of ancient Near Eastern texts "parallel" to biblical literature. The classic compilation in our day was published originally in 1950 (and since much enlarged in repeated editions) by James Pritchard as *Ancient Near Eastern Texts Relating to the Old Testament* (usually represented by the acronym *ANET*). The material has now become so vast that more recent publications have tended to focus upon particular areas, like Walter Beyerlin's *Near Eastern Religious Texts Relating to the Old Testament* (abbreviated as *NERT*; English edition, 1978) and Victor Matthews' and Don Benjamin's *Old Testament Parallels: Laws and Stories from the Ancient Near East* (1991).

BIBLICAL PARALLELS
WITH NONBIBLICAL
LITERATURE

Before going further, we should comment on the terms "parallel" and "relating to" as used above. To say that two stories are parallel implies that they can legitimately be printed in adjacent columns as a means of calling attention to detailed similarities between them. What it does not imply is anything about the nature of that relationship: "there it is—make of it what you will" is the only implication. At one extreme the resemblance might be purely accidental, and at the other it might have resulted from deliberate copying. The laws of chance make accidental resemblance less and less likely as the burden of evidence mounts, so that in the end it cannot be taken seriously as a explanation; and "copying" or plagiarism is a category that is not relevant here, for reasons we shall explain at the end of this chapter.

We are going to dismiss both of these possibilities, therefore, in favor of an intermediate position. What we are looking at in these parallel ancient texts might best be called "sharing of tradition"—a sharing that extended over centuries, originating probably in oral literature now lost to us forever, and reflecting similar ways of looking at the world, ways generated in overlapping or interdependent cultures. Over the several thousand years we have evidence for in the Near East, dynasties came and went, languages evolved, populations grew and died away—but all within a geographically circumscribed area where it would have been impossible for a given culture to develop without external influence. In what remains of this chapter, we shall discuss the major categories of tradition common to the Bible— the Old Testament in particular—and the literature of the ancient Near East.

NARRATIVES

Perhaps the most famous biblical/non-biblical parallel is that involving the flood story, discovered by George Smith. The basic structure of the story is given in the Sumerian myth "The Deluge," written down, as indicated earlier, in the late third millennium B.C.E. From the badly broken cuneiform tablet containing this story, it appears that the gods have decided to send a flood to destroy "the seed of mankind." Ziusudra, a particularly pious man who is attentive to revelation from the gods, is chosen to survive the flood. After he builds a "huge boat," the flood sweeps over the land for seven days and seven nights. Finally Utu, the sun god, appears, whereupon Ziusudra sacrifices an ox and is rewarded for his obedience with eternal life (his name means "life of long days"). In the early second millennium B.C.E., this story was incorporated into the Akkadian "Atrahasis epic" with certain details added, such as including the survivor's family among the passengers in the boat. Shortly thereafter the story was integrated into the even longer "Gilgamesh epic," in which form it was spread throughout the Near East.

Until archaeology and the recovery of ancient languages made it possible to go behind biblical narratives, there was no way for a reader of, say, Genesis 8:6–12 to know that the author was drawing upon an older narrative tradition for details in his story. Notice the parallels in the folllowing passages:

GENESIS 8:6–12:

At the end of forty days Noah opened the window he had made in the ark and released a raven, which flew back and forth as it waited for the waters to dry up on earth. He then released a dove, to see whether the waters were receding from the surface of the earth. But the dove, finding nowhere to perch, returned to him in the ark, for there was water over the whole surface of the earth; putting out his hand he took hold of it and brought it back into the ark with him. After waiting seven more days, he again released the dove from the ark. In the evening, the dove came back to him and there in its beak was a freshly-picked olive leaf! So Noah realised that the waters were receding from the earth. After waiting seven more days, he released the dove, and now it returned to him no more.

GILGAMESH TABLET XI, 145–54:

When the seventh day arrived,
I sent forth and set free a dove.
The dove went forth, but came back;
Since no resting-place for it was visible, she turned round.
Then I sent forth and set free a swallow.
The swallow went forth, but came back;
Since no resting-place for it was visible, she turned round.
Then I sent forth and set free a raven.
The raven went forth and, seeing that the waters had diminished,
He eats, circles, caws, and turns not round. (*ANET*, 1955, 94–95)

Since the detail about sending out birds from the ark is found in none of the earlier narratives except the Gilgamesh epic, we know that this is the version adapted for the Hebrew Bible, where all the key elements of the tradition are found: (1) deciding to send a flood to wipe out life on earth, (2) selecting a worthy man to survive, (3) building a boat, (4) riding out the storm in the boat, and (5) offering sacrifice on dry land at the end. While each version of the story is distinctive, most noticeably in the name given to its central human character, the two clearly belong to a shared tradition. There are many unique details in the biblical story of the deluge, but the basic story is one held in common with a number of earlier cultures in the Near East.

The use of a shared tradition, and especially its adaptation to the new use, is perhaps best shown in the creation story of Genesis 1. This is a reworking of the Babylonian creation story "Enuma Elish," sometimes called the "Babylonian Genesis." The first part of this work contains the Babylonian version of the victory over the River (Apsu) and the Sea (Tiamat). Tiamat is personified as a dragon that is resolved to take vengeance on the gods for killing her mate, Apsu. While the other deities cower before the threat of the mighty dragon, the god Marduk vows that he will defeat Tiamat, but on one condition: that the others acknowledge him as supreme being. They agree, he proceeds to defeat Tiamat, "split[s] her like a shellfish into two parts," and from this division creates heaven and earth (*ANET*, 1955, 67).* It is clear that a very ancient tradition existed concerning the need for the creator to conquer an adversary, who then became in a sense the material basis for the creation. But the Hebrew authors of Genesis 1, writing in the relatively sophisticated times of the sixth or fifth century B.C.E. and with a strong commitment to monotheism, edited out of the story any reference to gods other than the single, creating deity and reduced the adversary to the rather ghostly *tehom*. We are entitled to say this much because elsewhere in the Hebrew Bible there is ample evidence of the older belief, for example Psalm 74:12–14:

> Yet, God, my king from the first,
> author of saving acts throughout the earth,
> by your power you split the sea in two,
> and smashed the heads of the monsters on the waters.
> You crushed Leviathan's heads. . . .

This passage closely resembles the wording in a Canaanite poem that speaks of Baal as one who will "smite Lotan [i.e., Leviathan], the serpent slant / Destroy the serpent tortuous, / Shalyat [i.e., tyrant] of the seven heads" (*ANET*, 1955, 138).

The tradition of the deity in opposition to a primeval monster, a sea serpent or a personification of the ocean, was almost certainly mediated to Hebrew writers by Canaanite literature. Other biblical instances can be found in Isaiah 27:1, Job 7:12 and 26:12–13, Psalm 77:16, and Amos 9:3.** In addition to sharing this element, the two creation traditions are parallel in the order of created elements they have in common: (1) light, (2) "firmament," (3) dry land, (4) lights in the heavens, and (5) human beings—followed by rest and celebration.

The parallels with ancient Near Eastern texts that we are consider-

---

*In Genesis the antagonist is a watery chaos, called in 1:2 "the deep," *tehom* in Hebrew. Hebrew *tehom* and Akkadian *tiamat* are from the same Semitic root word. And the Hebrew word is used in Genesis without a definite article, which suggests that it was at one time understood as a proper name (that is, "and darkness covered the surface of Deep").

**The tradition was plastic in the sense that it could also be transferred from an initial conflict to a final one, as in Isaiah 27:1–2: "That day Yahweh will punish, / with his unyielding sword, / massive and strong, / Leviathan the fleeing serpent, / Leviathan the coiling serpent; / he will kill that dragon that lives in the sea."

ing are not limited to stories of the activities of deities: human beings are also involved. One of the most famous parallels of the latter kind is that between the story of Joseph and Potiphar's wife (Genesis 39) and an Egyptian antecedent known as the "Story of Two Brothers," the earliest manuscript of which dates from the mid-thirteenth century B.C.E. In the biblical story, Potiphar's wife importunes the handsome young Hebrew slave to have intercourse with her; he refuses out of a sense of loyalty to her husband, his employer; she accuses him of rape; he is imprisoned. The Egyptian story has the same characters (differently named, of course) and the same action, except that the young man is not imprisoned: rather, he "proves" his innocence by taking a knife and cutting off his own penis—a detail obviously unsuitable to the Hebrew writer, who was telling the story of a young man who was to become father of two of the tribes of Israel.

LEGAL FORMS

One of the principal reasons for honoring the ancient Near East as the cradle of civilization is that, without being in the least democratic, these nations attempted to promulgate the rule of law. Law codes—coherent sets of rules governing conduct, with specified penalties for disobedience—can be traced as far back as the two Sumerian codes, "Ur-Nammu" (about 2050 B.C.E.) and "Lipit-Ishtar" (about 1900), and the Babylonian "Eshnunna Code" (about 1900) and the famous "Code of Hammurabi" (about 1750). Later codes were the Hittite (sixteenth century B.C.E.), Assyrian (fifteenth century), and neo-Babylonian (seventh century). Curiously, no extensive law codes from Egypt have been preserved—whether by chance or because of something in the Egyptian temperament that made law codes uncongenial we do not know.

The Hebrew Bible contains several law codes, along with many passages of some other nature that presuppose the existence of such codes (for example, Isaiah 50:1 on the selling of children). The principal ones are found in Exodus 21–33 and Leviticus 17–26. The former is now known among scholars as the "Book of the Covenant" and the latter as the "Holiness Code." In addition, most of the book of Deuteronomy, excluding the prologue in chapters 1–11, is a legal code. These biblical codes are part of a shared tradition. Like their Mesopotamian predecessors, they are generally concerned with cultic activity, social justice, property (including slaves), sexual relations, and family matters.

The regulation that we recognize today as most "biblical" is surely the so-called *lex talionis* or law of retaliation: "life for life, eye for eye, tooth for tooth, hand for hand, foot for foot, burn for burn, wound for wound, stroke for stroke" (Exod. 21:23–24). But the quid-pro-quo form of the biblical law was merely following precedent. The *lex talionis* goes as far back as the Code of Hammurabi in the eighteenth century B.C.E.:

> If a seignior [i.e., commoner] has destroyed the eye of a member of the aristocracy, they shall destroy his eye. If he has broken another seignior's bone, they shall break his bone. . . . If it [i.e., the work of a builder] has caused the death of a son of the owner of the house, they shall put the son of that builder to death. (*ANET*, 1955, 175, 176)

The *lex talionis*, it should be pointed out, does not necessarily require death or mutilation: The codes that preceded that of Hammurabi were more lenient, stipulating that one could compensate for causing a personal injury by paying a fine.

Occasionally the resemblance between a prescription in the Bible and one in an earlier legal corpus is so close as to suggest actual borrowing. Compare "If anyone's ox injures anyone else's ox causing its death, the owners will sell and share the money for it; they will also share the dead animal" (Exod. 21:35) with this statement in the laws of Eshnunna: "If an ox gores another ox and causes its death, both ox owners shall divide among themselves the price of the live ox and also the equivalent of the dead ox" (*ANET*, 1955, 163).

In the ancient world the most fundamental legal form was the covenant, a solemn agreement between the ruler and his subjects; upon it all other laws were based. Unlike the U.S. Constitution, the covenant (or "suzerainty treaty," to give it the technical label) was imposed from above, but it did require the ritual assent of those whom it governed and it did remind them of the benefits that would accrue to those who observed it. In Hittite treaties, for example, we find a stereotyped structure: (1) preamble—identification of the sovereign; (2) historical prologue—why the king and his vassals are in this relationship; (3) stipulations—what is expected of the vassals; and (4) blessings and curses—what will happen if the covenant is observed or not observed. The covenant between Yahweh and Israel, given to Moses, is quite in line with this tradition. In Exodus 20 we find: (1) "I am Yahweh"; (2) I "brought you out of Egypt, where you lived as slaves"; and (3) "You shall have no other gods to rival me." The fourth element, blessings and curses, became detached from the Exodus context and shows up in Leviticus 26 and Deuteronomy 28.

Here we encounter a potential for confusion arising out of terminology, since the discussion in this chapter is about the parallels between Hebrew and earlier Near Eastern literature. That is parallelism in a general sense. But we must also speak (as we did in chapter 2) of a more particular kind of parallelism, that between two successive lines of Hebrew poetry. This basic literary device had been developed in earlier cultures and was adapted by Hebrew writers to meet their own needs. In the hands of prophetic poets like the man called "Deutero-Isaiah" (see chapter 8) and some of the contributors to the Psalms, this form served as the vehicle for great poetry and was manipulated with subtlety and force. But it was not original with them: The ground rules had been established many centuries earlier in Mesopotamia. At least a thousand years before any of the Psalms could have been written, an anonymous Sumerian poem described the descent of the goddess Inanna to the underworld this way:

> From the "great above" she set her mind toward the "great below,"
> The goddess, from the "great above" she set her mind towards the "great below,"
> Inanna, from the "great above" she set her mind toward the "great below."

And later in the same poem, with less rigidity of form:

> O father Enlil, let not thy daughter be put to death in the nether world,
> Let not thy good metal be ground with the dust of the nether world,
> Let not thy good lapis lazuli be broken up into the stone of the stoneworker,
> Let not thy boxwood be cut up into the wood of the wood-worker,
> Let not thy maid Inanna be put to death in the nether world.
>   (*ANET*, 1955, 53–54)

Canaanite poetry, too, employs this device of literal repetition, as in the following lines from "The Legend of King Keret" (dating from the fourteenth century B.C.E.):

> Muster the people and let it come forth,
>   The host of the troops of the people.
> Yea, let come forth the assembled multitude,
>   Thy troops, a mighty force:
> Three hundred myriads;
>   Serfs without number,
>   Peasants without counting. (*ANET*, 1955, 143)

The poetic technique here, parallel repetition, with its effect of reinforcing meaning, is precisely the one that shaped the poetry of the Hebrew Bible. And it was from their Near Eastern neighbors that Hebrew writers learned the technique.

HISTORIES

The historical writing in the Old Testament will be fully explored in chapter 4. This writing is distinct in many ways, not least in that its major connected narratives, like those in the Pentateuch, in Samuel–Kings, and in Chronicles, are (as far as we know) without precedent in the ancient Near East. Yet the authors in all these cases employed materials and techniques and had purposes that closely parallel those of earlier historians in other cultures.

One of the material sources of history in those days was annals: simple bookkeeping records, systematically compiled, of what happened when—since after the invention of writing it had become possible to support fallible human memory with something more permanent. The impetus and means for keeping such records typically came from royal functionaries conducting the business of government, which would have been in chaos without some reliable way of recording and dating events of public importance. (The vanity of kings, who wished to have their exploits immortalized, was an important factor, too.) Once Israel became a nation, the writing of annals was undertaken there as well. For example, that the author or authors of the so-called "Deuteronomic History" in Samuel–Kings had annals to draw upon is perfectly clear from passages such as this: "The rest of the history of Solomon, his entire career, his wisdom, is this not recorded in the Book of the Annals of Solomon?" (1 Kings 11:41). Elsewhere in 1 Kings are references to "the Book of the Annals of the Kings of Israel" (14:19) and "the Book of the Annals of the Kings of Judah" (14:29). How such annals might have been compiled and used is shown in the book of Esther. Early in the story Mordecai learns of a plot to assassinate King Ahasuerus and so informs the king, whereupon "the two conspirators were sent to the gallows, and the incident was recorded in the Annals, in the royal presence" (2:23). Later in the story a reference to these same annals becomes the means of rescuing Mordecai and his people and turning the tables on the wicked Haman.

A rich source of annals parallel to those in the Hebrew Bible can be found in the Babylonian Chronicle Series, dealing with important events (battles, deaths, uprisings, and so on) taking place between the Babylonians and their enemies from the time of Nabu-nasir in the eighth century to the third century B.C.E. This series is particularly interesting because it contains some objective history, including the occasional defeat and humiliation of the people who kept the chronicle. In a terse style it dates events with great precision (as is done in Haggai 1:1 and 2:1 and many other places in the Old Testament). For example:

> Fourth year of Mushezib-Marduk [King of Babylon]: In the month of Nisanu, the 15th day, Menanu, king of Elam, suffered a stroke, his mouth was paralyzed, he was unable to speak. In the month of Kislimu, the 1st day, the city [i.e., Babylon] was seized, Mushezib-Marduk was made a prisoner and brought to Assyria. Four years was Mushezib-Marduk king in Babylon. (*ANET*, 1955, 302)

When Israel split into two kingdoms after the death of Solomon, annals for the two were kept separately. Thus when the historian came to write the narrative that we see beginning in 1 Kings 14, he had to divide his attention between the two kingdoms, writing about first one and then the other. The result has been called "synchronized history," because its chronology works by reference back and forth between regnal periods. For example: "In the third year of Asa king of Judah, Baasha son of Ahijah became king of Israel at Tirzah for twenty-four years" (1 Kings 15:33). This technique is paralleled exactly in a document from about the fifteenth century B.C.E. that synchronizes the relations of Assyria and Babylon (described in A. K. Grayson's *Assyrian and Babylonian Chronicles* [1975]).

Along with official records, there is evidence that biblical writers drew upon popular histories—expanded narratives about an incident reflecting on the life and career of a great leader. The story of David's successful rise to power as told in 2 Samuel 1–5 is such a narrative. Ancient Hittite literature, in particular, has parallels to this kind of writing, such as in the story told by Mursilis about his father, King Suppiluliumas, and the request of the Egyptian queen (widow of Tutankhamun) to marry one of the great king's sons (*ANET*, 1955, 319).

The overarching principle that embraces and binds together all the elements of historical writing just described is ideology, and in this respect the biblical writers were exactly like their predecessors. As the next chapter will show in detail, none of those writers was ever interested in the past simply for its own sake. Their chief aim was to use the past to teach a religious lesson, specifically, that proper obedience to the deity leads to success in earthly affairs, with its corollary about disobedience. In the Deuteronomic History the conquest of the two Israelite kingdoms by foreign powers is attributed directly to this disobedience. Such a view is quite in the mainstream of ancient Near Eastern belief. On the Moabite Stone (an engraved basalt slab somewhat like the Rosetta Stone) from the ninth century B.C.E., King Mesha of Moab attributes his success and failure in dealings with Israel to his deity Chemosh. Mesha writes that "Chemosh caused me to triumph over all my adversaries" and that Israel was able to humble Moab for a time because "Chemosh was angry at his land." If a plague struck the land, this showed the anger of the deity, who then had to be appeased: thus the "Plague Prayer of Mursilis II," a Hittite document, and the efforts of David described in 2 Samuel 21 to remove the famine that had befallen Israel because of Yahweh's displeasure. From the Mesopotamian world we have a document, now called the "Weidner Chronicle," written around 1100 B.C.E., that purports to describe the history of Babylon during the third millennium B.C.E.; in this document the writer makes this traditional connection, contending that those rulers of Babylon who worshipped the deity Marduk and took care to provide fish for his temple succeeded, whereas those who neglected Marduk and his fish supply fared badly. While the nutritional requirements of deities might vary, it was universally believed that they had to be provisioned through sacrifice, which was a form of worship, and would retaliate if neglected. An ancient historian would have felt duty-bound in his writing to point out how events were determined by this relationship between humans and their deity—a kind of ideological commitment that modern historians go to great lengths to avoid.

WISDOM LITERATURE

Ideology and poetic form combine in the biblical genre called "wisdom literature," which will be explored at length in chapter 9, and both have their roots in shared traditions of the ancient Near East. Indeed, some of the Old Testament proverbs seem to be direct translations from an Egyptian collection of wisdom sayings known as "The Teaching of Amen-em-Opet." Consider these three sets of passages:

(1) Give thy ears, hear what is said,
Give thy heart to understand them.
To put them in thy heart is worth while. . . .
They shall be a mooring-stake for thy tongue. (*ANET*, 1955, 421–22)

Give ear, listen to the sayings of the sages,
  and apply your heart to what I know,
for it will be a delight to keep them deep within you
  to have them all ready on your lips. (Proverbs 22:17–18)

(2) Guard thyself against robbing the oppressed
And against overbearing the disabled. (*ANET*, 1955, 422)

Do not despoil the weak, for he is weak,
  and do not oppress the poor at the gate. . . . (Proverbs 22:22)

(3) Do not carry off the landmark at the boundaries of the arable
land. . . .
Nor encroach upon the boundaries of a widow. (*ANET*, 1955, 422)

Do not displace the ancient boundary-stone,
  or encroach on orphans' lands. (Proverbs 23:10)

These quotations reflect the practical side of the wisdom tradition, in which learning to play by the rules was a guarantee of success in life—a belief common to all ancient Near Eastern cultures. But there was also a contrasting side in which life was seen as a puzzle, perhaps a meaningless one, and in which someone who played by the rules might as easily fail as succeed. The biblical book of Proverbs exemplifies the former, Job and Ecclesiastes the latter. We do not have to look only to the book of Job to find someone puzzled by the apparent injustice of life: In Sumerian literature a document that modern scholars call "The Sumerian Job" has the sufferer complain:

Tears, laments, distress and pain abide in me,
suffering overpowers me like one doomed to nothing but tears,
the evil spirit of destiny holds me in its hand and takes away my
  breath of life,
the malignant demon of sickness bathes in my body. . . .
How long will you continue to leave me without guidance? (*NERT*
141)

Just as biblical Job desperately seeks to understand his unjust treatment, or even to get an answer to his questions, so the writer of a Babylonian wisdom piece ponders openly the mystery of reward and punishment:

I wish I knew that these things would be pleasing to one's god!
What is good for oneself may be offense to one's god,
What in one's own heart seems despicable may be proper to one's
  god.
Who can know the will of the gods in heaven?

He who was alive yesterday is dead today. . . . .
I am perplexed at these things; I have not been able to understand
  their significance. (*ANET*, 1975, 152)

Another Babylonian wisdom document, "The Babylonian Theodicy," is arranged like Job as a dialogue, here between the "Sufferer"
and the "Friend." It has the writer pondering the topsy-turvy nature
of reward and punishment and questioning the value of service to
his deity:

I have looked around in the world, but things are turned around.
The god does not impede the way of even a demon.
A father tows his boat along the canal,
While his son lies in bed. . . .
What has it profited me that I have bowed down to my god?
  (*ANET*, 1975, 165)

Job, too, wonders:

Why do the wicked still live on,
   their power increasing with their age?
They see their posterity assured,
   and their offspring secure before their eyes. (Job 21:7–8)

It is clear, then, that wisdom writing in the Bible incorporates a
tradition much older than its actual date of composition and that just
like biblical narrational, legal, poetic, and historical writing, it can
best be appreciated when seen in the broader context of the ancient
Near East.

PROPHECY

Chapter 8 will deal with prophets and
prophecy in detail. Here the point to
be made is that the biblical figure of
the prophet, who passes moral judgment on his times, and the form of
biblical prophetic writings, which are used to promulgate his message, owe a great deal to traditions shared with older Near Eastern
cultures. For example, from Egyptian literature, perhaps as early as
2300 B.C.E., we have the writings of a prophet who looks around him,
like the biblical Amos, and sees signs of disintegration everywhere
("The Admonitions of Ipu-Wer"). Nathan-like, this ancient prophet
rebukes the mighty Pharaoh: "Authority, Perception, and Justice are
with thee, but it is a confusion which thou wouldst set throughout
the land" (*ANET*, 1955, 443). An early second-millennium Egyptian
work, "The Protests of the Eloquent Peasant," denounces social injus-

tice in the manner of Amos or Isaiah (*ANET*, 1955, 407–410). And prophets in all of the ancient Near Eastern cultures, as in the Israelite, were allowed the freedom to rebuke those high in authority because the prophets were viewed as mediums of the deity in communicating his divine will. When the Hittite king Mursilis wants to know why his land is engulfed in plague, he sends for "the oracle," who apparently (the meaning of the text is not quite clear) proceeds to interpret two ancient tables for the king that explain the reasons for this disaster. An almost exact parallel to the latter story is found in 2 Kings 22, involving King Josiah, "the Book of the Law," and the prophetess Huldah.

Besides condemning social injustice and conveying the divine will to the king, the prophet of the ancient Near East might also claim to foretell events in a fashion familiar to us from the Bible. One of the more striking prognostications in the *Uruk Prophecy* promises that a ruler will emerge who will take dominion over the earth. In almost Davidic terms (as in Isaiah 11) the text promises that the ruler's kingdom will be established forever. Some twelfth-century prophecies from Mesopotamia suggest an apocalyptic mentality predating Jewish apocalypticism by as much as a thousand years. They predict that after a coming time of troubles, peace and righteousness will emerge triumphant. Note how "biblical" these representative passages sound:

> During his [i.e. the evil ruler's] kingship
> battle and slaughter
> will not cease.
> Under his rule brothers will consume one another,
> the people will sell their children
> for money.
> All the lands will be thrown into confusion. (*NERT* 120)

But a benevolent king is also in the picture:

> . . . This ruler will be powerful and [will have no] rivals.
> He will take care of the city, he will gather together those who are scattered.
> At the same time he will make the temple of Egalmach and the (other) sanctuaries splendid with precious stones.
> He will bring together and consolidate the scattered land.
> The door of heaven will constantly be open. (*NERT* 121)

Passages like these, together with the literature mentioned above concerning the prophet as critic and mouthpiece for a deity, demonstrate that prophets and their literature were an integral part of the total culture of the ancient Near East and were certainly not unique to Israel.

THE ETHICS OF LITERARY BORROWING

A final matter concerning the parallels between ancient Near Eastern and biblical literature must be considered. Did biblical writers merely *copy* existing literature? Were they literary pirates? The answer is quite simply, no. There are indeed biblical passages that closely resemble passages in older Near Eastern literature. But in the ancient world it was a perfectly acceptable literary activity to reproduce parts of an older literary work (quite often combined with still another such work) in a new context, with minor textual differences. In a world where anonymous authorship was the norm, literary works were regarded as common property; the concept of individual authorship simply did not exist. To reproduce already existing literary material did not carry the taint of plagiarism that modern readers attach to such a process. As we said at the beginning of this chapter, evidence of the process has always been recognized *within* the Bible, where even wholesale reproduction of earlier material occurs, as, for example, in the use of Samuel–Kings in Chronicles, of Ezra–Nehemiah in I Esdras, and of Mark in Matthew and again in Luke. It should not then be surprising to us that there are also in the Bible an immense number of passages with almost word-for-word counterparts in earlier Near Eastern texts.

SUGGESTED FURTHER READING

Bertil Albrektson, *History and the Gods: An Essay on the Idea of Historical Events as Divine Manifestations in the Ancient Near East and in Israel* (Lund: Gleerup, 1967).

Walter Beyerlin, ed., *Near Eastern Religious Texts Relating to the Old Testament*, trans. John Bowden (Philadelphia: Westminister Press, 1978).

W. W. Hallo, B. W. Jones, and G. L. Mattingly, eds., *The Bible in the Light of Cuneiform Literature*. Scripture in Context III (Lewiston, N.Y.: E. Mellen, 1990).

William G. Hupper, *An Index to English Periodical Literature on the Old Testament and Ancient Near Eastern Studies* (Metuchen, N.J.: Scarecrow Press, 1987).

Dorothy Irvin, *Mytharion: The Comparison of Tales from the Old Testament and the Ancient Near East* (Neukirchen-Vluyn: Neukirchener Verlag, 1978).

Victor Matthews and Don C. Benjamin, eds., *Old Testament Parallels: Laws and Stories from the Ancient Near East* (New York: Paulist Press, 1991).

James B. Pritchard, ed., *Ancient Near Eastern Texts Relating to the Old Testament*, 4th ed. (Princeton: Princeton University Press, 1975).

Ephraim Speiser, *The Anchor Bible: Genesis* (Garden City, N.Y.: Doubleday, 1964).

The Interpreter's Dictionary of the Bible, ed. George A. Buttrick et al. (Nashville: Abingdon Press, 1962). See articles on Assyria and Babylonia, Inscriptions.

The Anchor Bible Dictionary, ed. David Noel Freedman et al. (New York: Doubleday, 1992). See articles on Archaeology, Syro-Palestinian and Biblical; Covenant.

IV

The Bible and History

Nothing about the Bible is so surprising to uninitiated readers as the extent to which it appears to be a history book. On picking up the Bible, first-time readers might well expect it to be a lofty discourse on the nature of God or a description of the universe and humanity's place in it. Or they might reasonably suppose it to be a set of rules for living a moral and satisfying life. Instead, what they find is an account—sometimes in the form of historical chronicle, sometimes in the form of story—of the birth and growth of a particular people, Israel; of the later decline and ill fortune of that people; and of their hopes for a better life during a long-continued time of trouble. More than half of both the Old and New Testaments takes the form of historical writing, and most of the remainder relates in one way or another to the events and persons presented in the historical portions.

To become aware of the extent to which the Bible is involved with history is to become aware of a central conviction of those who wrote the Old and New Testaments: namely, that their deity—the God of ancient Israel and of Judaism and of Christianity—was a deity who stepped into human history and arranged events according to his own plan for humankind. In doing so he both accomplished his purposes and revealed himself to his people. The writers of the biblical books set forth no abstract descriptions of this deity; rather, they showed what he was and is through telling what he *did* in dealing with his people Israel over the centuries. Even when those writers wished to instruct their readers in how to conduct themselves in life, they did so (except in the "wisdom" books, such as Proverbs) within a historical framework, whether through the codes of law and principles of behavior delivered to individuals in the story or through the reports of divine judgment on the actions of individuals and nations.

To study the Bible, then, is necessarily to study history—and a very specific history. In the view of the writers of the Jewish scriptures, that history began at the point at which Yahweh chose as his special agent the man Abram (later called Abraham) and promised that Abram's descendants would someday become a great nation. The history continued as the earliest of those descendants went into bondage in Egypt, multiplied vastly in numbers, and were brought out of Egypt centuries later by Yahweh's power and led back to the land promised to Abram. They settled into nationhood in Palestine; produced their great kings, David and Solomon; built their Temple at Jerusalem; held the land over the centuries until their enemies defeated and deported them; and languished in exile until a small contingent was allowed to return and rebuild the Temple and commit themselves anew to Yahweh. For writers of Christian biblical literature, the relevant history included all of the above plus the events of the birth and ministry of Jesus and the founding and early growth of the Church. The time span covered in the combined Jewish–Christian account would be about two thousand years, from the period of the patriarchs (shortly after the beginning of the second millennium B.C.E.)* to the end of the Apostle Paul's career (about 60 C.E.).

Because biblical writers so often expressed themselves in the historical mode—that is, because they presented their religious message in the form of narratives about events and people—we as students of the Bible are faced with a specific obligation. We must become familiar with the sweep of Hebrew history as these writers conceived of it, beginning with the age of the patriarchs, proceeding on to the bondage in Egypt, then to the Exodus, and so on down to the time of the return from the Babylonian Captivity and the restoration of national

---

*There is a lively scholarly debate in our time as to when the patriarchs lived—if, that is, they actually lived at all. But if we begin with some datable event in later Hebrew history and work backward, using the Bible's own figures, we arrive at a date of not long after 2000 B.C.E. for the beginning of the patriarchal age. This will do for our present purpose, which is simply to understand what span of time biblical history professes to cover.

life and worship in Judea in the fifth century B.C.E. Even those biblical writers who do not directly report some part of this history make
constant reference to it. As we pointed out in chapter 1, over and over
again biblical characters—some of whom, like the outlaw Jephthah in
Judges 11, would seem to have no credentials as historians—lovingly
recite the great events of the national history down to their own
times. Most works in the Jewish scriptures were composed, or put
into their final form, near or shortly after the end of the classic period
of Hebrew history; but as a way of guaranteeing the authenticity of
their messages, the authors attached their compositions to notable persons who lived much earlier in that history. Thus Deuteronomy is
said to have been the work of Moses, many of the psalms the work of
David, Ecclesiastes the work of Solomon, and so on. And various
late writers also dipped into the ancient history at crucial points and
composed stories that expanded upon that history. Thus the book of
Ruth, written perhaps as late as the fifth or fourth century B.C.E., is
set far back in the time of the Judges (seven or eight centuries earlier)
and shows in close detail how the family that would produce King
David came into existence.

LIMITATIONS ON
READING THE BIBLE AS
HISTORY

Our intention thus far has been to
make a clear case concerning the centrality of history to the Bible. But if
we have succeeded in that effort, we
must immediately turn about and set
some serious limitations on the relationship between history and the
Bible. Although the Bible is intimately bound up with history, it cannot properly be read as a history book. This may seem at first a surprising statement, given all that we have said earlier, but there are
several compelling reasons for making it.

The first and most obvious reason is that, although the Bible is a
long book, it is not nearly long enough to cover a history that
stretches over two thousand years. A recently published work on the
American Civil War fills four volumes comprising some nineteen hundred pages; if the two millennia of biblical history were covered in the
same detail, sixteen hundred volumes and more than three quarters of
a million pages would be required! Plainly, the biblical writers were
highly selective with respect to the items they chose to dwell on. In
some places in the Bible, scores and even hundreds of years are passed
over in a single sentence or given no notice at all.

In addition to the degree of its selectivity, the particular nature of
that selectivity constitutes a major reason why the Bible cannot be
read as a book of history—not, at least, as a book of history in the

modern sense. What we require today of our historians is the highest level of objectivity possible in what they write. As the first chapter of this book has argued, objectivity in dealing with the past was of no interest to the writers of the Bible. Perhaps they could not even have conceived of such a thing. When they told their stories of the past, they did so not for the sake of the past but for the sake of the present—their present, of course. That is, they selected material concerning the past and shaped it according to what they felt were the needs of their own present-day audience. This can be taken as an axiom applicable to almost all of the historical portions of the Bible. In the following paragraphs we shall survey the Old and New Testaments with this axiom in mind, asking of those parts of the Bible concerned with the past just what the writers selected for treatment and how they shaped it for the profit of their particular audience in their own time.

HISTORY IN THE
PENTATEUCH

The first division of the Hebrew scriptures, which we call the Pentateuch ("five scrolls") or Torah ("teaching"), was often called the Law by Jews and Christians alike—and certainly there is a great deal of legalistic matter in it. But those five books from Genesis through Deuteronomy also present a sweeping account of Yahweh's dealings with his chosen people Israel. This account is constructed out of stories of Israel's ancient heroes and covers (exclusive of its preface, Genesis 1–11, which is set in the dateless past) the first seven hundred years of Israel's existence (about 1950–1250 B.C.E.). After those prefatory chapters have pictured the creation of the world and humankind and the spread of humankind upon the earth, the account gets down to cases in telling how Yahweh selected Abram through whom to create a special people. The account follows Abram's descendants into slavery and out of it, all according to Yahweh's design; shows them being welded into a nation with a covenant relationship with Yahweh; and brings them at last to the border of the land that they had been promised as their own. This extended narrative was artfully shaped by its final compilers at some time in the fifth century B.C.E. into what has been called a "salvation history." Its audience was the postexilic Jewish community in Jerusalem and Judea, a community aware of how pathetically its own circumstances compared with those of the great days of David, Solomon, and their successors. It was for the encouragement of that community that the Pentateuch was intended, showing, as it did, the miraculous foundation of the nation Israel, its rescue from bondage in Egypt, and its passage to the Promised Land—all by the gracious good will of its deity, Yahweh. How could any people despair, no matter what their present circumstances, when their beginnings had been so promising?

HISTORY IN THE FORMER PROPHETS

The second division of the Hebrew Bible, the Prophets, is divided into two parts called the Former Prophets and Latter Prophets. (This terminology is explained in chapter 6.) The Former Prophets is an account of Israel's past that embraces the approximately seven-hundred-year period (1250–550 B.C.E.) from the conquest of Palestine to the destruction of Jerusalem and the exile of its inhabitants. This particular account, which is frequently referred to as the Deuteronomic History (because it reflects attitudes pointedly set forth in the book of Deuteronomy), is, like the Pentateuch, a history book designed with a specifically religious purpose. Its elements were chosen and arranged and given emphasis to prove a point: namely, that when the people of Israel were faithful to their deity and observed his statutes, they prospered; but when they gave their allegiance to alien gods, they suffered at the hands of their enemies. A prediction that this would be the case for the Israelites was put into the mouth of Moses at the end of Deuteronomy. One of the ancient compilers of the Deuteronomic History inserted into the work (in Judges 2) an editorial comment that the whole course of events in the several hundred years between the Conquest and the beginning of the Monarchy constituted a series of cycles: Prosperity in Israel would characteristically lead to religious laxity and infidelity to Yahweh, who would thus allow Israel's enemies to defeat and enslave them; in their anguish the Israelites would turn back to Yahweh, who would then raise up a military leader (a "judge") to deliver his people and restore their security and prosperity. That, in turn, would set the stage for another cycle to begin.

This is the thesis of the Deuteronomic historians, and it firmly controls the shape of the story told in Judges. The thesis could not always be rigidly applied in the books of Samuel and Kings, however, mainly because the ancient materials available to the writers and compilers of those books were of considerable length and had their own literary integrity: Thus the materials could not be so easily manipulated in the service of the thesis as the short accounts in Judges could. We read, for example, that King Solomon's seven hundred wives and three hundred concubines "swayed his heart to other gods" (1 Kings 11:3–4) and led him into particularly loathsome forms of idolatry. According to the terms of the thesis, Solomon and his household and

the entire nation during this time should therefore have been griev-
ously punished. Instead, we are surprised to find, Solomon is simply
told that after his death his son will rule over only one tribe (Judah),
not all twelve. Some ancient history obviously had a life of its own
and had to be permitted to tell its own story.

But even so, the faithfulness-brings-reward/apostasy-brings-punish-
ment thesis can be seen to be the controlling idea of the Deuteronomic
History as a complete literary whole. By the end of the account, at
the conclusion of 2 Kings, destruction has come upon Judah; and it
has come specifically because Judah has wallowed in idolatrous wickedness under the leadership of the evil king Manasseh, who "misled
them into doing worse things than the nations whom Yahweh had
destroyed for the Israelites" (2 Kings 21:9). Except for the great Solomon and his great father David (who was, among other things, an
adulterer and a murderer), the kings of Israel and Judah are judged
one after another by the Deuteronomic historians according to
whether they jealously guarded the worship of Yahweh or permitted
the worship of other deities.

And not merely are these kings judged, they are allotted space in
the history in relation to their religious performance, not in relation
to their historical importance. We know from extrabiblical records
that, in terms of power and international influence, Omri was probably the most notable of all rulers of the northern kingdom, Israel. But
because Omri did not destroy Israel's system of idol worship and "was
worse than all his predecessors" (1 Kings 16:25), the Deuteronomic
historians found it inappropriate to chronicle Omri's accomplishments
and dismissed him with a short and negative notice. Jeroboam II,
another northern king, was given little space in the record even
though there was great prosperity during his reign and even though
the Deuteronomic historians themselves had to grant that he expanded the northern borders as broadly as King David had done long
before. The trouble was, as one might guess, that Jeroboam II "did
what was wrong" in the eyes of Yahweh. His actual stature mattered
little to religious historians intent on showing that soon after this
king's time the northern kingdom was obliterated as a result of its
collective unfaithfulness to Yahweh. Admittedly, a few wicked kings,
like Ahab and Manasseh, are given considerable attention in the Former Prophets, but that is only because it accorded with the Deuteronomic historians' bias to do so. Ahab is the wicked foil against whom
the great prophets Elijah and Elisha shine all the more brightly. And
Manasseh is so bad that he provides Yahweh with a final justification
for abandoning the kingdom of Judah to destruction and exile.

The Latter Prophets, the second part of the second division of the Hebrew Bible, comprises the writings that go under the names of individual prophets—Isaiah, Jeremiah, Ezekiel, and so on. These writings will be discussed at length in chapter 8, but here we can note briefly the large part that history plays in them. Even a quick perusal of this section of the Bible will make a reader aware of how thoroughly prophecy is involved with history. Amos, Micah, Isaiah, and the other "writing" prophets lived in times when momentous events were taking place in Israel and Judah and the nations surrounding them. The prophets saw the hand of Yahweh in those events and undertook to explain to their fellow Israelites the real significance of what was happening. The prophets' message was ultimately religious, but the occasion for their message and the stuff of which it was created lay in the history of their own times. The prophets shared with the Deuteronomic historians the view that uncorrected evil in national life would lead to national destruction. The historians had used that formulation to explain why destruction had come to Israel in the past; the prophets used it to explain why destruction could come in the near future. The practical effect they both hoped to achieve from their history lessons was a change in the lives of those individuals who were their contemporary audiences. History was only incidental to that religious end.

The third and final division of the Hebrew Bible, the Writings, contains a variety of kinds of books. But with the exception of Job, all of those books attach themselves to the history of Israel in some way or other. Several of them go further than that and relate what has the appearance of history. The books that most obviously do this are Chronicles, Ezra, and Nehemiah, which together form a continuous review of Israel's history from the time of David in the early tenth century B.C.E. down to the reestablishment of the worship of Yahweh in Jerusalem after the Exile in the fifth century B.C.E. Much of this survey was drawn from what the Deuteronomic historians had written in the books of Samuel and Kings; but it was made to serve an even more narrow religious purpose than the work of the Deuteronomic historians did. The Chronicler (as the final author–editor of Chronicles–Ezra–Nehemiah is traditionally called) undertook to demonstrate the legitimacy of Israelite worship as it was conducted at the Temple in Jerusalem during the Chronicler's own time. He did so by establishing a firm connection between the Temple (particularly its ritual and music, a special interest of the Chronicler) and King David. The "history" that is recounted here concerns David (the perfect king, whose acts of adultery and murder are ignored), the line of kings that descended from David, and the Temple that David made preparations to build. The work of the Chronicler is thus highly selective and extremely biased; it was not in the least designed to relate history objectively.

Though the other books in the Writings are not cast in the form of historical chronicles, some of them do tell stories that elaborate on the history of Israel. The book of Ruth relates the story of a widow from Moab (a neighboring nation) who journeys to Israel, marries a man from the tribe of Judah, and produces the line from which King David will ultimately come. The book is a skilfully told short story, quite attractive in itself; but the purpose for which it was written may have been to persuade Israelites to be less chauvinistic and more tolerant toward such neighbors as the Moabites. The book of Esther is another short story. It tells how a young Jewish woman saved Persian Jews of the fifth century from destruction at the hands of their overlords by interceding in their behalf with the Persian king. Like the book of Ruth, the book of Esther is a fine piece of narrative. It may have been written to give encouragement to the Jews living under foreign domination after the Exile or, even more specifically, to make a case for the Feast of Purim (referred to in the book's penultimate chapter) as worthy of becoming a major holiday in Judaism. But given the exaggerations in the book and its air of the fabulous, its purpose was obviously not to relate genuine history.

The same is true of another work among the Writings that has some appearance of historicity, the book of Daniel, which we shall discuss at length in chapter 10. Here we need only say that the book is about—and professes to have been written by—a young Jew, Daniel, who lived at the Babylonian court during the Exile of the sixth century. Actually, it was written in the second century B.C.E. by someone living in or around Jerusalem. His purpose was to encourage his fellow Jews to remain faithful to their religion at a time of terrible religious persecution. To do so, he harked back to a time of persecution four hundred years earlier and showed how his hero remained faithful despite every pressure that could be brought to bear on him. As in the previous instances, we see the stuff of history made use of in a literary work, not for its own sake but for decidedly religious purposes.

HISTORY IN THE APOCRYPHA

What is true of books with historical content in the canonical Jewish Bible is true of a number of books in the Apocrypha as well. Written long after the times in Israelite history in which their action is set, these books were intended primarily to encourage their contemporary Jewish audiences to be faithful to their God as the ancient heroes and heroines had been. The Apocryphal work that comes closest to being a genuine history book is 1 Maccabees, which reports how the members of a courageous Jewish family in the middle of the second century B.C.E. battled to overthrow the forces of the hated Antiochus IV Epiphanes and reestablish an independent Jewish nation. The book was written, perhaps fifty years after the events it recounts, by a patriotic and pious Jew; consequently it cannot be expected to be an unbiased piece of objective history. But it comes closer to that modern ideal than any other writing in the entire body of Jewish canonical or apocryphal works.

HISTORY IN THE NEW TESTAMENT

The New Testament is every bit as historically oriented as the Jewish Bible and the Apocrypha. More than half of its total—the four Gospels and the book of Acts—presents itself as a chronicle of events in the life of Jesus and in the first years of existence of the Christian church. Nevertheless, what is true of this sort of material in the earlier writings is true of it in the New Testament as well: Events of the past are set forth not to provide an objective account of the past but to serve the needs of some specific contemporary audience for whom each of the authors had a particular concern. The writer of the gospel of Luke informs the person to whom he is addressing his remarks that his intention is "to write an ordered account for you, Theophilus, so that your Excellency may learn how well founded the teaching is that you have received" (Luke 1:3–4). We know that the writer of the gospel of Luke had read the gospel of Mark because he drew heavily on it for his own work. Why then did he not simply send Theophilus a copy of Mark's gospel? The reason is that he felt that his particular audience required an account containing a different combination of events and shaped by a different editorial emphasis from what he found in Mark. It is for this same reason that the gospels of Matthew and John were written and that they differ from one another as well as from the gospels of Mark and Luke. A further discussion of the way the past—what happened forty, fifty, or sixty years earlier—was made to serve the gospel writers' present purposes will be found in chapter 13.

VALUE OF THE BIBLE
TO HISTORIANS

We have been urging the case that the purpose of the writers of biblical history was not to give objective accounts of the past but to meet the needs of their contemporary audiences. If that is so, is what those writers had to say of little actual value to modern historians in their attempt to reconstruct the past? Not at all. Precisely because biblical writings were designed to meet the religious needs of their contemporary audiences, the Bible is a primary source of information about what those religious needs were. Thus historians of religion have been aided in tracing the development of Judaism and Christianity by observing the sequence of contemporary issues to which the biblical writers responded. They have learned from biblical writings that religious issues important in one place—Jerusalem, for example—were not necessarily important elsewhere at the same time—in Alexandria, for example.

But the Bible has an even more direct historical usefulness than that. For the bulk of the long story it has to tell, the Bible is a unique source. No other ancient records exist for most of what the Bible reports about the past. It is surprising but true that no event in the Bible before the ninth century B.C.E. can be confirmed from outside sources. The very first item that can be so confirmed is the battle referred to in 2 Kings 3:5, in which Mesha, king of Moab, freed himself from Israelite rule; this battle is described on the famous Moabite Stone unearthed by archaeologists in the 1860s. But for the bulk of the notable stories out of Hebrew history—the accounts of Abraham and the other patriarchs, of Joseph as overlord of Egypt; of the Hebrews' bondage in and escape from Egypt, of the conquest of Palestine, and of the great glories of the reigns of Solomon and David—for all of these stories and for much else the Bible is our only source. With respect to the Christian story, the case is the same. There is secular confirmation that the various Roman rulers mentioned in the New Testament did, in fact, hold office. But except for a few incidental references to the Christian movement in late-first-century and early-second-century Latin literature, there is no independent evidence for events reported in the gospels and the book of Acts.

That the Bible is the only source for so much of what it reports means that its narrative of events is particularly valuable to historians of the ancient Near East. They cannot simply ignore the Bible on the grounds that its account has a strong religious bias. They must take great pains to extract from this unique account every possible kernel of solid fact. Some parts of the biblical text are not very promising in this respect, of course. Historians interested in knowing what really happened cannot find much help in the stories of the Garden of Eden or the Tower of Babel. They will find something of value in the accounts of the patriarchs, though those stories cannot be accepted at face value as records of events. Of considerably more worth as a historical source is the account in the books of Samuel and Kings about David's accession to the throne, of the later division of his realm into Judah and Israel, and of the subsequent fate of the two kingdoms down to the time of their destruction. Now we have said earlier that all the material in Samuel and Kings was employed by the Deuteronomic historians in the service of a specific thesis, and thus such material is highly biased. But it is also true that this material derived ultimately from records kept by court and temple scribes. Modern secular historians would of course prefer to have the original scribal records; in lieu of those, however, the account in Samuel and Kings is extremely valuable and yields a considerable amount of trustworthy information to historians accustomed to dealing with editorially biased writings from the ancient world.

Historians, then, will find "real" history recorded at places in the Bible. And the general reader of the Bible will find a historical tradition that is constantly being elaborated on and alluded to. But we must never forget that, so far as the biblical writers were concerned, history was only a means to a greater end, not an end in itself. In their conception, the truth of an event was not in the fact of its happening but in what it signified. To expect the Bible to tell us "what really happened" is to expect something that its writers never designed it to tell.

SUGGESTED FURTHER READING

Gosta W. Ahlstrom, *Who Were the Israelites?* (Winona Lake, Ind.: Eisenbrauns, 1986).

Philip R. Davies, *In Search of "Ancient Israel"* (Sheffield, Eng.: JSOT Press, 1992).

Diana Vikander Edelman, ed., *The Fabric of History: Text, Artifact and Israel's Past* (Sheffield, Eng.: JSOT Press, 1991).

Baruch Halpern, *The First Historians: The Hebrew Bible and History* (San Francisco: Harper & Row, 1988).

E. Theodore Mullen, Jr., *Narrative History and Ethnic Boundaries: The Deuteronomistic*

Historian and the Creation of Israelite National Identity (Atlanta: Scholars Press, 1993).

George W. Ramsey, *The Quest for the Historical Israel* (Atlanta: John Knox Press, 1981).

Klaas A. D. Smelik, *Converting the Past: Studies in Ancient Israelite and Moabite Historiography* (Leiden: E. J. Brill, 1992).

H. Tadmor and M. Weinfeld, eds., *History, Historiography, and Interpretation: Studies in Biblical and Cuneiform Literatures* (Jerusalem: Magnes Press, 1984).

John Van Seters, *In Search of History: Historiography in the Ancient World and the Origins of Biblical History* (New Haven: Yale University Press, 1983).

The Anchor Bible Dictionary, ed. David Noel Freedman et al. (New York: Doubleday, 1992). See article on Historiography.

V

The Physical Setting of the Bible

We asserted at the beginning of this book that the Bible was written by real people who lived in actual historical times. Chapter 4 explores in a general way the fascinating and important relationships of biblical writing to that human history. But time is not the only dimension within which human history exists: It exists also in space. The writers of the Bible lived some *where*, and the events they wrote about were always conceived of as occurring in a place. Not only the obviously historical writing of the Bible but even its poetic and visionary passages come to us dressed in the images of contemporary life as these writers knew it through their own surroundings. These surroundings—the physical context of biblical literature—were the world of what is traditionally called the Near East. Its reality can be documented extensively without having to rely on the biblical accounts. Moreover, in spite of the changes caused by the passing of years and the succession of human cultures imposed on them, the Bible lands are still there waiting to be brought back to life with a little careful study—and waiting to contribute that life to the words we read on the page.

The physical context of biblical literature is not as unfamiliar to most readers as are the events of biblical history, because it is much easier to picture a place than it is to picture a time. We have all experienced the biblical world at second hand through some visual form such as color prints stuck on the Sunday school wall or illustrations in gift-edition Bibles or (especially nowadays) historical movies recycled through television. It is open to question, though, whether such representations make the biblical world more real or less. The actors inhabiting these scenes wear what appear to be bathrobes and move stiffly, visibly burdened with the importance of the roles they are taking. Background details—a date palm or two, laden donkeys moving past—indicate that we are vaguely in the Near East. There are houses fronting the dusty streets, but nobody seems to live inside them. No work is being done. Even the rocks seem curiously weightless; they are probably made of Styrofoam and will disappear into the inventory of the property department once we turn off the TV set. All the energy and disorder of real life have been replaced by the artificial calm of the tableau vivant.

PALESTINE

As a first step in restoring reality to the biblical lands, let us consider the physical geography of Palestine. At its eastern end the Mediterranean Sea, elsewhere highly irregular, smooths out into approximately the shape of a rectangle. At this eastern end the sea is bordered on the north by the Anatolian (Turkish) peninsula, on the south by the upper corner of the African continent (Libya and Egypt), and on the east by some four hundred miles of gently curving coastline that connects these two landmasses more or less at right angles. The particular area we are concerned with, Palestine, is a skinny trapezoid occupying the lower third of this eastern coast. The traditional northern limit of our area, at the narrow end of the trapezoid, is the city of Dan, at the foot of snow-capped Mount Hermon in Lebanon; its southern limit, at the wide end, is the Negev, an arid region lying between the ancient city of Beer-sheba and the Sinai desert. Its eastern and western limits are the most definite ones, being respectively the Jordan valley and the Mediterranean Sea. On the east, running straight up and down the map, is a familiar series: the Sea (really a lake) of Galilee, the Jordan River, and the Dead Sea. From this valley westward to the Mediterranean coast lay the territory of the kingdoms of Israel and Judah, the biblical Promised Land. At various times it included portions of the Transjordan, the territory to the east of the river. But as the climax of the Exodus story makes clear, the true destiny of the Israelite people was always held to lie west of the river, and if they had not taken possession of this latter area, the Covenant would have come to nothing.

By our own standards it is a very small place: not more than 100 miles across at its widest point and fewer than 25 miles at its narrowest. Dan and Beer-sheba, the traditional extremes of the other axis, are only 150 miles apart. Its area is about 10,000 square miles—roughly the same as that of the state of Vermont or about a fourth that of the state of Ohio.

Our surprise is caused mainly by the fact that to the writers of the biblical text it was indeed a large place, and, in reading the Bible, we tend unconsciously to accept their point of view. This small corner of the eastern Mediterranean—we have to keep reminding ourselves that it takes up only the lower third of that coast—practically speaking was the whole world to them. Even had they known more about the rest of the world than they did, it would not have found any prominence in their writings. The land granted by Yahweh to the descendants of Abraham was all that really mattered. Foreign countries appear in the Old Testament only as military allies or enemies of the Israelites or as the habitat of alien gods; otherwise, not the slightest interest is shown in them. In the New Testament, of course, the horizon broadens to take in other lands of the Mediterranean, but again they are presented solely in the context of the writers' immediate interest, which is the missionary activity of the Church.

The effect of such a limited perspective is to magnify everything presented: Seen so close up, this crowded stage indeed begins to take on global dimensions. Thus one of our first tasks here will be to substitute a more realistic scale. In the course of doing so we can also learn how the physical characteristics of this stage themselves entered into the drama and helped determine the course of events.

The area we are considering is conventionally called "Palestine," a word derived directly from "Philistine." There is some irony in this because the Philistines were the great enemies of the Israelite nation in its early years. Eventually they disappeared from the land, but they left their name behind. The name was known and used as far back as the time of the Greek historian Herodotus (fifth century B.C.E.). It was taken up by the Romans as an official designation, became standard in Christian writings, and so came down to our time.

The oldest and probably the original name for Palestine was "Canaan." Its inhabitants are repeatedly characterized in the Old Testament by a stereotyped formula that contains six or seven names in various combinations, for example, "the Canaanites, the Hittites, the Amorites, the Perizzites, the Hivites, and the Jebusites" (Exod. 3:8). Except for the Jebusites, who are known to have been the occupants of the site of Jerusalem, these peoples cannot be identified with any certainty. It is customary now to call them all Canaanites. They were a mixed people, though mostly Semitic, with a well-developed culture. Their social structure was feudal-aristocratic, and they lived by farming, animal husbandry, and trading. Because their land was broken up into many small kingdoms with only local influence—city-states really—and because it was subject to frequent interference from its more powerful neighbors, the Canaanites never achieved a strong national identity. They were submerged by the Israelites in an occupation that began (according to the usual view) in the late thirteenth century B.C.E.

The Israelites did succeed in building a nation in Canaan, but they were able to unify the land politically for only a comparatively short period. In spite of the traditions of the Covenant, the Exodus, and the Conquest, regional loyalties proved stronger than national ones.* Ordinary Israelites knew more and cared more about their immediate neighborhood than about the land as a whole—and for very good reason: The way of life they followed was determined absolutely by the nature of the place they lived in, and Palestine (as we shall resume calling it) is divided geographically into regions that are quite different from one another. Within any of these regions, given its characteristic topography, soil type, rainfall, and temperature range, inhabitants had little choice. Whether they grew wheat or barley, tended sheep, or cultivated vines and olives depended on where they found themselves. It may well be for this reason that the Israelites themselves never used a comprehensive name, like Palestine or Canaan, for their land.

THE NATURAL REGIONS
OF PALESTINE

To begin our study of the geography of Palestine, let us take a position at the bottom or southern end of the map and look toward the north. We share the point of view of Moses and the Israelites camped at Kadesh-barnea, although we no longer need to send out spies to tell us what the land is like. The immediate prospect is not encouraging, for we are looking first at the Negev, an arid steppe where the annual rainfall of less than eight inches makes grow-

---

*We should remember that the only evidence of these traditions is furnished by biblical writers, whose view of the Israelite past was to a large extent an idealized one.

ing crops possible but chancy. At least this is true of the northwestern sector, below Gaza, which is relatively flat and is composed of wind-deposited soil. According to archaeological evidence, this area during some periods in its history contained a number of small cities and was regularly cultivated; but during other periods, perhaps because of minor shifts in climate, it lost its permanent population and reverted to the shepherds who have always been able to subsist here. The southeastern section, however, is true desert, with less than four inches of rain annually. It is a dramatic landscape of barren cliffs and valleys, a stony desert, not a sandy one, and furrowed by wadis (dry gullies that carry water only in times of storm). The traditional name of this southeastern sector is the Wilderness of Zin.

Looking northward from Beer-sheba, with the Negev behind us, we find more geographical variety and on the whole a less hostile environment. Figure 5 shows that the land here can be divided into four regions, lying parallel to one another and oriented north to south. We shall discuss them in order, beginning on the west with (1) the coastal plain. During the early history of Israel, the coastal plain was occupied by Philistines, who lived in the cities of Gath, Gaza, Ashkelon, Ashdod, and Ekron. Beyond the sand dunes that edge the Mediterranean shore, the land is flat, easy to cultivate, and well suited to growing barley. East of the coastal plain the land rises to an area of (2) limestone hills, the Shephelah. In early biblical times this was covered with scrubby forest and was not much settled or cultivated. Its main value was as a buffer zone between Philistia and Judah, and it was much fought over for this reason. Eastward, rising now to more than three thousand feet in some places, is (3) the hill country of Judah, composed of a limestone inherently more fertile than that of the Shephelah. Farming here was made difficult by the rugged terrain and, in earlier days, by the forests that covered the hills. But this region receives more than twenty inches of rain a year, and the soil can support grain fields, olive groves, and vineyards. The hill country of Judah had another advantage for the Israelites in that it was easy to defend, with many opportunities for ambush and many hiding places for guerillas—hence it was the last part of ancient Israel to be conquered by a foreign army. Between the Judean hills and the Dead Sea on the east lies (4) the desolate Wilderness of Judah. From Jerusalem, near its western edge, the land plunges downward almost four thousand feet in twelve miles, becoming more wild and forbidding as it goes. The lower part of this region gets as little as two inches of rainfall annually.

A simple phenomenon accounts for the difference in rainfall, and

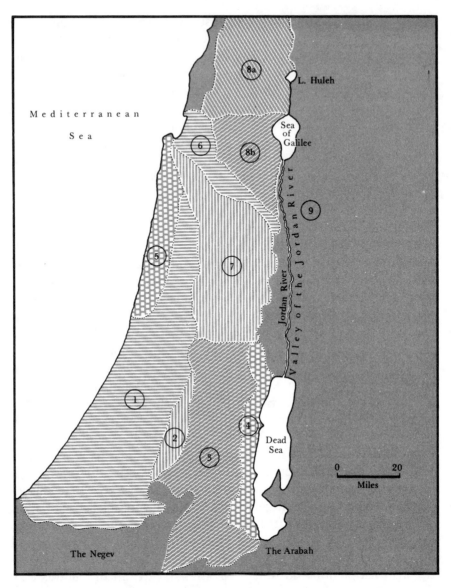

FIGURE 5. NATURAL REGIONS OF PALESTINE

Adapted from the *Oxford Bible Atlas*, 3rd ed., pp. 48–49.

it is the same one that created the deserts of the American West. In Palestine the prevailing winds in winter move eastward off the Mediterranean, carrying moisture-laden air with them. When this air strikes the hill country it is forced upward, cools below the dew point, and releases its moisture as rain. The higher and more abrupt the ascent, the more water is wrung from the clouds. But on the leeward side of the hills the air descends, heats up, and dries out. This leeward area lies in what is called the "rain shadow." Typically, the transition zone between the arable land and the rain shadow is very narrow. What rain does fall in rare thundershowers to the east of the crest of the hills in Judah is localized and quickly runs off through wadis or evaporates in the burning sunshine.

Moving northward from these four regions, we find (5) the Plain of Sharon, which is very different from the Philistine coast. In Old Testament times it was swampy and choked with vegetation, so difficult of passage that the Israelites never attempted to settle it. At the northern end of this plain the headland of Carmel juts into the sea, interrupting the smooth curve of the coastline. Mount Carmel itself is not very high, but its slopes are steep; in Old Testament times it was thickly wooded. It was strategically important, also, in that it blocked north–south movement except at a few easily defended passes. Immediately north of the Carmel ridge is (6) the Plain of Esdraelon, a valley about twenty miles in length connecting with the narrower Valley of Jezreel and eventually with that of the Jordan River. The soil here is alluvial and fertile, some of the best in all Palestine. This diagonal slash across the map marks the northern boundary of (7) Samaria, the southern being the Judean hills. Samaria was always well populated because its rolling hills offered no serious obstacles to cultivation, and its soil, of Eocene limestone, was easy to plow. On the other hand, Samaria was not so easy to defend as the hill country of Judah, and it lacked a natural fortress like Jerusalem.

The northernmost region of Palestine is Galilee, which extends from Samaria up to the border of Lebanon. It is divided into (8a) Upper or Northern Galilee and (8b) Lower or Southern Galilee because the topography of the two areas is quite different. Lower Galilee is a region of rolling hills, much like Samaria, and is well adapted to agriculture. Upper Galilee, however, is rugged and contains the highest mountain in all Palestine, rising to nearly four thousand feet. In Old Testament times both parts of Galilee were heavily wooded, especially the northern. Because Galilee was sparsely settled and remote from the centers of power in the southern provinces, it played little part in the history of Israel until the time of the Roman occupation. On the eastern border of Galilee, below the city of Dan, was Lake Huleh, surrounded by marshes and feeding into the Sea of Galilee. In the early years of the modern state of Israel, Huleh was drained and the marshland converted to agriculture.

From the foot of Mount Hermon in Lebanon, whose melting

snows begin the streams that form the Jordan River, we can look southward at the most remarkable physical feature of all Palestine: a rift valley, much of it below sea level, that stretches for more than 250 miles down to the Gulf of Aqaba. This rift valley contains the Dead Sea, the lowest spot on the surface of the earth. Geologists now believe that phenomena like this (and the even greater rift valley system of Africa with which it is related) are caused by the movement of tectonic plates, sections of the earth's solid crust that float on top of the plastic rock forming its upper mantle. As they spread apart from one another over aeons, these plates created our present continents. In this case the plate carrying the Arabian peninsula is thought to be moving slowly northeastward, pulling apart the shores of the Red Sea. In the process this plate has created the Jordan rift, where the earth's crust split along parallel lines and the section between these splits (or faults) dropped down to form the Jordan valley.

The valley begins at the Huleh basin, but its first major feature is the so-called Sea of Galilee, some seven by thirteen miles in area and already 695 feet below sea level. Its waters are relatively pure and abound in fish. From this lake the Jordan River begins its often meandering journey southward. The southernmost part of this valley is bare and inhospitable, the relentless sun and heavy air making its dryness even more severe, but it does contain some springs or oases, the most famous of these being at Jericho. On both sides of the desert valley is a narrow band of green, the Zor, a dense thicket of tropical vegetation that in ancient days was the habitat of dangerous wild animals. The Jordan empties into the Dead Sea. But this water, with the runoff from storms in the immediate area, has been just enough to balance the rate of evaporation from the surface; thus the level of the Dead Sea—which has no outlet—until recently remained fairly constant at about thirteen hundred feet below the level of the Mediterranean. The sea is "dead" because a concentration of dissolved salts— always present in natural water—has built up over thousands of years. The Sea today, about ten miles wide and fifty miles long, is merely the remnant of a much larger brackish lake that in prehistoric times filled most of the Jordan valley. The accumulated sediments from that period are visible now in the grayish-white marl soil of the Dead Sea shore.

The southern end of the Dead Sea is quite shallow, and according to a once-popular theory was believed to cover the sites of Sodom and Gomorrah, the infamous "cities of the plain" (Gen. 13:12). In any case, the rift valley here continues as dry land, the Arabah, down to the Gulf of Aqaba. To the west of this valley is the Wilderness of Zin and the Sinai desert, to the east the plateau of Edom.

Remarkable as the Jordan rift valley is, we must once again pause to remind ourselves of its size. The entire watercourse, from the top of the Sea of Galilee to the lower end of the Dead Sea, would fit comfortably inside the long dimension of Lake Ontario, the smallest of the five Great Lakes.

We need speak only briefly of the Transjordan, the area east of the river, because it was essentially foreign territory to the Israelites. The one region that they occupied for any length of time was (9) Gilead, which lies across the Jordan opposite Samaria but at a greater height so that its mountains extract further rainfall from the winter storms. It is well adapted to growing the traditional crops and was heavily forested in ancient times. Gilead is split east to west across its middle by the Jabbok River, which empties into the Jordan. North of Gilead lie the fertile plains of Bashan, south of it the vast tableland that dominates the rest of the Transjordanian valley. In the Dead Sea region, especially, the contrast between the two sides of the Jordan is striking: The tawny limestone hills of the Judean wilderness are abruptly replaced by the red sandstone of the Jordanian escarpment that faces them. Because of its height, the western Transjordan gets some further rain after storms have crossed the valley, but this effect is short-lived and soon the land returns to steppe and desert. The ancient kingdom of Moab lay across the Dead Sea from Judah, the kingdom of Edom to the south of Moab on a high plateau rising from the Arabah.

Within the major geographical regions we have described are a multitude of minor ones. Local variation is the dominant fact of life in this land. During Old Testament times this heterogeneity was also enforced by the lack of good roads—indeed, there were no roads at all by modern standards, only tracks of beaten earth winding across the countryside. The oldest and most important of these was the route from Egypt to Syria, which ran along the Mediterranean coast, skirted the plain of Sharon, crossed below Mount Carmel through the pass at Megiddo, and then continued toward Damascus between Lake Huleh and the Sea of Galilee. Another north-south route connected the Plain of Esdraelon with Beer-sheba along the crests of the hills of Samaria and Judah. This was paralleled on the eastern side of the Jordan by the so-called King's Highway. Several traditional routes crossed Palestine from west to east as well, but these were always secondary to the major corridors carrying travel through Palestine between nations to the north and south. This continual travel, with the economic and cultural interchange that accompanied it, made Palestine part of what we might call the international community of the Near East. Although at first relatively poor and overshadowed by its stronger neighbors, it was not isolated from them, and it was certainly not an undiscovered land waiting to be seized as a prize by the first major military expedition willing to tackle it, as the Exodus/Conquest story encourages one to believe.

SOIL AND CLIMATE

The exposed rock of Palestine is almost all sedimentary, either chalk or limestone. The chief exception is the eastern portion of Lower Galilee, where volcanic action has left behind a large area of basalt. This rock has been weathered into various kinds of soil, some more fertile than others. Most of this soil is to be found in the valleys and the coastal plain, where it was deposited over the years after being washed down from the hills.

The climate of Palestine is of the type known as Mediterranean, which is characterized by hot summers with no rain and cool wet winters, very much like that of the southern California coast. In effect, there are only two seasons: a five-month dry season (May through September) and a five-month wet one (October through April), with short transition periods between. During June, July, and August it never rains, in May and September very seldom. The wettest month is January. As a consequence, crops are grown in the winter and the land lies idle or awaits harvest in the summer. This inversion of the seasons from the major North American pattern is not the only peculiarity: The nature of the rainfall is also different from what most of us are accustomed to. The rain is brought by cyclonic storms moving eastward across the Mediterranean that dump their water on the land in widely separated and often intense showers. The concentration of rainfall into just a few days of the year (Jerusalem and London have about the same annual amount, but the average number of rainy days per year in the two cities is fifty and two hundred, respectively) means that relief from drought can be abrupt and sometimes quite spectacular, as it is reported to have been after the defeat of the prophets of Baal by Elijah (1 Kings 18). But it also means that the rainwater has little time to sink into the soil and runs off in muddy torrents through the ordinarily dry watercourses. A sudden storm of this type is credited with catching Sisera and his chariots by surprise and sweeping them away at the Torrent of Kishon (Judg. 4).

Rainfall is distributed over the country in a regular pattern: It is least in the south, at the edge of the Sinai desert, and increases as one goes north toward Lebanon. There is also a temporary increase from west to east because the hill county, as we have noted, drives the moisture-laden winds upward and wrings more water out of them.

By our standards most of Palestine is arid. Although averaging the rainfall in a land with so many regional differences would produce a meaningless figure, we can take Jerusalem as being close to the middle of the scale. Its twenty-plus inches a year is about half as much as that received by Washington, D.C., or New York City. Because of the many hours of sunshine annually, there is a high rate of evaporation from the soil. So critical is the need for moisture that a significant portion of the supply for growing plants in the summer is provided by the dew that collects on the ground at night. Farmers must take advantage of even marginal differences caused by the pattern of rainfall and evaporation; for example, in the hill country the western and northern slopes will always be, to some extent, moister than the eastern and southern slopes.

Techniques for obtaining and conserving water were developed through hard necessity by the early inhabitants of Palestine. Cisterns were dug in the ground to hold the runoff from rainstorms, as were wells where the water table was close enough to the surface. The importance of wells in Old Testament times is shown by the many biblical stories concerning their naming and the events associated with them. Local irrigation was possible in the immediate vicinity of springs, such as the one at Jericho, which allowed human settlement in the otherwise hostile Jordan valley; but the Israelites did not develop irrigation on any large scale because the only effective source of water available to them was rainfall or rain-fed streams. In this respect the contrast with both Egypt and Mesopotamia could not have been greater, as the author of Deuteronomy 11:10–11 recognized, for he quotes Moses as warning the Israelites that the land they will enter "is not like the country of Egypt from which you have come, where, having done your sowing, you had to water the seed by foot, as though in a vegetable garden." No, Palestine is "a country of hills and valleys watered by the rain of heaven." Hence the greatest possible punishment for the Israelites, if they disobey Yahweh, is to have him "shut the heavens" (Deut. 11:17), which in that land would have been tantamount to a sentence of death.

NATURAL RESOURCES

At the beginning of the Israelite period, large portions of Palestine were still covered by forest, especially in the highlands. This is reflected in the account in 2 Samuel 18 of the battle on the heights of Gilead between the forces of David and Absalom. The tenth-century forest might not have been so impressive to us, for whom the word "forest" means stands of towering trees, but it made

up in density what it lacked in vertical size; and there is no doubt
that clearing it for agriculture was a slow and laborious process. The
process—which had begun long before the Israelite occupation and
continued long after it—was immensely aided by the pasturing of
sheep and goats (especially the latter) because trees cannot regrow on
land where these animals graze. Once bare, the fields remained bare.
When the forest cover was gone, there being no foliage to break the
force of the rain and no root systems to hold the soil together, erosion
accelerated tremendously and vast quantities of irreplaceable soil
washed forever down into the seas. The starkness of Mediterranean
lands, which many find so beautiful, is not natural to them but is the
result of centuries of misuse by their human inhabitants.

Could Palestine ever have been a land "flowing with milk and
honey" (Exod. 3:8)? It might well have seemed attractive in the imaginations of homeless migrants camped across the Jordan after a long
sojourn in the desert—and the literary traditions embodying that period of Israelite history strenuously perpetuated the idea. Everything
depends on one's standard of comparison. Palestine is at one end of
the "Fertile Crescent," a band of arable land stretching in an arc from
southern Mesopotamia up the Tigris and Euphrates valleys, across
Syria, and down the Mediterranean coast. But Palestine is the least
fertile portion of the crescent; it was more desirable in ancient times
for its strategic location—being the corridor between Egypt and the
nations to the north and east—than for its natural resources, which
are insignificant. The only large deposits of copper ore were in the
Arabah, but they were never exploited by the Israelites. There was
no iron, gold, silver, or tin. For timber of any size, one had to send
to Lebanon, as Solomon did in building the Temple. Stone for building, of course, was abundant. The produce of the land was mainly
grain, olive oil, wine, and wool. Although the traditional crafts were
practiced, there was little or no industry as we understand it. The
most important item of manufacture was pottery, for clay vessels
were essential to almost every activity of daily life. Lacking anything
of great value to offer in international trade, the Israelites could become wealthy only through conquest—extending their borders and
drawing on the output of other peoples through expropriation or taxa-
tion—something that happened only a few times in their history.

The best opportunity for economic development, it might seem,
was one they never took: commerce by sea. With the Mediterranean
always at their door, the Israelites stubbornly remained a landlocked
people. They were effectively shut off from the coast at first by the
Philistines, but the warfare between the two had more to do with the
Philistines' attempts to expand toward the east than with any desire

of the Israelites to gain access to the sea. Although the Palestinian coast has no natural harbors south of Carmel, this need not have been a permanent obstacle.* The Israelites were content to let others, Phoenicians and Egyptians, conduct their merchant shipping for them, almost as though they believed the Covenant language in its narrowest sense as a promise of land and nothing further. It is clear from their writings in the Old Testament that the sea was always alien to them and had no significant part to play in their thought.

URBAN LIFE

Israelites entering the Promised Land in the thirteenth century B.C.E., as the traditional story has it, would have found the Canaanites living in fortified cities, each with a claim to be a "kingdom," mostly in the better agricultural areas outside the hill country. The location of these cities was not accidental. They had to be near good farmland and a dependable water supply, and they had to be defensible against attack. When possible, a city would be built on a hill that dominated the surrounding countryside. It was then ringed with a thick masonry wall in which were one or two heavy gates that could be closed at night. This fortification posed something of a problem because the spring that provided water would normally have been below the hill and outside the city wall, where it could not be reached in time of siege. A great deal of ingenuity was given, over the years, to devising ways of bringing such water within the walls through tunnels and shafts, the most famous example being Hezekiah's Tunnel in Jerusalem (2 Kings 20:20), which connected the spring of Gihon outside the eastern city wall to a protected reservoir on the western side through over 1,700 feet of rock, following an existing fissure—all the excavation of which was done by men working with hammers and chisels.

These fortified cities were quite tiny compared with anything that we would call a city. With just a few exceptions, no Canaanite or Israelite city before the Roman period occupied more area than that of an American university football stadium; most villages were hardly bigger than the playing field itself. King David's Jerusalem is estimated to have measured about three hundred by thirteen hundred feet. Inside the city walls, houses would be crammed together according to no particular pattern, leaving room for passages but not streets. Before the Greek period there were no public buildings of the

---

*The port of Ezion-geber (modern Elath) was supposedly used by Solomon (see 1 Kings 9:26–27), but it could be reached only by a long overland journey through difficult country and thus could hardly have been a major outlet.

kind that we take for granted, provided by the municipal government. The chief area for the conduct of business was the city gate, which was open during the daytime and had benches within it to accommodate those who wanted to meet friends, strike bargains, or just survey the passing scene. Such uses are shown at several points in the Old Testament narrative, for example in Ruth 4, where Boaz conducts negotiations at the city gate with the next of kin for Ruth's hand in marriage, and in 2 Samuel 15, where Absalom stations himself at the city gate to intercept passersby and curry their favor for himself as against his father, King David (see also Psalm 69:12 and Proverbs 31:23).

Obviously, the density of the urban population was quite high; because of their restricted areas, however, most cities in the Old Testament period had only a few thousand inhabitants. The combined population of Israel and Judah in the first half of the eighth century B.C.E., before the Assyrian and Babylonian conquests, was about one million.

Probably the bulk of this population did not live in the cities, which functioned primarily as administrative centers, storehouses, and military garrisons, but in satellite (or "daughter") villages scattered through the countryside. The people's daily work would be conducted outside the walls, but they could seek shelter inside in time of trouble. Until very recently the archaeological record of rural life was quite scanty. This record is now being filled in, but most of the evidence of life in ancient Palestine is still what has been turned up in excavating the sites where cities were built. Once chosen, these sites tended to remain as human settlements; many of them have histories that go back centuries before the Israelite period. Wars and conquests provided only temporary interruptions. If a conquest involved the nearly complete destruction of the physical city, as was common, the conquerors simply smoothed down the debris and built their own structures on top of it. Thus the "tell" grew, a flat-topped hill that everywhere in the Near East marks the place of ancient occupation. Inside the tell, preserved in strata like the layers of a cake, is the record of these successive stages.

DAILY LIFE IN PALESTINE   What we have been saying about life in Palestine so far applies down to the beginning of the period of Greek influence, in the third century B.C.E. And even though the pace of civilization must have quickened thereafter, especially under Roman rule, the ancient rhythms of life persisted because they were virtually independent of politics: People dug and sowed and harvested in Palestine as they had been doing for centuries, obeying the actual ruler of the land, which was nature herself.

The beginning of the growing season would be announced by the early rains toward the end of October. As soon as the soil was moist enough to be worked, it was sown by hand with wheat or barley. After sowing, the soil was plowed with a pointed stick tipped with metal, which made furrows in the soil but did not turn it over. The grain grew during the rainy months of winter. The first crop for harvest was the barley, some time in March. Barley was grown especially on the southern coastal plain because it will tolerate less fertile soil and a drier climate than wheat. But barley is less desirable as food, so wherever conditions permitted, wheat was grown instead. The wheat harvest would begin in April and run into July. Toward the end of the summer the grapes would ripen and the making of wine would begin. The picking of olives to be pressed for oil could be delayed into the autumn until all the grapes were safely in. Olives and grapes were the specialties of Judah: Both the olive tree and the grapevine have deep root systems that are not affected by the drought of summer, and both adapt well to growing on hillsides.

But however patterned the agricultural year was, farmers could never take their success for granted. Disaster was always just around the corner. The autumn rains might be delayed a month or more, so that the grain could not be planted in time to mature during the winter. Or the dreaded sirocco, the hot wind that blows up from the southeastern desert during the transition between seasons, might last too long and wither the crop before it could be harvested. Besides various plant diseases for which no preventives or remedies were known, there was always the threat of an outbreak of locusts. The vines had to be guarded because animals were fond of ripe grapes. Of the livestock, sheep were especially vulnerable to predation and accident because they had to be constantly out on the range searching for food and water. If farmers got through the season safely and had a successful harvest, it was cause for rejoicing and offering thanks to the supernatural power that had so favored them—for no one in that time and place believed that natural events just happen. Hence religious festivals were closely integrated with the agricultural year and were direct expressions of the people's ties to the land that gave them their livelihood.

The public or institutionalized religion, whatever it might have been at any given time, was in the custody of a priestly caste that lived and functioned at various shrines, like Shiloh, Shechem, Bethel, and Jerusalem—places with a historic claim to being sacred. At the more important of these there would have been a building, a temple, consecrated to the god or goddess and supported by contributions from people who desired the favor of that deity. It was believed to be a locus of the deity's power and in certain respects his or her actual home. The temple was not designed as a place for people to gather as a congregation, and nothing at all took place inside it that resembled what we would call a church service. Indeed, most persons never saw a temple from one year to the next; for not only did going to one require a journey on foot that might take several days, but few could afford the animal or perhaps even the vegetable sacrifice they would have been expected to bring.

The actual religion was private and domestic, constituted of beliefs and practices that had been shaped by centuries of rehearsal and that remained largely unaffected by events on the national scale (or the campaigns of religious reformers). It retained a considerable heritage of primitive animism: the belief that all objects are inhabited by spirits. And though the biblical authors went to great lengths to present Yahwism as a religion that abhorred the worship of idols (or even the creation of idols), archaeological evidence has by now made it clear that small clay and metal images of the gods, including that of Yahweh himself, were abundant. The sense that some deity was watching over one's daily activities was brought home, very literally, by the provision of a niche or ledge on one wall of the house where an image would sit and where it could receive the small attentions that wove it into the fabric of domestic life. Very likely the story in Genesis 31 of the household idols accurately reflects their importance, incidentally making it clear that they were part of the family inheritance, incorporating as it were the history of that family's relationship with the divine.

The center of daily life in ancient Palestine was, of course, the family. For the great majority of persons, home was a tiny stone structure with two rooms and an earthen floor. There was nothing that we would dignify with the name of furniture. Food was prepared separately for each meal, served in earthenware vessels, and eaten with the fingers. The diet was simple and almost entirely vegetarian. Flour for bread was ground in the home, and most bread was in the form of flat, unleavened cakes. Grains, whether baked as bread or made into porridge, were the mainstay, varied with garden produce such as onions, lentils, and melons—all washed down with home-made wine. The ordinary family would have meat to eat only on special occasions. Domestic fowl were unknown in Palestine until the Greek period.

In thinking about this society, we have to imagine one without social security, organized medical care, elected officials, postal service, public transportation, coinage, shopping malls, grocery stores, or piped water—to name only some of the things we expect and depend on in our own society today. Rich people, of course, lived as luxuriously as possible and largely escaped the effects of these limitations. But the rest could not. Within their horizons were only the inescapable and apparently eternal facts of daily hard labor, taxes, and armed vigilance, with hope for an old age sustained by numerous and loyal children if life had not been cut short before then by disease, accident, or war.

The basic dependence of the common people on the land transcended differences imposed by the peculiarities of geographical region. Not that we should minimize the importance of these regional distinctions. For the shepherds of the Negev, living in tents and moving their flocks from pasture to pasture, had little in common with the orchardists of Judah, carefully tending their olive trees and grapevines on a few acres of stony ground and seldom out of sight of their native village. And what did either of them know of the wheat growers in the plain of Esdraelon or the fishermen of Galilee? They all might have been—in a sense they all were—living in different worlds.

The centrality of human ties to the small geographical areas in which most spent their entire lives is manifested in the belief that the gods they worshipped were indigenous and local. Yahweh was the god of Israel—that is, of the land of Israel as well as of the people—as Chemosh was the god of Moab and Milcom the god of Ammon. (On this important matter, see Joshua 22:19, Judges 11:23–24, 1 Samuel 26:19–20, and 1 Kings 11:5–10.) The idea of a universal deity did not exist for most of the biblical period; and every religion, including Yahwism, at least implicitly acknowledged the existence of other gods (who, however, tended to be impotent outside the boundaries of their own realms). Under these circumstances it was inevitable that a covenant that Israel entered into with Yahweh would focus on possession of the land, the "promised" land, and that this possession would ratify the exclusiveness of the relationship with their deity in a very material way. The biblical picture of the promised land is a strongly idealized one (for example, in the allocation of areas in Palestine to the different tribes, or the fixation upon "Mount Zion," a place that owed more to the imagination of the prophets than to the topography of Jerusalem). Yet without question the real, physical land was the basis of the Israelite identity. One of the fateful ironies of history is that it took the loss of this land in the Assyrian and Babylonian conquests to remind its inhabitants how vital it was to their national identity.

SUGGESTED FURTHER READING

Yohanan Aharoni and Michael Avi-Yonah, *The Macmillan Bible Atlas*, rev. ed. (New York: Macmillan, 1977).

Denis Baly, *Basic Biblical Geography* (Philadelphia: Fortress Press, 1987).

G. S. P. Freeman-Grenville, *Historical Atlas of the Middle East* (New York: Simon & Schuster, 1993).

Herbert G. May, *Oxford Bible Atlas*, 3rd ed., rev. by John Day (New York: Oxford University Press, 1984).

Martin Noth, *The Old Testament World* (Philadelphia: Fortress Press, 1966).

James B. Pritchard, ed., *The Harper Atlas of the Bible* (New York: Harper & Row, 1987).

John W. Rogerson and Philip Davies, *The Old Testament World* (Cambridge: Cambridge University Press, 1989).

A. S. Van der Woude, ed., *The World of the Bible: Bible Handbook*, vol. 2, trans. Sierd Woudstra (Grand Rapids: William B. Eerdmans Publishing Co., 1986).

The Interpreter's Dictionary of the Bible, ed. George A. Buttrick et al. (Nashville: Abingdon Press, 1962). See articles on Agriculture, Canaanites, City, Palestine, Water, and Water Works. Supplement, 1976: See articles on Agriculture; Forest; Israel, Social and Economic Development of; Nomadism; Prehistory in the Ancient Near East; Tribes; Tribes, Territories of.

The Anchor Bible Dictionary, ed. David Noel Freedman et al. (New York: Doubleday, 1992). See articles on Agriculture; Geography and the Bible (Palestine); House, Israelite.

VI

The Formation of the Canon

Much of the preliminary work in the study of the Bible as literature involves the removal of misapprehensions that have grown up around the Bible because of its sacredness in the eyes of believers. Basic to all these misapprehensions is the view that the Bible is a single, complete, and integral document, unchanged and unchanging, which transcends the conditions of life on earth. We are countering that view here with the conclusions of modern biblical scholarship as well as with a certain amount of ordinary common sense. Although the writing of the Bible may well have been under divine inspiration and although it may be regarded in its entirety as God's revelation, God did not put a single word of it on paper (still less did he dictate the language of the King James Version). The writing of the Bible was done by human beings, acting in history—in human history.

But if the books of the Bible represent the writing efforts of many different persons, widely separated in time, what brought these writings together as they now exist? And what excluded all other writings produced during this period? How is it that these books—and only these books—found their place in the collection? This is our next major issue.

During the many thousands of years that it took the human race to expand over this planet and develop the arts of civilization, the Bible did not exist at all. It came into existence during comparatively recent times in a tiny country at the eastern end of the Mediterranean Sea as a part of the national experience of a specific people known as Hebrews or Israelites. But this did not happen immediately: For a good part of their own history, these people, too, had no Bible. The man revered as their ancestor, Abraham, had no Bible; their leader Moses had no Bible; their great kings David and Solomon had no Bible. What the Israelites did have were miscellaneous collections of writings, mostly without a well-defined identity or status. The writing of these documents began in the early years of the monarchical period (from about 1000 B.C.E. onward), although much of what was written down had been composed earlier and transmitted orally. The documents contained legends of heroes, old patriotic poems, accounts of theophanies, divine legislation, explanations of how things originated (including the story of Adam and Eve), genealogies, court records, and especially stories from the period of the Exodus and the Conquest of Canaan. They were, one might say, the archives of the national history. Many of the stories in these documents were still being transmitted by word of mouth in the monarchical period, and only priests and court officials would have had access to the written copies or would even have known of their existence. It is unlikely that the documents played any part in the actual worship of the people.

This is a long way from having a Bible. Obviously, many steps still had to be taken before these writings became a part of a book of faith with a fixed and unalterable content. No step in the process was more important than that of determining which written materials were sacred scriptures and which were not—in other words, determining the canon.

The English word "canon" is a direct descendant, through Greek and Latin, of a Semitic word meaning "reed" (*kaneh* in Hebrew). Because a reed is long, thin, and straight, it could be used for measuring, as we now use a yardstick; thus the word for reed came to denote a measuring rod and then by metaphorical extension a rule, standard, or norm. It still has these latter meanings. By another extension "canon" is now applied to the thing measured as well as to the standard of measurement. Its use to designate the list of genuine and authoritative books that make up the only proper contents of the Bible comes from the early period of the Christian church, and that is the sense in which we shall use the term here. Not all believers accept the same list, but every Bible in existence today is canonical to some group and contains nothing that the group does not consider genuine and authoritative.

We begin our study of the history of the canon with a present fact: the order of books of the Jewish Bible.

TORAH [also Pentateuch, Law]
    Genesis
    Exodus
    Leviticus
    Numbers
    Deuteronomy
PROPHETS
    Former Prophets:
      Joshua
      Judges
      1–2 Samuel
      1–2 Kings
    Latter Prophets:
      Isaiah
      Jeremiah
      Ezekiel
      The Book of the Twelve (Hosea, Joel, Amos, Obadiah, Jonah, Micah, Nahum, Habakkuk, Zephaniah, Haggai, Zechariah, Malachi)
WRITINGS [also Hagiographa]
    Psalms
    Proverbs
    Job
    The Five Scrolls (The Song of Songs, Ruth, Lamentations, Ecclesiastes, Esther)
    Daniel
    Ezra-Nehemiah
    1–2 Chronicles*

The chief alternative name for the first division is Pentateuch, a Greek word meaning "five scrolls" that indicates the physical form of this division. The Hebrew word *torah*, on the other hand, indicates the division's substance or basic nature: It is the "teaching" that Yahweh gave to Moses on Mount Sinai. ("Teaching" is a better translation

---

*The titles of biblical books and of the collection as a whole always begin with a capital letter, but in normal use they are not italicized or put within quotation marks.

of *torah* than "Law" because the Torah contains much more than just legislation.) No other books of the Bible are considered to have been thus directly given by the deity. The Torah is the foundation of the Jewish faith and is ritually read through from beginning to end, chapter by chapter, in Jewish congregations during the calendar year. The Christian attitude toward the Torah is somewhat different, but all Bibles, Jewish and Christian, begin with these five books in exactly this order.

The second division, the Prophets, includes four books that might more logically be classified as history; but they are considered prophetic because they deal with that period of Israel's history within which the great prophets lived—a period that, in a religious view, can be called the Age of the Prophets. The third division, the Writings, is obviously a catchall, containing poetry, moral tales, wisdom writing, a theological drama, historical chronicles, and an apocalypse. These are the leftovers: The other two divisions had already been completed before any of these became a candidate for inclusion.

Although the order of the Torah never varied, old manuscripts and printed versions show that there was variation in the order of books in the two divisions following it, especially in the Writings. These need not concern us here, but it should be noted that the order of books in the Christian Old Testament is considerably different from the Jewish one. The Protestant Old Testament canon is identical in content to the Jewish canon, but it is split into thirty-nine rather than twenty-four books* and arranged according to literary categories: history (Joshua through Esther), poetry and wisdom (Job through the Song of Songs), and prophecy (Isaiah through Malachi). The Catholic Old Testament is somewhat longer because it contains certain books not included in the Jewish canon (as we shall explain in chapter 12); but it is ordered on the same principle, the source of which is the ancient Greek version of the Hebrew Bible, the Septuagint.

The whole collection is sometimes referred to by Jews as the "Tanakh," an artificial term formed by combining the initial letters of the three Hebrew words designating the three divisions. But this is a modern device; in ancient times the collection had no title. The term "Bible" (from the Greek *ta biblia*, "the books") came into use during the early part of the Christian era. Needless to say, to Jews there is no

---

*In the Jewish Bible 1–2 Samuel, 1–2 Kings, and 1–2 Chronicles are each considered a single book in two parts. Ezra and Nehemiah are combined, and The Book of the Twelve is counted as one book; but the individual pieces in The Five Scrolls count as five books. In addition to the five books in the Torah, then, there are eight in the Prophets and eleven in the Writings, producing a total of twenty-four.

"Old" Testament, because for them there never has been any other—
certainly not a "New" one, although Jews will often use the phrase,
in conformity with general practice.

MAKING THE JEWISH
CANON

As we have already implied, the
three-fold structure of the Jewish Bible represents the actual order in
which its divisions became canonical:
first the Torah, then the Prophets,
finally the Writings. The first datable event in the history of the
Jewish canon is recorded as having taken place in 622 B.C.E. during
the reign of Josiah, King of Judah. A century earlier the stronghold
of Samaria had fallen to the invading Assyrians; with that event the
northern kingdom, Israel, had come to an end. The tiny kingdom of
Judah was all that remained of the land promised originally to the
descendants of Abraham. Shortly after Josiah's reign, Jerusalem, too,
would fall, and Solomon's Temple would be destroyed. Meanwhile,
there was a period of relative prosperity and stability for Judah. The
Assyrian threat had receded, and Josiah had been able to take advantage of the situation to extend the sphere of his control back into the
territory of what had been the northern kingdom. In addition to being
a politically successful leader, the young king was a reformer, with an
apparently genuine interest in purifying and strengthening the cult in
Judah.* One of the actions he took toward this end was to order repairs on the Temple in Jerusalem.

During the course of these repairs a scroll was found—perhaps
buried under rubble somewhere or stuffed into an obscure cubby-
hole—that, when examined, seemed to contain a body of ancient cultic legislation. This was brought to the king and read out to him.
With astonishment and dismay, he recognized that, according to this
book, the nation was and long had been in gross disobedience of the
cultic laws and that a sweeping religious reform was needed to purge
Judah of its apostasy. This he proceeded to carry out with great zeal.

The story is justifiably a famous one. It is told without much art,
and details are missing that a modern reader would very much like to
have, but that is probably because the writer of the account saw it in a
different light than we do and wrote accordingly. One of the principal
unanswered questions for us is what exactly was in "the Book of the
Law" (2 Kings 22:8), as the writer calls it, using the Hebrew word
torah. Certainly not the whole of what we now call the Torah. Most

---

*Scholars use the term "cult" in a neutral and strictly descriptive way to refer to
any system of religious practices supported by traditional beliefs.

likely it contained a portion of the present book of Deuteronomy, perhaps the central section, chapters 12–28; and it may well have been the somber eloquence of the curses in chapter 28 that moved the king to tear his garments in mourning.

Scholarly opinion does not take too seriously the possibility that the scroll had been hidden in the Temple ever since the days of Solomon. We can only guess about its actual origin. The document may have been composed in relatively recent times—though incorporating older materials—by refugees from the northern kingdom after its fall in 722 B.C.E. or, perhaps, by dissident priests working underground during the days of Manasseh, Josiah's grandfather. For them to write it as an address by Moses was not an act of literary deception. The authors were no doubt perfectly sincere in believing that Moses either did say or should have said these things (the distinction between "did" and "should have" is an invention of the modern mind and did not exist in the ancient world). Obviously, the authenticity of "the Book of the Law" was not questioned in 622 B.C.E. Josiah's purpose in consulting the prophetess Huldah was not to find out whether the scroll was genuine but to find out what he should do in response to it.

The surprising thing about the story—if it records an actual event—is that as late as 622 even the priests in the Temple apparently did not know the Mosaic law! The story does not suggest that those who found the scroll saw it as an addition to some code of law that they already possessed. For them it was something unique. In other words, up to that time there had been no authoritative scriptural text to which they were accustomed to turn for guidance, nor had they felt the need for such a text. We conclude that the scroll found in the Temple became the very first portion of canonical scripture—but because it was also the only piece of scripture ever to have been made canonical on the spot, it can tell us nothing about the process by which the rest of the Jewish Bible became canonical.

What happened next in the history of the canon is obscured by the catastrophe that overwhelmed Jerusalem in 587 B.C.E. when the final act of national independence was played out. We may easily imagine, though, that the besieged remnant in Jerusalem, especially the Temple officials, did everything possible to save their precious written records from the holocaust. These scrolls would have been of no interest to the conquering Babylonians, who had so much of real value to plunder; so the long file of sorrowing captives that made its way slowly into exile carried with it a substantial portion of the future Jewish canon. By this time, scholars believe, the older historical records had been compiled and edited into a single document, and the history of Israel and Judah had been rewritten and brought up to date in the spirit of the original Deuteronomic authors. Scholars disagree on when the final additions to this work, the Priestly writings, were made—whether just before the Babylonian Exile, during the stay in Babylonia, or shortly after the return from exile in 538 B.C.E.—but in any case the Torah certainly existed in its present form and was considered canonical by 400 B.C.E. At some time not far from this date, as the story is told in Nehemiah, "the Book of the Law of Moses which Yahweh had prescribed for Israel" (Neh. 8:1) was read to the people in a solemn assembly by the scribe Ezra. Out of the very considerable body of written materials that had survived the Exile, the Torah had been fashioned to stand alone as the supreme document of emerging Judaism.

We cannot call the process a speedy one. More than five centuries had passed since the oldest portions of what became the Torah were written down. And we are entitled to suspect that, without the destruction of the Temple and the Babylonian captivity, the process would never have taken place. A part of Deuteronomy was all that King Josiah needed—for centuries not even that had been needed. Only when it seemed to the Judahites that their faith was in danger of disappearing altogether did they feel obliged to fix its traditions permanently as canonical documents. This operation of salvage and reconstruction was a direct response to national crisis. Its product, a book, became ultimately a far more potent object than even the Temple of Solomon.

That the Torah should have been separated off from the rest of the religious literature existing in 400 B.C.E. is something of a historical accident. Indeed, its unity is artificial: To make the collection end with the death of Moses, the book of Deuteronomy had to be separated from a series of further historical narratives that Deuteronomy had already spawned (beginning with the present book of Joshua). In fact, the death of Moses may originally have been narrated at the end of what is now Numbers. Whatever had to be done to make a unit out of the Torah, its being frequently copied and widely distributed after 400 B.C.E. prevented alterations either in its content or its form. Once issued as a unit, it became canonical in the most effective possible way: in the minds and hearts of the devout. As we say now, the canon of the Torah was closed.

But the process of canon formation continued. By no means all of the extant religious literature of the Jews had been included in the Torah. Remaining were documents narrating the history of Israel after the death of Moses and collections of written prophecy attributed to individuals like Isaiah and Jeremiah. This material was full of moral lessons, it spoke of the glorious past of the nation, and, with postexilic additions, it offered hope of a glorious future. It was part of the bone and sinew of Israel's national identity. With the precedent of the Torah to guide them, pious Jews collected these documents and edited them into the books of the Former and Latter Prophets, which gained recognition as canonical at some time not later than 200 B.C.E.

The formalizing of the collection of Prophets merely ratified an opinion already current: that these books deserved honor second only to the Torah. This is shown by the comparatively short time between the two events. Just as the works constituting the Prophets had been in existence when the Torah became canonical, other works, known later as the Writings, were already in existence when the Prophets became canonical. Each event had been an act of selection in which a certain group of books, felt to have an inherent unity and a prior claim to recognition, was picked out of the larger number then available, leaving the remainder for the future to decide. The leftovers were a mixed bag, indeed. Their status is indicated by the preface to Ecclesiasticus, written about 132 B.C.E., which three times specifically mentions the Law (Torah), Prophets, and "the other writers" or "the other books" (language implying that the last group is still undefined).

It might have taken many centuries more for a Hebrew canon to be completed had it not been for two further crises in Jewish history. The first of these was the Jewish rebellion against the Romans in Palestine, led by the "Zealots." In 70 C.E., during the final phase of this rebellion, Roman soldiers commanded by Titus surrounded Jerusalem, placed it under siege, captured it, and burned the Temple. During the siege a Jewish elder, Rabbi Johanan ben Zakkai, managed to escape from the city through the Roman lines for the purpose of setting up a center for Judaism at Jamnia (or Jabneh), a village near the coast west of Jerusalem and out of the battle area. Here he was joined by other rabbis, and together they established a council to replace the now-extinct Sanhedrin, the ruling Jewish Council in Jerusalem. The Sanhedrin had had a degree of political authority and was dominated by the Sadducees, but the council at Jamnia had only religious authority and was composed of Pharisees. (For a discussion of the Pharisees and Sadducees, see chapter 11.) Its self-assigned role was a complex and immensely important one: to determine how Judaism could survive without the Temple. In some ways the situation was analogous to that during the Babylonian Conquest five centuries earlier. Again, the cultic center of Judaism was overwhelmed, again the Temple was destroyed, again the national faith was in grave jeopardy.

One of the responses to this situation was the closing of the canon of Jewish scriptures by fixing the limit of the third division, the Writings. The Pharisees, whose movement had been hospitable to these books, now saw the danger to their orthodoxy in leaving the canon unsettled. It was time to retrench, to purify, to define—in short, to fix once and for all the contents of the third division. That much is certain. But we do not know whether there was any systematic consideration of canonical questions at Jamnia or, indeed, whether there was a real "council" in a formal sense, with power to make decisions. No minutes were kept. We do know that there were sharp differences of opinion on the acceptability of Ecclesiastes and The Song of Songs. In the end both were accepted, perhaps because they had been ascribed to Solomon. The supposed date of the writing of a book would have been crucial because the belief was already current that there had been no prophets in Israel since the death of Ezra, and thus no books written after that time could be inspired. But all we can say for sure is that no book not already popular could ever have made its way into the canon, regardless of other considerations. Hence what the council at Jamnia did for the Writings was essentially to recognize a canon that had already been made in an informal way through the esteem of generations of the faithful. By 100 C.E. (a convenient round number that is accurate enough for our purpose) the canon of Jewish scriptures was closed for all time and Judaism became, as the phrase goes, "a religion of the book."

The action at Jamnia left a number of books outside the canon. These were miscellaneous works in Hebrew, Greek, and Aramaic dating from the previous several centuries—parts of the extensive library of Jewish religious writings. We should not think of them as having been deliberately excluded from the canon but rather as having failed to get in: For political or doctrinal or other reasons none of them had attracted a sufficient following. In chapter 12 we discuss these "outside" books in detail.

THE CHRISTIAN CANON

The books collected into the canon of the Jewish scriptures were all written in the Hebrew language (except for portions of the book of Daniel, which are in Aramaic). Since the latter part of the third century B.C.E. there had existed a translation of these (and some other) books into Greek, the Septuagint, which had been undertaken for the use of Greek-speaking Jews living outside of Palestine, who could no longer understand the original scriptural language. This translation was very popular. But it was never considered for canonization at Jamnia—indeed, the process at Jamnia was directed *against* the Septuagint. Not only was the Septuagint a translation but it was based on a Hebrew text that differed at some points from that used in Palestine, and it included a number of books felt to be of doubtful authority. Above all, it was now being used by the Christians to support their claims about the Messiah and thus to convert Jews.

Although the Septuagint was the Bible of the early Christians, who believed in it just as much as the Jews of Palestine believed in their own Hebrew text, it could not have satisfied Christian needs permanently. It spoke to them of the coming of the Messiah, but it did not bear direct witness to Jesus' life and teaching, nor did it address the issues that Christians now faced in the practice of their religion. We must not think, however, that in this situation the early Christians sat down consciously to produce scripture. They did not say to themselves, "We had better get busy and make a covenant book of our own to add to the book of the old covenant." They simply wrote. The first two centuries c.e. were a period of furious literary activity. The notion of a new canonical collection did not arise until everything that is now in it—and more—had already been written. In the end only a small portion of this literary output did become canonical. The rest either perished or remained forever "outside."

The earliest of the Christian writers known to us is Paul, whose letters date from the last decade of his missionary activity, roughly from 50 to 60 c.e. Paul's letters were written to deal with particular situations in the missionary field, but they also included instructions on faith and conduct that would have been relevant to Christians anywhere. They were preserved by the addressees for the first reason, copied and collected by others for the second. Because letters do not come together spontaneously, it is reasonable to suppose that some admirer of Paul's writings took the initiative a generation after the apostle's death and sought out such letters as still existed in order to publish them together. In the interval many letters must have disappeared. The Pauline tradition caught on quickly; by the middle of the second century, a Pauline canon of sorts existed (see, for evidence of this, 2 Peter 3:15–16), comprising thirteen letters (or fourteen, with the disputed book of Hebrews). To this Pauline canon the seven "catholic" or "general" letters (James; 1–2 Peter; 1–2–3 John; and Jude) were added during the second century: works obviously imitative of Paul, but bearing the names of apostolic writers.

The four gospels were not written until after Paul's death. That the latter part of the first century was a fluid and formative period in Christian writing is shown by the fact that each of the gospel writers presented his own account of the ministry of Jesus, on the assumption that his particular gospel would be adequate by itself to the needs of his audience. And because Matthew and Luke, writing in 80–90 C.E., saw fit to modify as well as to use the gospel of Mark, it is clear that to them Mark was not canonical scripture. Between this period and the middle of the second century, however, the situation changed rapidly. Within a few years of the issuing of the fourth gospel, someone obtained copies of all four and combined them into one book. The usefulness of their different perspectives on the life of Jesus, or, perhaps, their sheer bulk as evidence when combined, could have motivated him; but whatever his motive, we should pause to acknowledge that the four-gospel tradition, so obvious to us, was once a revolutionary idea. As we saw in the case of the Torah, when books are published together as a unit, they tend to stay together and to resist additions. When the influential churchman Irenaeus, writing about 180 C.E., vigorously championed the four-gospel tradition and proved in a way satisfactory to himself why there could not be more than four gospels, he was dealing with a canonical unit already well established in the Church, if not necessarily taken for granted in it (hence the need for defense).

The four gospels became canonical at a time when there was only a rough consensus about the status of other Christian writings. The next most stable element was the central group of Pauline letters. The book of Acts, which continued Luke's gospel, became canonical independently of that gospel during the latter half of the second century, having been separated from it for reasons that are not clear to us now. There were strong and continued differences of opinion in the Christian world on the acceptability of Hebrews and Revelation, and many powerful churchmen sponsored such popular (and now apocryphal) works as The Teaching of the Twelve Apostles, The Shepherd of Hermas, and the Letter of Barnabas. Thus it would be more accurate to speak of New Testament canons rather than of a canon prior to about 400 C.E., when the influence of Jerome's Vulgate began to prove decisive and the present New Testament canon of twenty-seven books (which is accepted by all Christian faiths) became fixed.

Again—not so dramatically as before, but no less certainly—a good deal of the impetus toward fixing the canon came from a religious crisis, this time the pressure of various heresies. These movements, now only historical curiosities, were once very real threats to the Church. One famous heretic, Marcion, actually issued a New Testament canon of his own in the middle of the second century. If the early Christians had seemed about to snatch the Jewish Bible away from the Jews, and the Jews retaliated, now the Christian heretics were doing much the same thing with Christian religious documents, and their challenge was met by the same kind of response.

The only principle that carried real weight in determining canonicity in the New Testament was whether the writer of a document was believed to have been an apostle, having known the living Christ, or, as in the case of Mark and Luke, one who had received the tradition directly from an apostle. The pseudonymous "catholic" letters, the originally anonymous gospels, and the book of Revelation were all justified in this way. This is indicative of the importance placed by the Church on living witness to the Incarnated Saviour. (Incidentally, it may explain why for at least a whole generation after the Crucifixion there were no gospels at all; who needed a written document if he or she had actually been in the Saviour's presence or knew someone who had?) The principle of apostolicity is firmly grounded in historical reality and differs quite markedly from the corresponding principle in the Jewish scriptures, that of prophetic inspiration.

But apostolicity, like inspiration, tends to be urged after the fact. At the stage when it becomes a question of deciding on explicit grounds whether a book should or should not be canonical, its status has already been determined by two different forces: time and consensus. How much time and how wide a consensus it is impossible to specify. A good deal of both, surely. But if generations of the faithful read and cherish a given document and find in it a source of spiritual strength, if it tells them what they want to hear, then all an ecclesiastical authority needs to do or can do is ratify this traditional choice. In a very real sense the Church, the body of believers, creates the writing that it wants to have. Individual writings do not force themselves onto the Church and gain admittance to the canon on their own merits without first having undergone this long process.

CAN THE BIBLICAL CANON EVER BE CHANGED?

The standards of time and consensus explain why the books that are in the Bible are in the Bible. But there are many books not in the Bible that meet these standards in that they are well loved and long used. Might not some of them be admitted? Why should not *The Pilgrim's Progress* be in the Bible? Surely its author, John Bunyan, was an inspired writer. It should not make any difference that he wrote in the seventeenth century rather than the first century, for God's truth never changes. Nor should it matter that he wrote in English: God's truth can be conveyed in the English language, as the Bible that we read testifies. *The Pilgrim's Progress* is not disqualified by being an allegorical fiction, for so, too, is some of the Bible. And there is no question as to the enormous popularity of Bunyan's work over the years among Christians. If *The Pilgrim's Progress* is too Protestant for some tastes, we could balance it with *The Confessions of Saint Augustine* or some other classic of Catholic faith. Having done this much, we could press on to urge the inclusion in the Old Testament of Kahlil Gibran's *The Prophet*, which is not less religious than The Song of Songs, or one of the great religious best-sellers of all time, Charles M. Sheldon's *In His Steps* (some thirty million copies printed since 1896). Our canon could then be brought up to date on a major social issue of the twentieth century by following the proposal of a group of African-American clergy in the United States that Martin Luther King, Jr.'s "Letter from Birmingham Jail—April 16, 1963" be added to the Bible because of its eloquent statement of the case for involving the Church in the struggle for civil rights. The possibilities are enormous. In none of the cited cases could any reasons be found for excluding such books that would not also exclude books already in the canon.

But the reasons have ceased to operate. Jews and Christians have decided that *this* is their Bible and that the canon is closed. Once closed, a canon never reopens. As time goes on, the possibility of change becomes smaller rather than larger: Unlike other works of mortal hands, the biblical canon is strengthened and upheld by age. By no effort of the imagination can one visualize the present Bible expanded to include another work, whatever its quality. For example, there is the Gospel of Thomas, a collection of sayings attributed to Jesus, discovered in Egypt during the 1940s. This collection preserves what some scholars believe to be authentic words of Jesus in a form closer to what he originally said than the versions given in the canonical gospels; but these sayings will never be added to the New Testament. It is not a question of our being unable to prove that they are authentic, for when it comes to that we are unable to *prove* that Matthew, Mark, Luke, and John are authentic. We may talk about divine inspiration as the great defining characteristic of all authentic scripture; in fact there is no way to tell simply by looking at a document whether it is inspired or not.

This discussion is not intended to suggest that there is any great pressure to add to the Bible. The vast majority of believers would be surprised merely to hear the question raised, so much do they take the Bible in its present form for granted. These believers are an enormous conservative force—a fact that has been tested from time to time with the issuance of new translations of the Bible or modernizations of language in church ritual. Invariably there are widespread and angry protests from persons who believe the innovations to be subversive of true religion. In addition, we should bear in mind that a canon is an *official* collection, sponsored by a religious establishment. Though the canon may have been created in the first place by the consensus of the faithful, the real interest in preserving it intact is that of the priesthood or rabbinate. *They* cannot afford to be casual about its contents, whatever other persons feel.

We should conclude by pointing out that if the Bible is throughout the word of God, it should in theory be useful and relevant in all parts to the same degree. But no one who uses the Bible treats it that way. Not all canonical books are in practice equally canonical. For all readers there is a canon within the canon, a list of favorite books or favorite passages to which they habitually return; correspondingly, there are books and passages that they never read or read only under duress. Few would deny that there is great weariness of the flesh in working one's way through the lists of names in Chronicles or the descriptions of sacrificial ritual in Leviticus; many Christians have as little to do with the book of Revelation as possible, preferring Paul's teaching on love to John's lurid visions of revenge; few readers can find religious inspiration in the The Song of Songs or in Esther; and most are not uplifted by the slaughter that accompanies the conquest of Canaan (Joshua 10–11) or the story of the Levite and the concubine (Judges 19); and so on. When using the Bible to provide proofs in religious argument, readers always turn to those passages that support their own point of view and ignore those that do not. The familiar maxim that even the Devil can quote scripture to his own purpose indicates the variety and multiplicity of this inexhaustible collection, for there is something here for all tastes. Most of us would vote to exclude some portions of it if the canon were to be submitted for ratification now.

But it will not be. Not only can no other books get into the Bible, none of the present canonical books can get out. That is what it means to say that the canon is closed.

SUGGESTED FURTHER READING

Peter R. Ackroyd, "The Old Testament in the Making," in *The Cambridge History of the Bible*, vol. 1 (Cambridge: Cambridge University Press, 1970), pp. 67–113.

G. W. Anderson, "Canonical and Non-Canonical," in *The Cambridge History of the Bible*, vol. 1 (Cambridge: Cambridge University Press, 1970), pp. 113–59.

James Barr, *Holy Scripture: Canon, Authority, Criticism* (Philadelphia: Westminster Press, 1983).

Hans Freiherr von Campenhausen, *The Formation of the Christian Bible*, trans. J. A. Baker (Philadelphia: Fortress Press, 1972).

C. F. Evans, "The New Testament in the Making," in *The Cambridge History of the Bible*, vol. 1 (Cambridge: Cambridge University Press, 1970), pp. 232–84.

R. M. Grant, "The New Testament Canon," in *The Cambridge History of the Bible*, vol. 1 (Cambridge: Cambridge University Press, 1970), pp. 284–308.

Frank Kermode, "The Canon," in *The Literary Guide to the Bible*, ed. Robert Alter and Frank Kermode (Cambridge, Mass.: Belknap Press, 1987), pp. 600–610.

Lee Martin McDonald, *The Formation of the Christian Biblical Canon* (Nashville: Abingdon Press, 1988).

James A. Sanders, *From Sacred Story to Sacred Text: Canon as Paradigm* (Philadelphia: Fortress Press, 1987).

K. Lawson Younger, Jr., W. W. Hallo, and Bernard Batto, *The Biblical Canon in Comparative Perspective*. Vol. 11 of *Ancient Near Eastern Texts and Studies* (Lewiston, N.Y.: E. Mellen Press, 1991).

The Interpreter's Dictionary of the Bible, ed. George A. Buttrick et al. (Nashville: Abingdon Press, 1962). See articles on Canon of the NT, Canon of the OT.

The Anchor Bible Dictionary, ed. David Noel Freedman et al. (New York: Doubleday, 1992). See article on Canon.

VII

The Composition of the Pentateuch

We have been able to reconstruct in a general fashion the process by which the Bible was put together. Although there is much disagreement among scholars about certain details in the process and although some of the crucial information about it is now lost beyond recall, few doubt that something very much like what we described in the preceding chapter did take place.

But when we turn our attention from the making of the canon to the actual composing of the texts that went into the canon, we run into sharp and fundamental disagreement. This centers primarily on the authorship of the Pentateuch. According to an ancient tradition that was already well established in Jesus' time, these five books—Genesis, Exodus, Leviticus, Numbers, Deuteronomy—were written by Moses. In fact, they were often called "the book of Moses." Conservative believers, both Christians and Jews, still hold to this tradition and defend it earnestly. Indeed, no other article of faith about the Bible (except perhaps that of its divine inspiration) is held more tenaciously, and probably nothing that we are proposing in this book will be rejected with greater heat by some of our readers than what follows in this chapter. For we are going to present a theory that the Pentateuch comes from a number of different sources—none directly attributable to Moses—and that it was composed over a period of many centuries by persons with various motives who worked with traditional records of their people, culling, patching, rewriting, and amplifying them. The work thus produced was not a coherent and unified composition from a single pen but the product of anonymous writers and editors, most of whom lived long after the days of its reputed author. This view of the composition of the Pentateuch is called the "documentary hypothesis" or the "documentary theory," or, sometimes, the "Graf-Wellhausen hypothesis," after two German scholars who played important roles in its development.

THE TRADITION OF
MOSAIC AUTHORSHIP

Before taking up the documentary theory, let us first look at the evidence for Mosaic authorship. Moses appears early in the book of Exodus; from that point onward, the Pentateuch is directly or indirectly the story of his career. From Exodus through Numbers this story is told in conventional narrative form: Moses is presented in the third person as an actor in the events recorded; and when he speaks, his words are part of a general dialogue. Deuteronomy, however, presents Moses more as author than as actor because almost the entire book is made up of speeches purported to have been delivered by him to the Israelites before their crossing of the Jordan. Such narrative as it contains, up to the last four chapters, comes to us through Moses' own words. But nothing is said in Deuteronomy to imply that Moses also composed the four books preceding it; in fact, in those books themselves no author is mentioned. Taken by itself, this evidence would never suggest that Moses wrote the whole Pentateuch, and since a number of other biblical books are generally agreed to be anonymous (Judges, for example), the reader may wonder why it is so important to tie the Pentateuch to Moses. If all this material is inspired anyway, does it matter which human hand transmitted the inspiration?

It does matter a great deal to those persons who are made nervous by any attempt to analyze the sources of the biblical text, which they interpret as an attempt to discredit the Bible. By their logic, the God who intended us to have the sacred text would not have allowed it to come into being haphazardly but would have entrusted its composition to one inspired person. This would only be consistent, they feel, with what the Bible itself shows us of God's providence working through human history. And who would be more appropriate for this task than Moses himself, the hero of the story, chosen by God to receive the Law and transmit it to the people? If more than two thousand years of tradition assert the Mosaic authorship of the Pentateuch, why should anyone be so presumptuous as to deny it? Surely the burden of proof rests, does it not, on those who would do so?

The majority of modern biblical scholars accept the burden willingly. And the endeavor they are engaged in is hardly new: Behind them are two centuries of biblical criticism by scholars (many of them devout persons, both Christian and Jewish) who did not shrink from asking searching questions. As a result, our understanding of the Bible has been enormously furthered in all kinds of ways. Such biblical criticism has given much of its attention to the investigation of sources. It is perfectly logical that source criticism (as it is technically termed) should begin by addressing itself to the Pentateuch; for not only does the Pentateuch contain the first five books of the Bible but the claim that these books are a unit composed entirely by one man, Moses, is the most ambitious claim that the traditional view of biblical authorship has to make. Hence it must be settled one way or another before further criticism can take place.*

THE EARLY QUESTIONING
OF MOSAIC AUTHORSHIP

The traditional view held sway without serious challenge until three centuries ago. The first important challenge to it came from the pen of the English philosopher Thomas Hobbes in 1651. In his *Leviathan* Hobbes used evidence in the Old Testament text itself to argue that Moses did not write any of the Pentateuch except the laws in Deuteronomy specifically attributed to him. From there Hobbes went on to consider the authorship and dates of composition of the rest of the Old Testament books, concluding that the historical books through Kings and Chronicles were written long after the events described in them and that the Old Testament text was "set forth in the form we have it in" by the scribe Ezra following the return from Babylonian captivity. Hobbes's bold attack was not fully documented, and it was only incidental to other matters in his long philosophical treatise. It remained for the Jewish philosopher Benedict Spinoza, in 1670, to make the first systematic attack on the tradition of Mosaic authorship. Reading the sacred text as closely and care-

---

*In ordinary speech to "criticize" usually means to find fault with. The term "criticism" as used in biblical scholarship is perfectly neutral and has no such connotations. Criticism here is simply informed study of the biblical text under one or more of its possible aspects, such as authorship, sources, redaction, textual transmission, literary forms, or intention. If anything, biblical criticism is a tribute to the importance of the Bible; a lesser book could never have commanded such lavish attention or justified the application of so much human learning.

fully as the rabbis did, Spinoza found abundant evidence pointing to an author who lived long after the time of Moses. His candidate for this role was also Ezra. Spinoza believed that Ezra, in writing the Pentateuch and the narratives following it, "merely collected the histories from various writers," that is, from extant documents, sometimes setting them down simply as they were, but did not live to finish the job of unifying them properly. Hence the text was left a kind of jumble, filled with repetitions, inconsistencies, and historical improbabilities.

It is important for us to remember that these early critics were reading the Bible very literally and were taking the words of the text at face value. When they found a problem, they did not attempt to solve it by interpretation (by looking for some partly obscured divine purpose underlying the words) but rather by the application of the same methods that one would use with any other text, on the assumption that the author or authors meant what they said. The most often cited evidence against Mosaic authorship was a statement in Genesis 12:6, "The Canaanites were in the country at the time." The crucial words are "at the time." The Promised Land was, indeed, occupied by Canaanites as Abraham entered it to make his new home, and at the time when Moses would have been writing this passage (before the Conquest), the land was *still* occupied by Canaanites. "At the time" implies that the Canaanites are no longer present. It would have been nonsensical for Moses to write this but perfectly proper for someone who lived in later times to do so. Similarly, the "to this day" or "still today" passages (for example, in Genesis 26:33 and 35:20 and in Deuteronomy 3:14 and 10:8) indicate a much later perspective than that of Moses.

In 1678 Richard Simon, a French Catholic priest, joined the argument. Part of his evidence was the difference in style that various passages in the Pentateuch show, thus indicating that they came from different documentary sources. But Simon's work, like those of his predecessors, did not go beyond substituting the anonymous authors of these documents for Moses. Source criticism still needed a way of identifying and separating the documents before anything else could be done with them.

The first step toward this end was taken by a French physician, Jean Astruc, who in 1753 published a book on Genesis and the beginning of Exodus. He reasoned that, as Moses could not have had personal knowledge of all the events he recorded in Genesis and Exodus, he must have depended on written sources handed down to him from the actual eyewitnesses. Astruc thought that two such "original accounts" could be identified and reconstructed by finding the passages in which the Hebrew deity is called "Yahweh" and those in which he is called "Elohim," assuming that the two names were not used indiscriminately or by chance but, instead, reflected the characteristic vocabularies of two different authors. That this particular difference could be significant had been pointed out by H. B. Witter in 1711, but Astruc went further by producing an actual analysis of the contents of Genesis according to his own scheme. Even though Astruc clung to the idea of Mosaic authorship, his work would prove to have revolutionary impact on biblical source criticism.

If the approach to the documentary sources by this means is valid, then what remains is (1) to strengthen the approach by adding further criteria; (2) to try to discover when, by whom, and why the various documents were written; and (3) to investigate the process of editing that brought the documents to their present state. Of course, this is quite a large order. Filling it was the work of criticism in the eighteenth and nineteenth centuries, a process much too long and complicated to be summarized here. We can only say that biblical criticism flourished during this period in an unprecedented way and that out of the attention lavished on the problem of sources has come the documentary theory that we are presenting. Not every detail in it is above dispute but the consensus of present-day scholars would heavily favor its general outlines.

EVIDENCE FOR A
DOCUMENTARY THEORY

We can begin with Astruc's discovery. Hebrew has a number of words for referring to the deity; the commonest are the two rendered in English as "Elohim" and "Yahweh." (For a full discussion of these terms, see the first appendix of this book, "The Name of Israel's God.") A reader of the opening chapters of Genesis cannot help noticing that the story of creation is presented twice: in 1–2:4a and again in 2:4b-3:24. Anyone reading the original Hebrew might also notice that the first creation story uses only "Elohim" in reference to the deity, whereas the second uses "Yahweh" or "Yahweh Elohim"—but never "Elohim" alone. Continued reading in Hebrew would reveal that a similar variation in terminology exists in the rest of Genesis and that the contexts in which the two terms appear have other characteristics, of style and content, peculiar to each. Perhaps, then, the difference in terminology is no accident: It may point to two different sources for the text.

This hypothesis explains certain obvious repetitions and contradictions. To cite just a few examples, keeping to Genesis, Noah is directed in 6:19 to take two of every kind of living creature into the Ark with him, but in 7:2 he is told to take seven pairs of every clean animal and one pair of those that are not clean. The purpose of the seven pairs of clean animals becomes evident after the Flood when Noah in gratitude sacrifices to God; the other account of the post-Flood activities makes no mention of a sacrifice. In Genesis 21:31 Abraham names a well Beer-sheba; in 26:33 the well is named all over again by his son Isaac. There are two parallel and separate accounts of God's offering the Covenant to Abraham, in 12:1–9 and 17:1–14. And in the story of Jacob and Esau in chapter 27, Jacob is given two separate means of deceiving his father: wearing Esau's clothes, which smell of the open country, and wearing goatskins on his arms to simulate Esau's hairy skin. Afterwards Jacob leaves home and travels to Harran, either to escape Esau's wrath at his mother's advice or to find a wife as his father commands. Still later, Jacob's name is changed to "Israel," but one source of the text ignores this and calls him "Jacob" to the end. In the story of Joseph there is a flat contradiction as to whether the Ishmaelites or the Midianites sell him to the Egyptians. The two sources of the story compete with one another in pushing either Judah or Reuben to the forefront of the brothers who remain at home. And when the brothers with their aged father finally make the trip down to Egypt, in one version it is because Joseph invited them to come down, in the other because the pharaoh did.

One point should be made very clear. We are not citing these problems to undermine the authority of scripture, as used to be the fashion when professional skeptics would lecture to audiences on "the mistakes of Moses." We are merely supplying some of the data on which the documentary theory rests. Efforts to reconcile contradictions or explain away problems have been made and will be made by persons who feel that the integrity of the text (which for them means its divine authority) must be preserved at all costs. The costs, however, tend to be rather high. Whenever there are contradictions or other problems, the documentary theory usually presents a more reasonable alternative, and it is accepted by a great many scholars who do not feel their faith threatened by the possibility that the Bible text, being a product of human history, experienced some adventures in reaching the point where it is now.

Let us now look at the two creation stories in some detail. The differences in their structure and content can best be seen in tabular form:

GENESIS 1–2:4a	GENESIS 2:4b–3:24
Creation is divided into days.	No days or other periods of time are mentioned.
Creation has a cosmic scope.	Creation has to do with the earth only.
Animals are created before man.	Man is created before animals.
Animals are part of a cosmic design (along with plants and everything else).	Animals are created for a special purpose: to keep man company.
Man is to rule the world.	Man is to have charge of Eden only and, presumably, is never to leave it.
Woman is created simultaneously with man.	Woman is created after (and from) man.
No names are given to creatures.	All creatures, including man and woman, are given names.
Only the deity speaks.	Four speakers engage in dialogue, one of them an animal.
The deity makes a day of the week holy.	The deity forbids eating the fruit of a tree.

It is clear, even in translation, that the first story employs much verbal repetition and is precisely and regularly organized, with the separate acts of creation carefully set in parallel form. It is austere, dignified, solemn in movement, almost ritualistic, as befits the theme that it unfolds. No word is used carelessly. Obviously, a literary artist of great skill wrote it. The second story is no less skilful. In contrast to the first, however, it is down-to-earth and appeals to the mind's eye with many vivid and concrete details. The deity creates Adam not by uttering a verbal command but by descending to the barren plain of earth, taking some clay, molding it into the figure of a man, and then breathing life into it. (The Hebrew verb here, *yatsar*, is the same one that would be used of a human potter molding or shaping a vessel.) The creator is anthropomorphically represented as one of the actors in a drama. A major purpose of the second story seems to be etiology: to explain how something got started. And it is an incomplete story, bringing man to the threshold of history with all of time before him, whereas the first account is complete and implies no sequel or further action.

The two voices that we hear in the opening chapters continue to be heard as we proceed through Genesis. The characteristics just noted still prevail. The first voice seems preoccupied with order and regulation and now and then produces a genealogical list that makes dull reading for us but must have seemed quite important to the writer. The deity of the first account is remote and abstract—powerful, but not distinct to the human imagination. At appropriate intervals he issues sweeping laws: for observing the sabbath, against eating blood, for circumcision of all males. The second voice, however, continues with its anthropomorphic presentation of the deity, and this writer's flair for the dramatic is revealed in a string of fascinating stories: Cain and Abel, the Tower of Babel, Noah drunk and naked in his tent, Abraham bargaining with God over Sodom and Gomorrah, Isaac and Rebecca, Jacob wrestling with a divine antagonist at the ford of the Jabbok.

THE DOCUMENTARY SOURCES

Biblical scholars attribute these stories to a source they call the "Yahwist," following Astruc's insight. The Yahwist source is designated by the letter J (according to the German system of spelling in which the sound of Y is represented by J). The other source came to be called the "Priestly" or P, because of its overriding interest in ritual legislation (this is much more apparent in the later books of the Pentateuch than in Genesis). A third source that can be traced in Genesis is now called the "Elohist" or E, even though the Priestly source also uses "Elohim" as the name of God. Finally, there is a fourth source, the "Deuteronomic" or D, which in the Pentateuch is found only in the book of Deuteronomy (we shall discuss this a little later).

While the four sources were being identified by nineteenth-century biblical scholars, the question naturally arose as to their order in time. Unless we know this or have a good hypothesis about it, we are at a dead end, for the Pentateuch remains just as mysteriously cut off from the stream of human history as it was when Moses was believed to be its author. The key to solving this problem was the realization that P, the first document encountered and the source of Genesis 1, was actually the last one to have been composed. It simply did not exist in the times of the kings of Israel and Judah and of the prophets before the sixth century. There is no other way to explain the absence of any mention of ritual legislation by the authors of the history of Israel in Samuel–Kings or by the preexilic prophets. There was, of course, a national cult that worshiped the deity with animal sacrifice, observed a sacred calendar, and had certain rules of conduct; but the historians and prophets never connect any of this with a written document or with written authority. Indeed, the prophets refer to the practice of sacrifice only to denounce it. This picture was considerably altered by the Priestly writers who, late in the process of forming the Pentateuch, took it on themselves to project back into the earlier history of Israel an idealized version of the national cult as they themselves then practiced it or wished it to be, thus justifying it by the authority of Moses during the Exodus when, it was believed, Israel as a covenanted people came into being.

It is interesting that the Priestly writers did not, apparently, attempt to avoid inconsistencies and duplications in their revision and completion of the history of Israel. Some of these problems they inherited with the JE document (discussed later), but some they created for themselves by their determination to retell theologically important episodes in their own way (such as the story of creation, the covenant with Noah, and the revelation of Yahweh's name at Sinai) and then to retain both versions in their text. We have difficulty understanding the psychology of this procedure because a modern writer would, above all, want to produce a harmonious and internally consistent text. But we must be careful not to impose our own literary values on texts composed over two millennia ago.

Of the two sources that the Priestly writers worked with, J is thought to be the older. In fact, many scholars believe that the Yahwist writing was the basic document and that it once stood independently. When we say "document," we mean a written record. The art of writing was surely not unknown to the early Israelites, and from the beginning of the Monarchy onward it must have been in common use. But many of the Yahwist stories, like the epics of Homer, are older than their written forms. In all cultures the earliest legends and histories circulated in oral form for generations before being written down. Hence there may once have been an epic poem that recounted the patriarchal history of the Israelites (beginning with the creation of the world) that is now lost as such and survives only in prose fragments in the J document. There is evidence that J originated in the southern part of Palestine, in the area that later became the kingdom of Judah. Separate parts of it undoubtedly continued to circulate by word of mouth (perhaps in storytelling sessions by professional entertainers) for some time after scribes committed it to writing. We must remember that the ancient Israelites had no Bible. They did, however, have a lively sense of national identity, and they were proud of their past, which in typical human fashion they enlarged to take in the whole scope of earthly history. Living in an age without the diversions we take for granted, they had plenty of time to listen to public speakers and plenty of incentive to do so. Many a quiet evening hour in a dusty Palestinian village in, say, 800 B.C.E., must have been spent listening with pleasure to a learned man recite the story of Abraham or the miraculous deliverance of the Exodus. The fact that the people had heard the stories many times before only deepened their pleasure at hearing them again.

Although the influence of the oral tradition is not absent from the other sources, it seems to be particularly prominent in J. The Yahwist author makes artful use of humor, irony, suspense, hyperbole, and concrete detail: all devices for appealing to an immediate audience. If we put ourselves in the place of this audience, understanding its attitudes and expectations, we can better appreciate the skill of the storyteller, as, for example, in the story of the courtship of Rebecca in Genesis 24, which is one of the finest products of the Yahwist author's narrative talent. We are carefully made to see, by many devices, that the long and risky journey to find a wife for Isaac, Sarah's only son and the ancestor of Israel-to-be, was entirely under the protection of Yahweh, just as his birth to a ninety-year-old woman had been a miracle. The little episode of Rebecca at the well in verses 15–27, which seems to be the climax of the quest, makes that point quite clearly. (It also gives its audience an exemplary picture of courteous behavior on the part of Rebecca.) The modern reader tends to lose interest in the story at this stage, feeling that the denouement has nothing more to offer. Not so the listeners to the ancient folktale, who looked forward to something still to come. Thus the narrator makes the servant retell the whole story of his search in verses 35–49—even though the reader already knows it—because everything still depends on the reaction of Rebecca's brother, who is listening to him. What will Laban say? Will Yahweh's purpose prevail? By having the servant retell the story—in what is for us suffocating detail—the Yahwist author kept his own audience in delighted suspense.

The Elohist material is not quite so easy to identify or to characterize. Much of what early scholars assigned to the E document has now been assigned to P, and what remains has so many gaps that some scholars now deny that there ever was such a document separate from J. The usual theory is that the Elohist material originated in the north, in what became the kingdom of Israel, also called "Ephraim" after the split with Judah. (It is just a coincidence, but a happy one, that the document which originated in Ephraim is called "E" and the one which originated in Judah is called "J.") The E material is thought to be slightly more recent than the J material. It enters the Pentateuch late, its first substantial contribution being Genesis 20, and it is responsible for the story of the testing of Abraham with the sacrifice of Isaac, portions of the story of Jacob and Esau, and about half of the story of Joseph. It is marked by a certain tact or reserve in the portrayal of the deity, who does not appear to humans in person but communicates through dreams and angels, and by an interest in prophets and seers. It, too, is very skilful, but it lacks the earthy appeal of the Yahwist material.

The D document is the one we can speak about with most confidence as a document, for almost all scholars agree that the "Book of the Law" discovered in the Temple and brought to King Josiah in 622 B.C.E. forms its basis. As pointed out in chapter 6, the D document is exceptional in that it was received at once as a piece of sacred writing, something that occurred only in time with the J and E materials. Furthermore, in contrast to J and E, the Deuteronomic document stimulated a school of like-minded writers to begin producing further material with the same tone and religious outlook, and thus to extend its influence widely over the early books of the Hebrew Bible. Though we said "school," there is in fact no evidence at all about the authorship of the D material, and in biblical criticism it is common to hear its authors reduced to one hypothetical figure, called the "Deuteronomist."

Deuteronomy is notable for its style, and on that ground alone it could easily be separated from the other documents. It is solemn, deliberate, eloquent, much given to certain verbal formulas. Again and again it hammers home the requirement of total obedience to Yahweh. It does this, of course, in what are supposedly the words of Moses. We have already noted in discussing the P document that these later writers read back into the past the practices they wished to see established or reinforced in their own times. The Deuteronomist did the same thing, as indeed he had to if he was to do anything at all, for who would pay attention to any laws that did not appear to come from Moses? There was simply no other authority. But after reminding ourselves not to judge the Deuteronomist by standards of literary conduct appropriate to our own day rather than to his, we find our sympathies strained when we discover that he is not only speaking as Moses but at some points contradicting Mosaic law! The opening sentences of the "Book of the Covenant" (an earlier law code that occupies Exodus 20:22–23:33) enjoin the Israelites to worship Yahweh at simple altars built of earth or unhewn stones, which may be scattered through the countryside. The Deuteronomist, on the other hand, wants all sacrifice moved to one place, Jerusalem, and at Solomon's Temple, which was anything but a simple altar. (The position of this passage, at the beginning of Deuteronomy 12, suggests that the writer was consciously confronting the doctrine that begins the Book of the Covenant.) We wonder how he dared to write this, and we wonder also how the redactor who later added it to the body of existing writings could have failed to see the contradiction. But we have to stop again and remind ourselves that the concept of an organized canonical body of scripture did not yet exist in these times. Certainly there was no person or institution that had the duty of harmonizing inconsistencies among the various documents. What the Deuteronomist did was to reinterpret Israel's covenant relationship with Yahweh according to his own vision of it—which he sincerely believed to be carrying out the authentic Mosaic tradition—and then leave his document to be added to the existing ones. Thus what he wrote—as the German scholar Otto Eissfeldt long ago pointed out— "neutralized" the older documents. The older ones would be either understood in a different sense or simply ignored. This process (without the writing of new documents) continues in our own time, for both Judaism and Christianity have unofficially neutralized considerable portions of the Old Testament by various means—one of which is by ceasing to pay any attention to them.

PUTTING THE
DOCUMENTS TOGETHER

We now need to review the documentary history of the Pentateuch because it has been presented piecemeal and a little out of sequence. The oral material that is thought to be the basis of J, the oldest document, probably originated in the period of the Judges as a product of the growing sense of national identity among the Israelite tribes. In the tenth century B.C.E., during the early years of the Monarchy, some unknown individual wrote these stories down in coherent narrative form to make the document. It was a patriotic undertaking and reflected the feeling of accomplished destiny that David and Solomon inspired in their people. About a hundred years after this time, in the ninth to eighth centuries B.C.E., somewhere in the north (which had become the separate kingdom of Israel after Solomon's death), a writer collected stories then circulating in his area about these same heroes of the past and created the E document. It may well have been less extensive and less complete than the J one. After the Assyrians conquered the north, refugees from Israel brought the E document down to Judah, where a redactor, himself a southerner, wove the two documents together into something that we call "JE." The J document was the basic one in this blend, E being used mainly to flesh out the story at certain points. This took place in the early seventh century. During or after the period of reform instituted by King Josiah in 622, the recently discovered Deuteronomic book was added to JE by another redactor to make what is called "JED."*

Both writing and redacting continued after the creation of the JED document, producing the Deuteronomic History (or DH), which comprises the books from Joshua through 2 Kings. The Deuteronomic History, as such, is not part of the Pentateuch, and because we discussed it at length in chapter 4, we will not say more of it here.

The final document, the Priestly one (P), was written during or very shortly after the Babylonian Exile and clearly reflects the dire need of the people to salvage what they could of their national past from this disaster. It was seen as—indeed, it was—their one final chance to set the record straight. Substantial additions were therefore made to JED, including the whole book of Leviticus: and the result is called "JEDP." A Priestly redactor may have been the one who did the actual combining, patching, and harmonizing of the various documents.

SOME ADDITIONAL
SOURCES

The formula JEDP is a neat one and certainly worth remembering, for it gives us the four major sources of the Pentateuch and the order in which they were composed. But we should also be aware that this is not the whole story: In reading the Pentateuch we encounter JEDP in a very different fashion, often out of sequence and sometimes so scrambled together as to make separating the sources impossible. Nor were these the only sources of the Pentateuch. Many scholars believe that there were independent sources for such units as the "Holiness Code" of Leviticus 17–26 and the Book of the Covenant mentioned earlier. Genesis 14 comes from somewhere quite outside the range of the other Abrahamic materials. The Song of Moses in Exodus 15, down through verse 12 (or perhaps only the refrain, which Miriam sings), is surely very much older than its context and was available to the compiler of Exodus from some kind of document. Bits and pieces of old material, such as the strange story of the proxy circumcision of Moses in Exodus 4:24–26, seem to have been inserted into the narratives by writers or redactors who knew

---

*The role of the redactors in the creation of biblical texts was discussed in chapter 1. Their importance is clear from what we have just presented here, for redaction was involved at every state of the process. Without redaction, the documents would have remained isolated and uncoordinated fragments and no doubt in time would have been lost altogether.

them from an existing source and wished to preserve them even if they did not quite understand what to make of them. These writers also had before them sets of official records that had been kept since the earliest days of the Monarchy. Joshua 10:13 refers to something called "the Book of the Just" (or alternatively "the Book of Jashar"); 1 Kings 11:41 refers to "the Book of the Annals of Solomon"; and in 1 and 2 Kings there are frequent references to "the Book of the Annals of the Kings of Israel" and "the Book of the Annals of the Kings of Judah." These records, now entirely lost, must have been well known at the time, for they are referred to in an almost casual way, with the implication that they can still be consulted should anyone want further information.

THE INTEGRITY OF THE
PENTATEUCH

As the history of the canon in chapter 6 made clear, the Pentateuch is an artificial unit. The JE narrative must originally have closed with the death of Moses at the end of what is now Numbers. This ending was picked up and transposed by the Deuteronomist or his redactor to the end of what is now Deuteronomy in order to allow for the insertion of the speeches of Moses, thus making our five books. But the book of Joshua continues the story in the Deuteronomic manner without interruption, completing it with the successful end of the Conquest and the death of Joshua. Hence many scholars prefer to talk of a "Hexateuch," a six-book unit. On the other hand, because Joshua was written by the Deuteronomic historian responsible for the narrative that extends through 2 Kings, there is no good reason to divorce his work from that narrative by putting it in a Hexateuch. Perhaps the break should be made earlier, reuniting both Deuteronomy and Joshua with the historical books following them, thus leaving a "Tetrateuch," or four-book unit, at the head of the Bible. But there is no real solution to the problem of larger structure when the components themselves are arbitrary. Whatever the other circumstances may have been, we should not forget that the Pentateuch is the unit that became canonical by 400 B.C.E. and that the death of Moses, wherever it occurred in the record, was and still would be the appropriate place for making this first division.

So the Pentateuch remains. The Documentary Theory has certainly not diminished its impressiveness as a human achievement. If anything, we respect it the more, now that we know something about its composition. In age and in magnitude it is like an immense mountain looming before us, a perpetual source of awe, inspiration, and challenge. To read it, as to climb to the top of the mountain, is a worthy achievement. But only to read it and put it down—to clamber over the surface of the mountain and then go home—hardly does it real justice. The mountain is surely more than an inexplicable interruption of the plain, a mere lump or swelling on the earth's surface remarkable only for its size and the way it hinders passage; so, too, the Pentateuch is more than just a week's reading assignment. Dig into it and one begins to find a vivid record of the past. Here is a series of layers of ancient sediment from a river delta or a shallow sea, squeezed and hardened into rock, folded and upthrust by forces in the earth's crust, carved by glaciers, and weathered into fantastic shapes. Over here is rock so changed by heat and pressure that we cannot tell anything about its origins. Here is an abrupt discontinuity that puzzles us. And here is an upwelling of hardened lava that suggests something about turbulent episodes in this mountain's otherwise gradual formation. The more we dig, the more carefully we study, the more we learn. Both the mountain and the Pentateuch survive our researches, and who is to say that the view from the summit is not improved by our knowledge of what had to take place before we could stand there?

SUGGESTED FURTHER READING

Antony F. Campbell and Mark A. O'Brien, *Sources of the Pentateuch: Texts, Introductions, Annotations* (Minneapolis: Fortress Press, 1993).

Otto Eissfeldt, "The Analysis of the Books of the Old Testament," in *The Old Testament: An Introduction*, 3rd ed., trans. Peter J. Ackroyd (New York: Harper & Row, 1976), pp. 155–241.

Peter F. Ellis, *The Yahwist: The Bible's First Theologian, with the Jerusalem Bible Text of the Yahwist Saga* (Notre Dame, Ind.: Fides Publishers, 1968).

Richard Elliot Friedman, *Who Wrote the Bible?* (Englewood Cliffs, N.J.: Prentice-Hall, 1987).

Alan W. Jenks, *The Elohist and North Israelite Traditions* (Missoula, Mont.: Scholars Press for the Society of Biblical Literature, 1977).

Philip Peter Jenson, *Graded Holiness: Key to the Priestly Conception of the World* (Sheffield, Eng.: JSOT Press, 1992).

R. W. L. Moberly, *The Old Testament of the Old Testament: Patriarchal Narratives and Mosaic Yahwism* (Minneapolis: Fortress Press, 1992).

Martin Noth, *A History of Pentateuchal Traditions*, trans. and intro. Bernhard W. Anderson (Englewood Cliffs, N.J.: Prentice-Hall, 1972).

Brian Peckham, *History and Prophecy: the Development of Late Judean Literary Traditions* (New York: Doubleday, 1993).

David Rosenberg and Harold Bloom, *The Book of J* (New York: Grove / Weidenfeld, 1990).

Jeffrey H. Tigay, ed., *Empirical Models for Biblical Criticism* (Philadelphia: University of Pennsylvania Press, 1985).

John Van Seters, *Prologue to History: The Yahwist as Historian in Genesis* (Louisville: Westminster/John Knox Press, 1992).

The Interpreter's Dictionary of the Bible, ed. George A. Buttrick et al. (Nashville: Abingdon Press, 1962). See articles on Biblical Criticism; Biblical Criticism, History of; Pentateuch.

The Anchor Bible Dictionary, ed. David Noel Freedman et al. (New York: Doubleday, 1992). See articles on Deuteronomistic History, Elohist, Priestly ("P") Source, Yahwist ("J") Source.

VIII

The Prophetic Writings

The prophetic writings of the Old Testament occur, not surprisingly, in that section of the Jewish Bible called "The Prophets." But, what is surprising, the prophets share that section with the historical books: Joshua, Judges, Samuel, and Kings. Why this is so was touched on in chapters 4 and 6. But we must now look more deeply into the relationship between the prophets and history, for a knowledge of that relationship is the key to reading this difficult body of material with sympathetic understanding.

In chapter 4 we saw that Yahweh, the God of Israel, was a deity who—however much he transcended the human sphere by virtue of his creative and sustaining functions in the universe—was thought of as being intimately involved in human history. This deity constantly interjected himself into human affairs: as a means, obviously, of controlling those affairs, but beyond that of revealing himself to his creatures. When Yahweh produced the sunshine and rain in the appropriate seasons, this was a gracious exercise of his power in behalf of his people and, furthermore, a revelation of his pleasure in their obedience. When Yahweh withheld those good things so that the crops withered and died, or were ruined at harvest time, this too was both an exercise of his power and a revelation—a revelation of his anger at some failing on the part of the individual or community involved.

It was one thing to know whether Yahweh approved of past actions, but it was quite another to determine what Yahweh wished for the present and future. In the complex give-and-take of human affairs, particularly those involving persons in high government office, how was the will of Yahweh to be precisely discerned? If an enemy threatened to invade the land, should Israel yield? Should it resist? Should it ally itself with some third party? If a new king was to be chosen because the previous one had been found unworthy and his family line had been eradicated, how was the new candidate to be recognized? If over time the worship of Yahweh had been altered through the introduction of elements from the worship of other gods, by what means were the details of this trespass to be found out? In situations where the point in question required a direct answer of yes or no—this as opposed to that—then some simple means of choosing between alternatives could be employed: for example, the casting of lots (the outcome of which was never, to the ancient mind, a matter of chance). But when the matter involved was more complex, when uncertainty prevailed and fear was rife, the faithful longed to hear the authoritative word of God. In that circumstance it was inevitable in ancient Israel that prophets should arise—persons who believed themselves to have a special relationship with Yahweh, one that enabled them to read his intentions and articulate his will and judgments.*

Throughout the long sweep of history covered in Joshua, Judges, Samuel, and Kings and concluding in Ezra-Nehemiah, Israel is represented as having a steady stream of spokesmen (and a few spokeswomen) for Yahweh. Theirs was the obligation to interpret events from Yahweh's point of view and either to threaten the people

---

*Some biblical prophets are represented as receiving the word they are to speak through spectacular means. Moses, whom Israel conceived to be the greatest of its prophets, is described as talking with Yahweh "face to face, as a man talks to his friend" (Exod. 33:11); lesser figures are given their messages in dreams or in visions or through angelic visitations. But just as commonly, no spectacular sort of divine communication is reported, and the prophets seem to have spoken out of their own personal awareness of violations of the ancient religious relations, of callousness in dealing with the downtrodden, of patent folly on the part of leaders of the nation. In a very real sense, any religious figure at any period in history (including our own) who calls fervently on others to return to the old standards of faith and behavior is playing the role of prophet, every bit as much as did those great spokesmen (and spokeswomen) of Yahweh in the Old Testament.

with judgment or to promise them good fortune—as circumstances required. But in that long history, stretching from the thirteenth to the fifth centuries B.C.E., one period above all others required the particular services of the prophets, for it was during that period—from the mid-eighth through the mid-sixth centuries—that Israel lost its independence as a nation and went into captivity. How could this have happened to a nation that believed itself to be Yahweh's own people, a nation descended from the great Abraham, a nation brought into being and given Yahweh's law by the great Moses, a nation once ruled by the great King David, who had been promised that his royal throne would be established forever? The prophets stood ready to explain precisely how it had happened, to warn of what further could happen, and to promise deliverance if Yahweh's conditions were met. So impressive to modern eyes are the words of the prophets of that era that the eighth through the sixth centuries B.C.E. are often called by biblical historians the "Age of Prophecy."

THE GREAT EIGHTH-CENTURY PROPHETS

That age can be said to have begun on a day in approximately 750 B.C.E. when Amos of Tekoa trudged into Samaria, the capital city of the northern kingdom, and began to denounce the apostate religion and the social injustice that he saw about him. The kingdom appeared to the outer eye to be enjoying great prosperity under King Jeroboam II, but the prophet could see that prosperity had been purchased at the expense of the poor and that organized religion was making no effort to remind the well-to-do of their humane obligations to their less fortunate neighbors. Amos warned that judgment must come upon a nation so casual about social injustice and bad religion. At about the same time Amos was speaking out, Hosea began to prophesy along these lines, also in the northern kingdom, and shortly thereafter Micah was preaching a similar message to Judah in the south. Mere religious observances (and it did not help that some of them were of pagan origin) were not sufficient to please Yahweh, these prophets insisted. Amos could have been speaking for all the prophets when he represented Yahweh as saying,

> I hate, I scorn your festivals,
> I take no pleasure
>   in your solemn assemblies.
> When you bring me burnt offerings . . .
> your oblations, I do not accept them

and I do not look
　at your communion sacrifices
　　of fat cattle.
Spare me the din of your chanting,
let me hear none of your strumming
　on lyres,
but let justice flow like water,
and uprightness
　　like a never-failing stream! (Amos 5:21–24)

The northern kingdom fell to the Assyrians in 722 B.C.E. To those observers who maintained their faith in the God of Israel, it would have seemed certain that judgment had come—just as the prophets had warned—because of the apostasy of the people of Israel. What of the southern kingdom? Was it faithful to Yahweh and exempt from the fate that had befallen its neighbor? Not according to the greatest of the prophets of the time, Isaiah (that is, Isaiah of Jerusalem—so called to distinguish this man, whose prophecies and activities are recorded in Isaiah 1–39, from the author of Isaiah 40–55 and from another individual or other individuals responsible for Isaiah 56–66). Isaiah was no humble country person like Amos but an adviser of kings and a man with knowledge of affairs well beyond his nation's borders. Amos, it is true, had denounced the surrounding states before focusing his attention on the northern kingdom, but Amos possessed nothing like the international outlook of Isaiah. For us to read Isaiah intelligently requires us to know something not only of the history of Assyria (the superpower that was a threat to the entire Near and Middle East and that Judah's fearful kings were so wary of) but also of such closer neighbors as Egypt, Syria, Moab, and Edom— all of which, according to Isaiah, were both the tools of Yahweh's justice and ultimately the objects of his wrath. Like Amos, Micah, and Hosea before him, however, Isaiah gave most of his attention to his own people, berating Judah for its idolatry, immorality, and callous disregard for the poor and powerless.

THE GREAT SIXTH-
CENTURY PROPHETS

In addition to the four great literary prophets of the eighth century B.C.E. (called "literary" because biblical books are named for them), there are three major figures who prophesied in the late seventh and the sixth centuries: Jeremiah, Ezekiel, and the author of Isaiah 40–55 (whom scholars call Deutero-Isaiah or Second Isaiah). At the outset of his prophetic career, Jeremiah stressed one of the major themes of the earlier prophets, the infidelity to Yahweh evident in the people's participation in pagan rites of worship, particularly those of the fertility cults. But the great work of Jeremiah's life was to offer counsel to his people through the unhappy years when the Babylonians moved against Judah and the city of Jerusalem, first deporting a portion of the population to Babylonia in 597 B.C.E., then in 587 burning the Temple and city and deporting additional citizens, and finally making a third deportation in 582. Jeremiah's constant message during all this was that the Babylonian victory was inevitable—it being a judgment of Yahweh on the nation's apostasy—and that the people must submit to it. Such a message was, of course, unpopular with the leaders of Judah, and Jeremiah suffered great hardship at their hands for speaking as he did. Those who had been carried into exile, he said, must settle into their new circumstances, work hard and prosper, and pray for the welfare of their captors, for their exile would be a long one and future generations of Israel must be provided for.

During part of the time that Jeremiah was active in Jerusalem, Ezekiel too was prophesying to the inhabitants of that doomed city, apparently while he himself was among the exiles in Babylonia. It appears that Ezekiel had been carried off to Babylonia during the first deportation of 597 and there, having been called to the prophetic office, communicated his message to Jerusalem in writing. In general that message was like Jeremiah's: Because Jerusalem had broken the Covenant and abandoned Yahweh, destruction was coming upon it. But Ezekiel's driving concern was to portray the utter wickedness of Israel's revolt against Yahweh, and thus he gave much less attention to the practical politics of Jerusalem's last days than Jeremiah did. And to a greater extent than Jeremiah, Ezekiel expressed confidence that in some future day Israel would be revived as a nation, would repopulate its ancient territory, and would exist thereafter as an ideal state enjoying a new covenant relationship with Yahweh. Ezekiel's vision of Israel's distant future was as heartening as his view of its present condition was grim.

The third great prophet of the sixth century B.C.E. was the nameless individual called Deutero-Isaiah, author of the bulk of chapters 40–55 of the book of Isaiah. This body of prophetic material, sometimes called "The Consolation of Israel," was apparently composed at about the time when permission for the exiled Israelites to return to Judah was granted in 538 B.C.E. by the king of Persia, Cyrus (Persia had by this time replaced Babylonia as the reigning superpower). The message of Deutero-Isaiah was largely a joyous one, celebrating the fact that there was about to be a new exodus from captivity, during which the Israelites, led by Yahweh, would cross the desert and return home in triumph. Israel had served its full term of bondage, not merely for its own sins but for those of all nations; now the time of its glorification as the faithful servant of Yahweh was at hand.

Deutero-Isaiah's joyous confidence was, alas, somewhat premature. A Jewish community did reestablish itself in Jerusalem not long after Cyrus published his decree permitting the return; but it was not until several decades had passed that a new building was completed to replace Solomon's Temple, and there was no major return of descendants of the exiles until the late fifth century B.C.E. in the time of Ezra.

We have remarked on the seven great figures of the Age of Prophecy: Amos, Micah, Hosea, and Isaiah in the eighth century and Jeremiah, Ezekiel, and Deutero-Isaiah in the sixth. There were other (to our eyes lesser) figures who prophesied shortly before, during, and just after the sixth century B.C.E. and whose words are now recorded (along with those of Amos, Micah, and Hosea) in what are termed the Minor Prophets or, in the Jewish scriptures, "The Book of the Twelve." All of these prophets, major and minor alike, have in common that they were caught up in the critical events of their own times, events they saw as proceeding according to an immutable principle: Unfaithfulness to Yahweh would necessarily bring destruction, but from that destruction a remnant would be saved through which Yahweh would reestablish his people forever in the land promised to their forefathers.

POETIC FORM OF THE
PROPHETIC BOOKS

Before considering the appeal of the prophetic books to later ages, including our own, we must note briefly two characteristics other than their concern with history, both having to do with form. The first is that these books consist largely of poetry, a fact much less evident in the King James Version than in any modern translation. In the Revised English Bible or the New Jerusalem Bible, for example, poetic passages are laid out on the page as poetry usually is, in lines of uneven length, and thus these passages do not have the appearance of prose that the King James Version gives them. This appropriate poetic arrangement makes immediately apparent the parallelism (discussed in chapter 2) that is a distinguishing feature of Hebrew poetry. This arrangement also predisposes a reader to accept poetic metaphor as simply that and not as something to be taken literally.

That there is so much poetry in the prophetic books raises an inevitable question: Were the ancient prophets actually poets? Did they deliver their prophecies in the poetic form in which they have been preserved? No one can say how the prophets spoke, of course, for we have no way of getting past the words on the page to the words that were spoken. What we do have are written versions of what the prophets are supposed to have said on certain specific occasions. They did, perhaps, speak in a poetic fashion, employing highly patterned and figurative language. Or perhaps this form was imposed on their words either by those who repeated them and passed them on to later generations or by those who first wrote them down.

STRUCTURE OF THE
PROPHETIC BOOKS

The other formal characteristic of the prophetic books that deserves our attention is their frequent lack of coherence. In most the material has no logical arrangement; we could as well start reading them in the middle as at the beginning. With few exceptions (Daniel and Haggai, most notably), the prophetic books are random collections of individual units, called "oracles," that can be classified into a few basic types. The prophetic message was, as we have seen, quite simple; it varied little from one occasion to the next, from one prophet to another. Typically, a prophet is represented as looking about him and seeing wrongdoing, most often bad religion (which involved fertility worship or other practices of Baalism and worshiping Yahweh in a superficial way) and social injustice (taking advantage of the poor, using dishonest weights and measures, accepting bribes to thwart justice in the courts). The prophet denounces the wrongdoing and threatens that, unless the sinners repent, Yahweh will send dreadful punishment on them, punishment that will culminate in the destruction of the nation. That fate, says the prophet, will bring about repentance at last; then Yahweh will relent and will restore a remnant of his people to a good life in their homeland.

Each point in this recurring situation, this prophetic paradigm, has its appropriate kind of oracle: There are oracles denouncing bad religion or social injustice, oracles calling for repentance, oracles announcing destruction, and oracles promising restoration. Each of these types occurs over and over again in the prophetic books and—to the confusion of the inexperienced reader—in every possible arrangement. Modern translations provide some help with this problem in that they print the individual oracles with spaces between them, so that a reader knows when to shift gears, so to speak: Now I am reading an oracle denouncing social injustice, now I am reading an oracle calling for repentance, and so on.

Given their poetic content, their general lack of coherence, and particularly their thoroughgoing concern with the history of the ancient Near East, the prophetic books are not easy to read—as any beginning student of the Bible can testify. Despite their difficulty, however, no other portion of the Old Testament, with the exception of Genesis and Exodus, receives more attention in our time than do the prophetic books. One need only glance at the religious announcements in any newspaper or on the bulletin boards of any college campus to see that this is so; prominently advertised will be sermons, lectures, and conferences on the prophetic writings. Why should they receive this attention? It is not, after all, as though they record the great foundational events of Judaism or Christianity such as we find in Exodus and the gospels. And it is not as though they present the basic principles of either of those faiths, as do the Pentateuch and the letters of Paul. Why are these writings that are tied to the history of long-past events in a far-off part of the globe of such interest to the modern world? Wherein lies their appeal?

APPEAL OF THE
PROPHETS FOR EARLY
JUDAISM

In accounting for that appeal, it is well to begin with the question of the appeal of the prophetic writings to a more ancient audience—to the Jews of those postexilic centuries during which the oracles of the old prophets were passed on in oral or written form, copied and recopied, and gathered into edited collections. Why should Jews of, say, the third century B.C.E. have accepted as sacred scripture, possessed of eternal significance, material that was thoroughly bound up with events of three or four hundred years earlier? Several answers to that question present themselves. First, in Jewish eyes the message delivered by the prophets as the very word of Yahweh did not cease to be authoritative simply because the time of its original application had passed. What an Amos or a Micah had spoken was forever precious in Judaism and could not be allowed to perish. The northern kingdom *had* gone down to destruction, as Amos and Hosea had warned, and the southern kingdom *had* been defeated, as Micah, Isaiah, and Jeremiah had warned. Prophecy could thus be studied as a means of accounting for the situation of Israel in the third century B.C.E., of discovering the ways of a just God, and of perceiving the nature of the relationship between Yahweh and his faithful human servants. Second, and ironically, the words of the prophets were precious not simply because much in them *had* been fulfilled but because much *had not*—not yet, at any rate. That is, those writings contained foretellings and promises that had not yet been satisfied and (the words of the prophets being the very words of God) that must therefore yet be satisfied. In those centuries following the Babylonian Exile, Israel no longer existed as a nation; but Jews had confidence that the old prophets' utopian visions of a renewed and restored Israel, ruled once again by a descendant of David, would in God's time be fulfilled, and until that time their words were to be studied and treasured.

Finally, and perhaps most significantly, even those words of the prophets that had been fulfilled had not thereby been exhausted of meaning. What the prophets had said about specific situations in their own times could be taken and applied to other, similar situations in later times. We see an instance of this in the book of Daniel, where Daniel is "studying the scriptures, counting over the number of years—as revealed by Yahweh to the prophet Jeremiah—that were to pass before the desolation of Jerusalem would come to an end, namely seventy years" (Dan. 9:2). While he is praying in anguish about his people's continuing captivity, an angel appears to him and says,

> Seventy weeks are decreed
> for your people and your holy city,
> for putting an end to transgression,
> for placing the seal on sin,
> for expiating crime,
> for introducing everlasting uprightness
> for setting the seal
>   on vision and on prophecy,
> for anointing the holy of holies. (Dan. 9:24)

The book of Daniel was written 400 years after Jeremiah's time, when Jerusalem, though it did not actually lie "in ruins," was suffering dreadful persecution under its Greek overlord, Antiochus IV Epiphanes. How might Jeremiah's predicted 70-year period in the sixth century B.C.E. be made use of in the second century B.C.E.? By the simple expedient of reinterpreting 70 years as 70 *weeks* of years—that is, 70 times 7, or 490 years, the end of the period (and thus the time of deliverance) could be extended down to the writer's own time in the mid-second century. Such a system of double application (which we can be confident was not intended by Jeremiah) enabled the prophecy to have both its original sense—in the sixth century, Jerusalem did lie in ruins for more or less 70 years—and a new sense immediately applicable to the second-century audience of the book of Daniel.

APPEAL OF THE
PROPHETS FOR
CHRISTIANITY

The earliest Christians, convinced that the crucified Jesus had been resurrected from the dead, searched the Jewish scriptures for passages that could be taken as applying to Jesus. Various events in biblical history were believed to foreshadow events in the New Testament story (this way of reading the Old Testament is called "typological" and will be discussed further in chapter 17), and likewise a great many remarks of the prophets were understood as references to Jesus. The writer of the gospel of Luke pictures the resurrected Christ as appearing to some of the disciples who do not immediately recognize him and who are puzzling over the events of the Crucifixion and Resurrection. In exasperation Christ exclaims: "You foolish men! So slow to believe all that the prophets have said! . . . Then, starting with Moses and going through all the prophets, he explained to them the passages throughout the scriptures that were about himself" (Luke 24:25, 27). Matthew's gospel repeatedly brings elements from the Jewish Bible to bear on the life of Jesus. Most notable, perhaps, is the application to the birth of Jesus of Isaiah 7:14, a passage that in context—as the author intended it—says simply that a young woman is going to bear a child and that, before the child is more than a few years old, the enemies currently advancing upon Judah will no longer be a threat. Matthew (who surely was not the first to do so) takes the word for "young woman" in the Hebrew text to mean specifically what we today mean by "virgin"—one who has had no sexual relations—and thus introduces a miraculous element into what was meant by the prophet Isaiah to be simply a way of measuring time.

What is striking about the New Testament writers' application of Old Testament passages to their own areas of interest is the way they ignore the context of those passages—and thus ignore one of the most important means of defining what the original author had in mind. This proof-text method (that is, the picking of a fragmentary unit out of a complicated larger statement) allows a passage to mean whatever the interpreter can find in it. Prophetic writings that were thoroughly involved in the history of the eighth or sixth centuries B.C.E. could thus be freed of history and made to apply to the one great New Testament story: the birth, ministry, death, and resurrection of Jesus.

APPEAL OF THE
PROPHETS FOR MODERN
TIMES

Having looked at why and how early Jews and Christians adapted the ancient prophets to their own circumstances, we can now return to the question we asked earlier: What accounts for the considerable appeal in our own time of the prophetic writings of the Bible? The answer is that those writings can be applied to the particular circumstances of our own time in the same way that they were applied to the crucial events of early Jewish and Christian times. For modern readers of the Bible, constantly assailed by news of threatening affairs at home and abroad, the appeal of the prophets lies in their general message of destruction for evildoers and (after a period of purifying tribulation) reward and endless bliss for the faithful. The great final acts of God against the wicked and in behalf of the righteous have not yet happened but—having been predicted by the prophets—they will surely yet happen, the faithful believe. Might this not well be in our own age, they ask, an age marked by worldwide wars and the awesome ability to blow ourselves into oblivion? To the weary inhabitants of such a world as we live in, the hope that the prophets hold out and the certainty with which they speak can be immensely satisfying.

But more than something so general as this in the prophetic books can be drawn upon. For the general message of these books is conveyed not in abstract terms but through the specific details of ancient Hebrew history, details that can be lifted out and examined and applied to our time ("After all, they're in the Bible, aren't they?") with little concern for their original significance or how they fit together into the larger historical situation in which the prophets were participants. Thus one can hear commentators on the Bible say that "the north" in prophecy is always a reference to Russia, "the east" always a reference to China, and any eagle always a reference to the United States. Predictions in prophetic writing of a return to their homeland of the Jews deported to Babylonia in the sixth century B.C.E. are made to apply to the founding of the modern state of Israel in 1948; the ideal temple that Ezekiel envisions as being built by the returned exiles is made to apply to a temple that will yet be built in the time of the Antichrist; the dreadful Day of Yahweh spoken of by Amos as the time when destruction will come upon the northern kingdom is made to apply to the final destruction of evil in our own near future; and so on and so on.

What should we think about this sort of interpretation of prophecy that ignores the actual historical circumstances to which it applies? It is certainly no recent invention, for, as we have seen, it was practiced by Jews before the Common Era and taken up with particular intensity by the earliest followers of Jesus. But even if its pedigree is long, interpretation of this kind is too easy a game to play. Interpreters look at the text with the preconception that it applies directly (not merely morally or by way of example) to the present day; they consider what associations its individual words and phrases create in their minds and what relations those words may have to recent news of the world. And from that they develop a statement of what the text "means" for us today (if not what it meant for its author some two and a half thousand years ago!). But how, one might ask, is anyone to determine whether the resulting interpretation is right or wrong, since every such interpretation will differ inevitably from every other one in significant respects? And will not an interpretation worked out in 1950, say, differ from one worked out forty or fifty years later, inasmuch as historical circumstances will be different from one decade to another? There is nothing else in religious literature that goes out of date quite so quickly as a book on prophecy written along these lines. Used-book shops are overstocked with fading volumes written to prove that one Russian premier or another was the Antichrist or that one war or another in the Near East was the prelude to the End.

Perhaps the best preventive against overinterpreting the prophets is to ponder a simple question: Is it likely that Amos or Micah or Isaiah or Ezekiel—standing in the dusty streets of Samaria or Jerusalem and addressing the curious citizens who crowded about them—were speaking of America and Europe and Asia in the nuclear age? Is it probable that any of what they said was intended to apply to specific events twenty-six hundred years in the future and half a world away? To claim that this is so is to claim a great deal, and it surely reveals more about ourselves and our own needs and anxieties than it will ever reveal about the prophets.

Anyone seriously approaching biblical prophecy for the first time should, in view of its historical basis, always make use of a modern translation that is accompanied by footnotes explaining the historical references and by maps showing those parts of the ancient world with which the text was originally concerned. Historical meaning may not be the whole meaning of biblical prophecy, of course, but it is certainly the foundation upon which all other genuine meaning must be built. Whatever religious view one brings to the reading of the prophets, one must be aware that the reference of those prophets is in the first instance always to circumstances and events of their own time. To ignore that fact is wilfully to cut oneself off from the surest means of understanding the sense of the prophets' words.

SUGGESTED FURTHER READING

John Barton, *Oracles of God: Perceptions of Ancient Prophecy in Israel after the Exile* (New York: Oxford University Press, 1988).

Terence Collins, *The Mantle of Elijah: The Redaction Criticism of the Prophetical Books* (Sheffield, Eng.: JSOT Press, 1993).

Philip R. Davies and David Clines, eds., *Among the Prophets: Language, Image and Structure in the Prophetic Writings* (Sheffield, Eng.: JSOT Press, 1993).

Bernhard Lang, *Monotheism and the Prophetic Minority: An Essay in Biblical History and Sociology*, The Social World of Biblical Antiquity Series, no. 1 (Sheffield, Eng.: The Almond Press, 1983).

David L. Petersen, ed., *Prophecy in Israel: Search for an Identity*, Issues in Religion and Theology, no. 10 (Philadelphia: Fortress Press, 1987).

John F. A. Sawyer, *Prophecy and the Prophets of the Old Testament* (New York: Oxford University Press, 1987).

The Interpreter's Dictionary of the Bible, ed. George A. Buttrick et al. (Nashville: Abingdon Press, 1962). See article on Prophet, Prophetism. Supplement, 1976: See articles on Prophecy in Ancient Israel, Prophecy in the Ancient Near East.

The Anchor Bible Dictionary, ed. David Noel Freedman et al. (New York: Doubleday, 1992). See articles on Prophecy (ANE), Prophecy (Pre-Exilic Hebrew), Prophecy (Post-Exilic Hebrew).

IX

The Wisdom Literature

The term "wisdom literature" designates three books in the canonical
Old Testament (Job, Proverbs, and Ecclesiastes), two books in the
Apocrypha (Ecclesiasticus [also called The Wisdom of Jesus ben Sira]
and The Wisdom of Solomon), and sometimes other, scattered portions of the Old Testament. Job, Proverbs, and Ecclesiastes are
among the most popular books of the Bible, and beginning students
will find them full of interest. But if such students are reading the
Bible straight through for the first time and have thoughtfully been
forming a sense—based on the Pentateuch, the books of history, and
Psalms—of just what the Old Testament is, they will find much in
the wisdom books that is puzzling and even astonishing. By the time
they have finished reading the Old Testament and look back again at
Job, Proverbs, and Ecclesiastes from the vantage point of the prophetic books, they will wonder even more at them.

For the wisdom writings are in some ways quite "unbiblical." The
book of Job tells of a man who is made to suffer horribly as a result
of a casual wager between Yahweh and "the satan"*; and it tells how

---

*In Hebrew the word usually translated "Satan" is literally "the satan," best
translated as "the adversary" (of human beings, not God).

the hero (ignorant of the wager but very much aware that he does not deserve his suffering) hurls questions and challenges at the Almighty that verge on blasphemy. Ecclesiastes goes beyond Job's questioning of God's justice. It presents the bleak prospect that humans can understand almost nothing about this world and about their place in it except that they will come to the same end—death and the grave— that animals do, and that the best way to occupy themselves in view of that end is merely to eat, drink, and be merry. And Proverbs, although it is not negative in spirit in the way the other two books are, and although it cheerfully advises that the first principle of wisdom is to have a proper respect for God, nevertheless seems remarkably this-worldly in its advice on how to conduct one's life. Certainly it pays scant attention to the proper form of religious worship or to the ecstasy and despair of a soul communing with God or to the great facts of Israel's relationship to its deity Yahweh.

If these three wisdom books are so unlike the rest of the Old Testament (someone has called them a "foreign body" in the Bible), how can their existence be accounted for? As is true with most of the other books in the Old Testament, next to nothing can be said about the individuals who composed them; certainly Solomon was not the author of either Proverbs or Ecclesiastes, as tradition maintains. But if we do not know who the authors were, we do know something about the group in ancient Israel to which the authors belonged—a group most simply called "wise men."

THE WISE MEN IN
ANCIENT ISRAEL

That there was a distinct class of individuals who claimed to possess particular wisdom and who were recognized for it is demonstrated at a number of points in Jewish scriptures, most notably in the words of the prophets. Jeremiah speaks of personal enemies who want to do away with him and quotes them as saying, ". . . let us concoct a plot against Jeremiah, for the Law will not perish for lack of priests, nor advice for lack of wise men, nor the word for lack of prophets" (Jer. 18:18). Ezekiel predicts at one point that there are bad times coming when people will anxiously attempt to learn the truth about what is happening: ". . . they will pester the prophet for a vision; the priest will be at a loss over the law and the elders on how to advise" (Ezek. 7:26). It seems justifiable to generalize from these and other such passages that there were, at the royal court of Judah in the late-seventh/early-sixth centuries B.C.E., counselors to the king and people whose ability and duties put them on a par professionally with the leading figures in the national religion. It seems also justifiable to project this situation back several centuries earlier to the time when King David and, particularly, his son Solomon brought the fragmented Israelite people together into a nation with a central government established in a fixed place. Such a government could function effectively—that is, collect taxes, administer justice, raise armies, and press work forces into service—only through the efforts of a bureaucracy distinguished by its ability to read, write, and count and headed by persons in the king's court capable of planning and administering large-scale programs.

Among the functions of the civil servants (sometimes called "scribes") in the central government would have been not merely the record keeping necessary for any governmental operation but also the writing of official annals of the reigning king and—in Solomon's time, we can guess—the producing of a national history that would portray Israel's past in a good light and serve to glorify the king. Another of the civil servants' functions would have been the handling of written correspondence between Israel's rulers and the governments of neighboring nations. And, finally, they would have been responsible for training a select few young men in the mysteries of the written word. At first such training would probably have been limited to one teacher and one student: a father tutoring a son, an old scribe instructing a young apprentice. But sooner or later in that situation the idea of formal schooling would have developed. This must not be thought of as a system of general public education, for the boys (not girls, of course) involved would have been an elite drawn from the upper classes. Judging from evidence about such schooling in other nations in antiquity and from what can be inferred from the wisdom books themselves, instruction would have been not merely in reading, writing, and counting but in practical morality: how to conduct oneself in the world of affairs, what ends and means are appropriate to men of consequence.

As the reach of formal education broadened over the centuries, we can assume that what was originally a small professional group of scribes would have developed into an educated class in Israel—still a relatively small and privileged caste but one embracing a broader spectrum of society than merely the scribal families. That caste would have constituted the readers of the literary works produced not only by the wisdom writers but also by Israel's poets and historians. That caste would thus have been the intended audience for the Deuteronomic History, Job, Ecclesiastes, Proverbs (in its final, edited form), and more popular books, such as Ruth, Jonah, and Esther.

Of course, it is one thing to say that there was a professional class of men in ancient Israel who performed scribal functions and taught the young; it is another thing to say that such a class constituted a genuine group that shared a set of values and a world view. But this is in fact what scholars hold to be true about those in the wisdom tradition. Some of the evidence in the matter is very persuasive, some less so. The major difficulty is that most of the evidence about the wise men is found in the wisdom books themselves, and yet those books are the very ones we are trying to understand in the first place—we want to know about the wise men so that we can know how to read the wisdom books. Thus our process of reasoning is circular, and we are in danger of merely proving in the end what we assumed at the beginning. How helpful it would be for scholars studying the wisdom literature if archaeologists were to unearth a series of documents that described the makeup of Israelite society every century or so from David's time onward and, in the process, defined a class in that society called "the wise," a class that performed such and such a function and held to these values and those views of God and man! Then we could study the wisdom literature in terms of this information and come to an understanding uncontaminated by circular reasoning.

PATTERNS OF THOUGHT
OF THE WISDOM WRITERS

Lacking any such contemporary sociological analyses of Israel, we are limited for information about the wise men to what they themselves wrote and to scattered references to them in the prophetic and historical writings. Yet by using that information carefully, scholars have been able to go a considerable way toward defining the basic patterns of thought of the wisdom school and the values their literary works were written to promote. As we survey the matter here, we can best begin by proceeding negatively, that is, by showing what the wise men were *not* concerned with or, at least, what they had no interest in writing about. We must stress that what will be said here applies only to the wise men of the century or two before and after the Babylonian Exile in the sixth century B.C.E., that is, to those who produced the books of Proverbs, Job, and Ecclesiastes. The wisdom writers of later centuries held views that differed from those of their predecessors in several important respects, as we shall see later when we consider the books of Ecclesiasticus and The Wisdom of Solomon.

First, the wise men in the centuries before, during, and after the Exile were little concerned with the cult, that is, the organized religion of Israel. There are only passing references to cultic observances in the three canonical wisdom books. At one point in Proverbs—and it is the only positive allusion to cultic matters in the entire book—the reader is advised to "Honour Yahweh with what goods you have / and with the first-fruits of all your produce" (Prov. 3:10); but later there is more characteristic advice: "To do what is upright and just / is more pleasing to Yahweh than sacrifice" (Prov. 21:3). In Ecclesiastes, if we ignore the several passages inserted by scribes and later editors in an effort to soften the book's harsh view of life, the author's only comment on religion is that those who participate in cultic activities should do so with full awareness of what they are doing and of what commitments they are making. In Job the only cultic activity referred to is found in the prose prologue and epilogue, where we are told that Job performs sacrifices and prays to Yahweh on behalf of his children and friends; but in the large poetic bulk of the book, where Job and his friends investigate in minutest detail the ways a man may sin and thus deserve his suffering, cultic matters are never once mentioned. In these three books, it is almost as though organized religion is taken for granted as something that has no bearing on the really profound questions of life.

Second, the wise men appear not to have been nationalistic in spirit. Their works exhibit none of the Deuteronomic historians' abhorrence of all things and all people non-Israelite, none of the prophets' distrust of other nations with whom Israel's leaders might wish to ally themselves. The wise men did not address themselves, as the Deuteronomists and the prophets did, to Israel as a people—they uttered no "Hear, O Israel"—but rather they spoke to individual human beings concerned with the nature of the world and with how one lives a satisfying life. That sort of concern was not simply Israelite, it was universal. Indeed, it was the topic of concern in the wisdom literature of neighboring nations, some of whose literature the sages of Israel would have known well through contact with their professional peers in those nations. In both Egypt and Mesopotamia, a great many works along the lines of the Israelite wisdom books were written, as we know from copies that have survived to the present. One of these, an Egyptian book of instruction addressed to a young man, was the source of a passage in Proverbs 22 and 23. Works that are quite like Ecclesiastes and Job and that may have been known to the authors of those biblical books were written—as we have discussed in chapter 3—in Egypt and Mesopotamia as far back as the beginning of the second millennium B.C.E. We may suppose that the wise men of Israel considered themselves to be in the tradition that produced such works and as belonging to an international body of sages.

Third, as the wise men had no concern for Israel's place among the nations, they had no concern for the nation's past—not for its birth through Yahweh's covenant with Abraham, or its rebirth through Yahweh's rescue of his people from Egypt and giving of the Law at Sinai, or the conquest of the land of Canaan, or the establishing of David on a throne that was to belong forever to him and his descendants. As we remarked earlier, it may have been members of the wisdom school in the tenth century B.C.E. who wrote the official history of Israel; but their successors several hundred years later who wrote Proverbs, Job, and Ecclesiastes made no reference to the facts of that history in their works. The later wise men paid their allegiance to David's son Solomon, of course, but that was not really a matter of history; they looked to Solomon solely because of his reputation for wisdom: He was the patron saint of the sages, so to speak, just as David was the patron saint of musicians.

Fourth, although the wise men assumed the existence of a deity who created and sustained the world, they had no sense of a personal relationship between believer and God. Thus their books reflect nothing of the kind of personal relationship with Yahweh that we find throughout the Psalms and the pronouncements of the prophets; and there is consequently no "Thus says Yahweh" to be reported back to their countrymen. Job does, admittedly, cry out to God in his anguish and does receive a response; but that response takes the form of an overwhelming series of questions intended to belittle Job—to crush him into insignificance by virtue of his being a mere man. In the book of Job and the other wisdom writings, no special truths are revealed from heaven; those of men's questions that cannot be answered from the observation of nature and of human society must remain forever without an answer.

The classic wisdom writers, then, had in several important respects a considerably different outlook from those who produced the Pentateuch, the historical writings, the prophetic books, and the Psalms. For the wise men, the world of created things and of human relationships was God's book, complete and unchanging. And it was observation of things as they are, not divine revelation, that would produce knowledge and, ultimately, with years of experience and contemplation, wisdom. What followed from having achieved such hard-earned wisdom was, of course, that its possessors could be expected to instruct the young concerning what was of value in the world and how to live the good life. Two of the wisdom books, Proverbs and Ecclesiastes, would have served this function directly; the book of Job would have done so indirectly by exposing common, but false, expectations about behavior and its consequences.

Although the wisdom writers did not derive their insight from religion and had little to say about it, we can suppose that few Israelite sages were antagonistic to the religious values of their society and that most were probably quite comfortable with them. Their advice to youth can be assumed to have coexisted easily enough with the traditional lore of the priests and the preaching of the prophets. Despite the fact that they were not religious enthusiasts, the wise men believed in the existence of God and his creation of an ordered universe. The majority of them would have accepted the ethical principles of traditional religion, the most basic being the principle that there is a necessary connection in God's ordered world between one's behavior and one's lot in life: If one does well, one will prosper; if one does ill, one will suffer. But if this basic ethical principle was accepted by most of the wisdom writers, it was not accepted by all. The three books before us—Proverbs, Job, and Ecclesiastes—can profitably be examined in terms of their authors' view of whether one's behavior determines one's lot in life.

BEHAVIOR AND ITS
CONSEQUENCES IN
PROVERBS

Proverbs contains material composed by a number of authors over a number of centuries, but in its final version it takes the form of a book of instruction for young men on the nature of the world and on the conduct required for success in the world. We are not surprised, therefore, to find that throughout the book a necessary connection between behavior and fortune in life is always assumed and frequently mentioned. The book's intended audience would have come away from it understanding that they would reap as they had sown, both in the moral realm (good produces good/evil produces evil) and in the practical realm (wise planning and effort lead to prosperity/carelessness and laziness lead to ruin). Consider these typical observations and warnings from Proverbs:

> No harm can come to the upright,
>   but the wicked are swamped by misfortunes. (12:21)

> In the way of uprightness is life,
>   the ways of the vengeful lead to death. (12:28)

> Hard work always yields its profit,
>   idle talk brings only want. (14:23).

> Idleness lulls to sleep,
>   the feckless soul will go hungry. (19:15)

Throughout Proverbs this simple principle of cause and effect can be seen as basic to its authors' thinking: Prudence and fairness will— indeed must—lead to success, because that is the way things work in this world and because God's watchful concern will guarantee that it is so. The perceptive young person will grasp this principle and employ it to shape the sort of happy ends he wishes to achieve.

This is all very well in theory, of course, but even those most committed to the principle cannot deny that it does not always work out that way in real life. Here, let's suppose, is an unhappy orphan who can scarcely be thought to have deserved to have lost his parents; here is a poor widow whose husband perished through no fault of hers; here is a godly merchant whose stock in trade has been suddenly destroyed by a fire caused by lightning. What is to be said to these who suffer through no fault of their own? Proverbs has remarkably little to say to them. When it faces up to the matter at all, the book's only attempt at a solution to the problem of undeserved suffering is to imply that there is no such thing: Sufferers must be guilty of sin that others cannot see and that even they themselves may be unaware of. Thus their suffering is to be understood as God's way of reproving and chastening them for their own good. Immediately following the passage cited earlier in which the reader is told to honor Yahweh with his wealth in order to achieve prosperity, we find this further advice:

> My child, do not scorn correction from Yahweh,
>   do not resent his reproof;
> for Yahweh reproves those he loves,
>   as a father the child whom he loves. (Prov. 3:11–12)

By explaining the suffering of the godly in terms of chastening, the sages of the book of Proverbs could maintain the validity of their cause-and-effect ethical principle and at the same time the justice of the deity who stood behind it.

BEHAVIOR AND ITS CONSEQUENCES IN JOB

The same principle is espoused by the friends who gather about Job in his time of trouble. They say to this apparently godly man who has suddenly lost his prosperity, his family, and his health:

> Can you recall anyone guiltless that perished?
> Where then have the honest been wiped out?

> I speak from experience: those who plough iniquity
> and sow disaster, reap just that. (Job 4:7–8)

His friends advise Job that he "not then scorn the lesson of Shaddai" (Job 5:17), the implication being that he is not so blameless as he appears or as he thinks he is. But that easy answer will not do for Job; he is innocent and he knows it. The truth seems plain to him— and a terrible thing it is to utter: God is simply not just!

> It is all one, and hence I boldly say:
> he destroys innocent and guilty alike.
> When a sudden deadly scourge descends,
> he laughs at the plight of the innocent. (Job 9:22–23)

Throughout his long debate with his "comforters," Job pleads that God appear (in court, as it were) and state plainly what wrong Job has committed that warrants such ill fortune. When, near the end of the book, the Almighty does speak to Job, it is in terms designed to make him understand that his reasoning about human suffering and God's justice—or lack of it—is utterly misinformed. Overwhelmed, Job confesses to having spoken foolishly about matters too great for him; but, astonishingly, no sooner has he done so than the deity says to one of the staunchly orthodox counselors, "I burn with anger against you and your two friends, for not having spoken correctly about me as my servant Job has done" (42:7). In view of this statement, are we to understand that Job's friends were wrong when they defended God's justice and that Job was right when he attacked it? Probably not. The author simply has gone as far in this poetic exploration of the insoluble problem of suffering as he can go. The old prose tale that got him into the topic now serves to get him out of it—but at the price of severe inconsistency.

BEHAVIOR AND ITS CONSEQUENCES IN ECCLESIASTES

Job argues that there is no relationship between the good or evil a man does and what happens to him in life. Where Job leaves off, the author of Ecclesiastes picks up and then pushes the argument to its ultimate conclusion. Not only are there no guarantees that doing good or bad will lead to good or bad consequences for the doer—there are no guarantees that *any* kind of action will have the consequence the doer intends or thinks he has a right to expect. The only certainties in this world are that natural processes will continue forever—sunrise, sunset, sunrise, sunset—and that death will follow life. All human speculation about cause and effect comes to nothing, "For the fate of human and the fate of animal is the same: as the one dies, so the other dies; both have the selfsame breath" (Eccles. 3:19). And death as this writer views it is certainly no realm where all wrongs will be set right, with rewards and punishments distributed according to how human beings have merited them by their actions in life. Death is an utter nothingness, as well for the good as for the bad, as well for humans as for animals.

What sort of behavior can the author of Ecclesiastes then recommend, seeing that there are no long-term satisfactions toward which one can work and no comfortable system of rewards and punishments in life or death? Well, some situations in life are obviously better than others, and one can reach for a few short-term satisfactions: It is better, all in all, to be wise than to be foolish; better to have food and drink than to lack them; better to be young than old; and, finally, better to be alive than dead. Cherish these few good things, entertain no great expectations, and in general live in the knowledge that soon you will be dead.

At this point it may appear to the reader that Ecclesiastes and Proverbs are so radically different from one another that they can scarcely be considered to belong to the same school of thought. Some scholars do, indeed, propose that there was an optimistic/pessimistic division in the wisdom tradition. But the simple truth is that any of us, thoughtfully contemplating the way the world goes, is capable of widely varying responses to what we witness and experience. Smelling the good air on a balmy spring day, enjoying the affection of dear friends, making progress toward the achievement of our life's goals—in such circumstances any of us can expansively advise our fellows to cheer up, be alive, think positively, and so conquer the world; after all, a negative attitude never gets anyone anywhere. But when the weather changes for the worse, friends are less than honorable, and expectations fail through no fault of our own, any of us is capable of turning pessimistic and ready to counsel all who will listen that a person just can't win in this world. The point is not that Proverbs is simply a wise man's advice on a good day and Ecclesiastes his advice on a bad day but that those two works represent the diametrically opposed outcomes of the same process: the philosophical investigation of human existence with unaided human reasoning. Life presents a wide spectrum of conditions, and the range of responses of those who examine it can be wide indeed. But as far as the wise men were concerned, it is worthwhile to do the examining of life, whatever one's response may be. As we said earlier, even the pessimistic author of Ecclesiastes must admit that the wisdom—the inquiring mind—that produces his melancholy is nevertheless a good thing to possess. An unexamined life is merely a fool's paradise.

THE RELIGIOUS ELEMENT
IN THE LATER WISDOM
WRITINGS

What we have been saying thus far applies to the three canonical wisdom books, composed over a span of several centuries in the middle of the first millennium B.C.E. There are two other books in the wisdom category, both in the Apocrypha. Ecclesiasticus (the similarity of its title to Ecclesiastes is unfortunate) was probably written about 180 B.C.E.; The Wisdom of Solomon (or simply Wisdom) is generally dated about 100 B.C.E. These two books, although recognizably very much in the wisdom tradition, share certain characteristics that set them off from the earlier three works we discussed. The most obvious of these characteristics is a specific commitment to Israel's deity and to the scriptural record of his mighty works in behalf of his elect nation. In the earlier books, wisdom is a nearly secular quality and one that knows no national boundaries; in these later books it has become a quality that is specifically religious and Israelite (or Jewish, as we can appropriately say of these works of the second century B.C.E.). In the early chapters of Proverbs, wisdom was personified as something of a goddess, who was somehow involved in the creation of the world and who summoned humankind to leave folly and become her companions. But now, in Ecclesiasticus, wisdom is identified as the very word of God that was spoken at the time of creation to bring all things into existence. That task accomplished, wisdom looked about for a place to make her home and was assigned by God to dwell in Israel—indeed, in Jerusalem, "the beloved city" (24:11). Wisdom is, furthermore, identified with the Law of Moses, "the Book of the Covenant of the Most High God" (24:23). Although Ecclesiasticus contains a great deal of the practical observation and advice that Proverbs has, it also contains blocks of material that sound very much like passages in Psalms—passages praising God's power, wisdom, and mercy—and a long section (chapters 44–49) reviewing Israel's history in terms of the great men who were champions of the faith. For all its practicality Ecclesiasticus is a thoroughly religious book in which the wisdom tradition is put into the service of orthodox belief. It has no room for the gloomy speculations and doubts of Ecclesiastes and Job and has considerably more of a godly orientation than Proverbs.

Just as Ecclesiasticus converts the sort of prudent counsel found in Proverbs into a more pious, spiritually satisfying form, The Wisdom of Solomon reworks the sort of philosophical and ethical skepticism found in Ecclesiastes into a strong defense of the Jewish view of God. Ostensibly addressed to "rulers on earth" by their fellow ruler King Solomon, the book was actually written by a Greek-speaking Jew to coreligionists (perhaps those living in Alexandria, Egypt) who were in danger of abandoning their monotheism in favor of other, more popular and supposedly more reasonable kinds of religion—or no religion at all. Against this the writer brings to bear Solomon himself and also wisdom personified as an attractive female figure, of whom Solomon says, "Wisdom I loved and searched for from my youth; / I resolved to have her as my bride, / I fell in love with her beauty" (Wisd. of Sol. 8:2). To everything that these potent figures represent are added several ideas from Greek thought that fit well with the wisdom tradition: the immortality of the soul and the preeminence of the spiritual over the physical, for example. Whereas the earlier wisdom writers surveyed God's creation and (on the basis of what they perceived) advised the young how to live, the author of The Wisdom of Solomon performs that survey in order to praise the God who did the creating. True wisdom brings one back to the God who is the origin of wisdom and through him to immortal life; false wisdom leads one to the folly of atheism or idolatry and ultimately to death. Like Ecclesiasticus, The Wisdom of Solomon ends with a long meditation on Israel's history as revealed in the sacred scriptures, thus tying the God who is the source and end of wisdom firmly to the people of Israel. The point made is that Jews should not allow themselves to be seduced by alien philosophy and religion; truth is theirs because the God of truth is theirs.

CANONIZATION OF THE WISDOM BOOKS

A curious irony may have occurred to the reader by this time: Of the five wisdom books we have been considering, the two most positively religious ones—Ecclesiasticus and The Wisdom of Solomon—were not admitted by ancient Judaism to its canon of scripture, whereas the earlier three books, despite the fact that they contained much that is questionable in religious eyes, were admitted. Why books like Ecclesiasticus and The Wisdom of Solomon were not included in the Jewish canon is a matter we touched on in chapter 6: They were recognized as relatively recent works written after the period of divine inspiration was thought by Jews to have ended. But why were the earlier three books admitted?

We have to suppose, first, that it was because Proverbs, Job, and Ecclesiastes had great appeal in ancient Judaism. Probably the appeal was due to the honesty with which those books deal with the hard questions of religion and ethics. Whatever the ideals of any system of belief, it must in some way face up to the problems of life as it is really lived and experienced; the writers of the three books in question did that for Judaism. Ideals must not, however, be put under excessive stress or cross-examined too relentlessly; and these three books, no matter what their appeal, would not have been found worthy during the last stages of the canonizing process had ways not been found earlier in the process to temper their frankness or at least to interpret their texts in line with traditional ideas.

Such ways were found. The biting skepticism of Ecclesiastes, for example, was rendered palatable in part by the simple device of interpreting as straightforward statements certain pious remarks that the author intended to be taken ironically. In addition, several pious rejoinders that scribes had written in the margins as they copied out the text gradually came to be understood as part of the text (despite the fact that these sentiments were out of keeping with everything surrounding them). Finally, Ecclesiastes had a postscript tacked onto it by a later sage who admired the work but felt that its rigor had to be alleviated a bit. Thus he advised readers of Ecclesiastes—with a degree of assurance that the author of the book himself was certainly incapable of achieving—to "fear God and keep his commandments, for that is the duty of everyone. For God will call all our deeds to judgement, all that is hidden, be it good or bad" (12:13–14).

When we consider this same sort of softening of the book of Job to make it more acceptable to orthodox tastes, we must note that the hero's harsh judgment of the justice of God was muted by the author himself by having Job admit, after God has responded to him so crushingly, that he has been in over his head and has spoken of matters too great for him to handle. And in the prose conclusion of the book, as we have seen, Job is praised by God and his friends are blamed; if God could approve of Job's seemingly blasphemous words, then (as orthodox readers of the book might have said to themselves) surely those words must not be so dangerously unorthodox as they sound. But just for good measure, changes were made in the text of the book in the centuries following its composition to make it even more acceptable to the godly. For example, in chapters 32–37 the speeches of Elihu, a fourth comforter who comes unannounced from nowhere, appear to be someone's addition to the text as yet another pious attempt to refute Job's assault on God's justice. And one particularly uncompromising statement of Job's, at 13:15, was changed to make it say the opposite of what the author wrote. The original sense of Job's words is reflected in the Revised Standard Version reading:

> Behold he will slay me; I have no hope;
> yet I will defend my ways to his face.

Early in the history of the text, a scandalized copyist slightly altered the Hebrew so as to produce this sense (New International Version):

> Though he slay me, yet will I hope in him:
> I will surely defend my ways to his face.*

Proverbs would at no point in its history have presented the difficulty for the pious that the book of Job and Ecclesiastes did, even in their edited form. The only thing questionable about Proverbs, as we have said earlier, is that much of its material is secular, this-worldly advice to young men on how the world goes and how to conduct oneself in it. The compiler of the book undertook to make this material more acceptable to the orthodox by casting it in a new light. As a prologue to the older collections of wisdom material he was gathering together into a single work, he composed a long introduction (Prov. 1–9) that presents wisdom as a divine quality and as the essential foundation of a godly life. With that identification established, even the utterly pragmatic pieces of advice that follow could be taken as appropriate by the godly.

THE LITERARY FORM OF
WISDOM LITERATURE

A final matter that requires discussion is the literary form in which biblical wisdom presents itself. The book of Job takes the form of a prose story interrupted in the middle by a poetic dialogue. The other four works take the form of something like a classroom lecture in which the instructors in wisdom undertake three tasks: to describe what they have observed in life, to advise their hearers/readers how to live, and to praise wisdom as a quality. The chief form of expression that the wise men employ is a series of two-part, self-contained statements shaped in the parallel structure that, as we said in chapter 2, is characteristic of Hebrew poetry. In some places the individual statements are related in content to one another and so form loose paragraphs; elsewhere, succeeding statements have no logical relationship to one another. It has often been observed that Proverbs is in places monotonous to read; those places are precisely the ones in which there is a long succession of disconnected bits of obser-

---

*For more on this verse, see chapter 16, where it is discussed as a problem for translators of the Bible.

vation and advice. Among the wisdom books Proverbs suffers most from this defect, The Wisdom of Solomon least.

A moment's thought will make us aware that the form described above is more appropriate to displaying a truth that has already been worked out than to demonstrating the process of reasoning that produces a truth. Each wisdom statement has behind it a lifetime—or a number of lifetimes—of observation and contemplation necessary to its formulation. It is a dignified form appropriate to the great experience and wisdom of the sage. It is a form suitable to the feeble capacity of humble learners sitting at the feet of a master, for the truth it expresses will strike them as inevitable and unarguable, and its brevity and parallelism will assist them to learn it by heart. This latter element was particularly important in a society with little access to writing materials and that depended on memory in training its young and passing on knowledge from one generation to another.

The self-contained units of thought of which we have been speaking can be referred to simply as "proverbs," which is also the title assigned to the best-known collection of them. Actually this English word is not completely appropriate for the thought units of wisdom literature: Few of these units take the form that we normally think of as a proverb. A proverb is more terse, more pithy than the typical statement of the wisdom writers; and although it is memorable by virtue of repetition of sounds, the true proverb is less elaborate and less artful—or to put it another way, more primitive in poetic effects—than most of what is found in the book of Proverbs. Consider, for example, "Haste makes waste," "Easy come, easy go," "A stitch in time saves nine," and so on. Where proverbs of this sort do occur in the book of Proverbs, they are usually only one element in a larger statement, serving as the first of two parts in a parallel structure. But although the term "proverb" may be an inexact designation for what we find most often in wisdom literature, it is well established by convention, and we shall use it to refer to the individual wisdom statements.

The parallel structure of Hebrew poetry makes it a particularly suitable literary vehicle for the moralizing that the wise men engaged in. In the typical two-part unit of Hebrew poetry, an idea or subject expressed in one part was in the other part (1) restated in other words (synonymous parallelism), (2) defined as being *like* something else (emblematic parallelism), (3) further developed (synthetic parallelism), or (4) set against an opposing idea (antithetic parallelism). (See the discussion on poetry and figure 4 in chapter 2.) All of these possibilities

served the moralist well. Having made an observation or stated a principle, he could profitably restate it for emphasis:

> Wounding strokes are good medicine for evil,
>   blows have an effect on the inmost self. (Prov. 20:30)

Or, wishing to define something in the moral realm, he could compare it to a down-to-earth element that his audience was familiar with:

> Cold water to a thirsty throat;
>   such is good news from a distant land. (Prov. 25:25)

> Base silver-plate on top of clay:
>   such are fervent lips and a wicked heart. (Prov. 26:23)

Synthetic parallelism served the moralist particularly well. For, as
a moralist, he believed in a necessary connection between deed or
attitude and its consequence; and synthetic parallelism allowed him to
display the relation between these two elements:

> To make a fortune with the help of a lying tongue:
>   such is the idle fantasy of those who look for death. (Prov. 21:6)

> Whoever strays far from the way of prudence
>   will rest in the assembly of shadows. (Prov. 21:16)

> Give a lad a training suitable to his character
>   and, even when old, he will not go back on it. (Prov. 22:6)

Antithetic parallelism was an ideal mode of expression for the
moralist, for he tended to see the world (and to offer advice) in terms
of paired opposites: good and evil, wise and foolish, divine and human, humble and proud, poor and rich, better and worse, and so on.
Consider the effectiveness of the form to the content in these instances:

> A wise child is a father's joy,
>   a foolish child a mother's grief.
> Treasures wickedly come by give no benefit,
>   but uprightness brings delivery from death.
> Yahweh does not let the upright go hungry,
>   but he thwarts the greed of the wicked.
> A slack hand brings poverty,
>   but the hand of the diligent brings wealth. (Prov. 10:1–4)

Of the thirty-two proverbs/verses in chapter 10 of Proverbs, from
which the example above was taken, all but one have the form of
antithetic parallelism. We remarked earlier on the monotony that can
result in reading a string of proverbs that have no logical connections

between them. This monotony is intensified when the proverbs in the string all employ antithetic parallelism, for a rocking motion is produced by the succession of paired opposites that tends to detract from the sense of what is being said.

THE APPEAL OF WISDOM
LITERATURE

The wisdom literature we have been discussing constitutes about 12 percent of the combined canonical Old Testament and Apocrypha. We could wish that the proportion were even higher. For, if this body of material lacks the great historical significance of the Law and Deuteronomic History and the elevated religious awareness of the Psalms and the Prophets, it has its own importance and appeal. If it teaches us little of the great forces at work in ancient Israel and early Judaism that made them permanently significant in the history of civilization, it nevertheless corrects the impression that there were only giants in the earth in those biblical days. Certainly wisdom puts us in touch with that part of Israelite/Jewish culture with which we can most readily identify. The individuals for whom it was written were people much like ourselves, people caught up in the complexities of organized society, anxious to get ahead and to avoid making a mess of things, trying to see meaningful patterns in the phenomena of daily existence. The issues to which the wisdom writers addressed themselves are timeless and universal; thus their books will always and everywhere be relevant. The problem of suffering can never be solved, but humankind must forever maintain the anguished attempt to solve it, and so we read the book of Job. The search for meaning in the constant flux of everyday events is unending, and so we read Ecclesiastes and The Wisdom of Solomon. The necessity for persevering, with some degree of equanimity, in the struggles of daily life demands that we use whatever good advice is available, and so we read Proverbs and Ecclesiasticus. Biblical wisdom literature is one of the world's great cultural storehouses, and the student unacquainted with it is missing some of the best that the Bible has to offer.

SUGGESTED FURTHER READING

Dianne Bergant, *What Are They Saying About Wisdom Literature?* (New York: Paulist Press, 1984).

Joseph Blenkinsopp, *Wisdom and Law in the Old Testament: The Ordering of Life in Israel and Early Judaism* (New York: Oxford University Press, 1983).

James L. Crenshaw, "The Wisdom Literature," in *The Hebrew Bible and Its Modern Interpreters*, ed. Douglas A. Knight and Gene M. Tucker (Philadelphia: Fortress Press, 1985), pp. 369–407.

James L. Crenshaw, *The Old Testament Wisdom: An Introduction* (Atlanta: John Knox Press, 1981).

John G. Gammie and Leo G. Perdue, eds., *The Sage in Israel and the Ancient Near East* (Winona Lake, Ind.: Eisenbrauns, 1990).

Burton L. Mack, *Wisdom and the Hebrew Epic: Ben Sira's Hymn in Praise of the Fathers* (Chicago: University of Chicago Press, 1985).

R. B. Y. Scott, *The Way of Wisdom* (New York: Macmillan, 1971).

John Mark Thompson, *The Form and Function of Proverbs in Ancient Israel* (The Hague: Mouton, 1974).

The Interpreter's Dictionary of the Bible, ed. George A. Buttrick et al. (Nashville: Abingdon Press, 1962). See articles on Ecclesiastes; Ecclesiasticus; Job, Book of; Wisdom; and Wisdom of Solomon. Supplement, 1976: See articles on Ecclesiastes, Wisdom in the OT.

The Anchor Bible Dictionary, ed. David Noel Freedman et al. (New York: Doubleday, 1992). See articles on Theodicy, Wisdom in the OT.

X

The Apocalyptic Literature

Two of the most popular books in the Bible are Daniel in the Old Testament and Revelation in the New Testament. Both are spectacular, full of color and action, and concerned in large part with the future—the future, that is, from the point of view of the ancient author or of a character in his account. Because of this element of futurity, Daniel and Revelation are often thought of as being among the prophetic books of the Bible and thus contributing to that large body of texts on which sermons, lectures, and books about the "last days" can be based. Daniel and Revelation are particularly useful to anyone with this sort of interest because, unlike any of the prophetic books except Ezekiel, they lay their information out on a time scale of sorts—first this will happen, then that—and so serve to provide a chronological framework into which the less focused remarks of the prophets can be fitted. But although there are similarities between the work of the prophets, on the one hand, and Daniel and Revelation, on the other, the two groups should not be allowed to fall together into one large class, for Daniel and Revelation were composed for a different purpose from that of prophecy and on a different set of assumptions.

The kind of literature these two books represent is called "apocalyptic"; a single instance is called an "apocalypse" (derived from the Greek word for "revelation" which appears in the first verse of the book of Revelation); and the system of thought they embody is called "apocalypticism." It is important for a serious student of the Bible to know the distinctive nature of apocalyptic writing and the typical circumstances in which it was produced. We can begin to deal with those matters by looking closely at the book of Daniel.

THE BOOK OF DANIEL

Daniel is a relatively brief work, containing but twelve chapters. The first six describe the wisdom and brave deeds of the young man Daniel and his three friends, all of them Jews living in Babylon during the Exile. In the seventh chapter there is an abrupt change: The hero Daniel is no longer a Joseph-like interpreter of the dreams and visions of others but a man who himself experiences dreams and visions and requires the assistance of angelic visitors to interpret them. In chapters 1–6, Daniel is spoken of in the third person: Daniel did this and Daniel did that; in chapters 7–12, Daniel speaks of himself in the first person: "I, Daniel, was deeply disturbed" (Dan. 7:15). The figure Daniel was a traditional folk hero among both the Israelites and their Canaanite neighbors; we can guess that a collection of tales about this Daniel (there are other tales like them among the Apocryphal writings) was taken over by the writer of the biblical book and used as an attractive introduction to the more important material in chapters 7–12. These later chapters comprise four separate apocalypses: In chapter 7 and again in chapter 8, Daniel has a vision that is interpreted for him by a heavenly being; in chapter 9, Daniel meditates on the scriptures and prays and receives a response from an angel; in chapters 10–12, he falls into a trance in which an angel describes events of the recent past, the near future, and the final days.

To even a casual reader it becomes apparent quickly that all of these four apocalypses (and for that matter the dream of Nebuchadnezzar in chapter 2, which Daniel interprets) have reference to the same set of circumstances. Although differing elements receive emphasis in each apocalypse, the subject of all of them is the succession of empires that held sway over Israel from the time of the Babylonian captivity in the sixth century B.C.E. onwards. The majority of modern scholars agree that the focus of attention is on the very end of that succession—that is, on the very time when the author of Daniel was composing chapters 7–12—a time in the second century B.C.E. when the land of Israel, having been ruled previously by the Babylonians,

Persians, and Greeks, was under the oppressive control of Antiochus IV Epiphanes, a descendant of one of the Greek generals who took control of Syria and Palestine on the death of Alexander the Great.

By means of the bizarre imagery of the apocalyptic visions, the author reviews history from the time of the Exile down to his own present moment. His purpose is to comfort and encourage his people by demonstrating that the horrors they are experiencing, even as he is writing, constitute the climax of all the troubles that Israel will ever have to endure at the hands of pagan oppressors: The next step can only be that God himself will intervene in their behalf. In the year 167 B.C.E., as part of a campaign to weld all the people of his captured territories into one, Antiochus set up an altar consecrated to Zeus in the Temple in Jerusalem and forced Jewish participation in pagan religious rites upon pain of death. The book of Daniel seems to have been composed shortly after that time, when it appeared that no human solution to the Jews' problem was possible and only God's intervention could save his people. As is evident from chapters 11 and 12, the writer moved from a precise recounting of events in the near past and present directly into a description of events "at the time of the end," when the heavenly host itself would come forth to destroy Antiochus and his army. That victory having been won, the author believed, living Jews and their resurrected dead would be brought to judgment and rewarded or punished eternally in accord with how they had behaved during the terrible persecution of those days.

COMPARISON OF DANIEL
AND THE PROPHETS

If we set the book of Daniel beside the prophetic writings, a most obvious point of comparison presents itself. Both the book of Daniel and the words of the individual prophets were composed in response to the immediate circumstances of their respective times. To know what Amos, for example, was referring to in his prophecy requires us to know something of the political, religious, and social circumstances of the northern kingdom in the mid-eighth century B.C.E. during the reign of Jeroboam II; to know what the author of Daniel intended in his work calls for a knowledge of events during the reign of Antiochus IV Epiphanes and of his campaign to stamp out Judaism. Both the prophet Amos and the author of Daniel perceived in looking about them that Israel was poised at a moment of dire crisis. Whatever use we may wish to make of their words today, we must be careful to take into account their concern for the contemporary situation.

But having perceived the above similarity between Daniel and the prophetic writings, we can also see that there is a crucial difference between them, specifically in the response that was expected from the audience for each. The prophet delivered his message in hopes of getting his hearers to change their wicked behavior and by that means to ward off the punishment that such behavior demanded. The author of Daniel, on the other hand, wrote not to persuade his audience to do something (he never called on them to repent) but rather to inform them that all that had happened was within the plan of God and that what was yet to happen had already been determined by God. The words of the prophets were, we suppose, originally spoken aloud to individuals or groups who needed the message of repentance; the words of the apocalypticist were originally written down in a document intended to be read by the faithful. The first readers of Daniel, those at whom the author specifically aimed his book, would have taken comfort from being told that their present suffering was within the plan of God and was about to come to an end, to be replaced by unending bliss.

There is profound significance in the distinction between the prophet's intention, on the one hand, and the apocalypticist's, on the other. The prophet, though he spoke for the deity about ultimate matters, nevertheless lived in *this* world and was concerned with real offenses committed against real people and with obvious violations of time-honored principles of religion and right conduct. In the prophet's view, punishment, when it came, would be administered by the armies of real enemies, and rewards would take the form of being allowed to live in peace on one's ancestral land. But the author of Daniel had given up on the real world. He offered neither practical steps for people in distress to take nor advice to the suffering on how to bear their lot. Having outlined the details of history to the present (his present, that is), he abandoned the real world and leapt to a completely superhuman agency for the solution of his people's problems. The spectacular element in the book of Daniel—God's sudden breaking into human history and terminating it—is a distinguishing feature of works in the apocalyptic mode. It does appear at places in the words of the prophets, but it is not central to the prophets' way of seeing things.

The author of Daniel proved to be both right and wrong in his predictions. The terrible reign of Antiochus IV Epiphanes did indeed come to an end, soon and abruptly; but it happened through the efforts of the Maccabees, a courageous family of Jews who led the effort to free the nation and cleanse and rededicate the Temple. And when Antiochus died it was of natural causes, in distant Persia, and not at the hands of the archangel Michael. But in spite of the fact that what was foretold did not come true, for reasons not entirely clear the book of Daniel came to be widely known among Jews in a relatively short time. Not long after its appearance, it began to be imitated; over the next three hundred years, several dozen works which still survive were composed under its influence. These extant apocalypses differ widely from one another in some respects; but a set of characteristics can be isolated for the class as a whole, most of which apply to any single instance of the form.

THE CHARACTERISTICS OF APOCALYPSE

First, the level of conflict in apocalypses is cosmic. Characters move easily between heaven and earth (and sometimes the underworld) and into the realm where angels war against angels and where immense earthly armies are struck down by divine power. There is an irony here, for all of this tremendous cosmic activity occurs in response to the fortunes of a minority people (whether Jews or early Christians) living in a not very notable corner of the earth. This suffering people is also, of course, the audience to which the author of an apocalypse addresses his words; and it is in their behalf, he says, pushed as they are to the uttermost limits of endurance by hostile forces, that the power of God will be unleashed and the present age will be brought to an end.

Second, as befits the cosmic stage on which it is played out, the drama of apocalypse presents two mighty opposites who must meet in mortal combat. Though Antiochus IV Epiphanes may strike us as being just one more of history's petty tyrants, he is represented in the book of Daniel in colossal terms. He is said to be a king who "will insult the Most High, / and torment / the holy ones of the Most High" (7:25); "growing more and more arrogant," he will think himself "greater than all the gods; he will utter incredible blasphemies against the God of gods . . ." (11:36). What is striking here is that the author conceives of the persecutor of his people as being so nearly omnipotent that only God himself can defeat him. Thus the universe this writer imagines is what we would call "dualistic"—that is, operating within it is a force for good and a force for evil, and those two are so closely matched that only with the greatest of difficulty will one ultimately overcome the other. The embodiment of the power of evil in a godlike figure, one who employs supernatural agents parallel to God's angels and who appears for a time to be invincible, is a typical feature of apocalyptic literature.

Third, despite the fact that an individual apocalypse is written in response to historical circumstances in its author's own time, this body of literature is less concerned with actual history than with the end of history or what we call "eschatology"—that is, the "last things," namely, the final stages in an individual human being's existence (death, resurrection, judgment, and reward and punishment) and the final events of the "present" age (the ultimate confrontation between good and evil, the triumph of God's cause, the consequences of that triumph for the good and for the evil, and the closing of this age and the beginning of the next). Although these writers did take up anew some of the historical and biographical matters of the Law and Prophets, they did so only for the purpose of romantic and sometimes sensational elaboration; thus we hear in apocalyptic works just what it was the serpent said to Eve while tempting her in the Garden of Eden, just how Potiphar's wife attempted to seduce Joseph, and how Isaiah was slain by being cut in two with a saw. In some apocalyptic works (as we have seen in Daniel) there is occasionally genuine history—specifically, of the events in the writer's own time that provided the terrible occasion for composing the book. But when the focus moves off that narrow "present" moment, history in apocalypses is dealt with in stylized units: in "weeks" of years, or successive kingdoms, or millennia. The most common historical scheme employed in this whole body of writing is a twofold one: There is the Present Age (which includes all past time) and the Age to Come. The dividing point between the two is set shortly after the time of writing, when the conflict between good and evil reaches its climax and God summarily sweeps away the old and brings in the new. So history is indeed *there* in the apocalypses, but a most peculiar sort of history it is.

Fourth, apocalyptic literature frequently takes the form of a report of a vision experienced by the speaker in the work (why we must say "speaker" instead of "author" will be discussed later). The vision that comes to the speaker—sometimes while waking, sometimes while sleeping—consists of concrete images that represent the intended meaning through vivid and usually fantastic allegory. If what is to be represented is the overthrow of one kingdom by another, for example, what the vision may well picture is one mythic beast attacking another. To derive abstract meaning from something that is presented in a largely visual way is no simple matter. What, after all, is one to make of a vision in which a male goat appears with one horn and that horn breaks off and is replaced by four horns; then out of one of those four horns a little horn grows that throws down armies and stars and treads on them? Small wonder that the man Daniel can make no sense of this until an angel appears to explain it to him (Dan. 8:19ff.): The goat represents the Greek kingdom and its horn represents the first

Greek king (Alexander the Great); the four horns are the four kingdoms into which the Greek realm split after Alexander's time; and the little horn growing on one of those horns is the most terrible king in that line, Antiochus IV Epiphanes; and so on. Plainly, the vision that comes to the "seer" and the interpretation of that vision by the heavenly being are two necessary sides of the revelation; the seer describes the vision and reports the interpretation provided to him. Behind it all, of course, is the writer, who to serve the need of his contemporary audience has created the seer, the vision, the interpreter, and the interpretation.

Finally, in all apocalyptic works except the book of Revelation, the one to whom the revelation comes is some great individual out of the past. The figure Daniel, as we have noted, was a celebrated wise man, one to whom folktales had attached over the centuries. The writer of the book that bears his name selected that particular figure as the one to receive and describe a series of apocalyptic visions. Other such figures in apocalyptic writings are Enoch, Isaiah, Baruch (the secretary of Jeremiah), Ezra, Peter, and Paul. It is a legitimate question to ask why a pseudonymous (that is, falsely named) author should have been so universally assigned to apocalyptic books: Why, if the real writer had something vitally important to say, did he cast his work in the form of a report from a notable person of the past? We shall explore this matter in some detail at the end of chapter 12, which concerns the Apocrypha and Pseudepigrapha, bodies of works containing a number of apocalypses. Here it will suffice to remark that (1) the author of the book of Daniel chose the potent figure of wise Daniel as his spokesperson in order to gain a hearing for his message of hope, and (2) the success of the book encouraged later imitation of its major literary features, including pseudonymity. In other words, what we are dealing with is a literary convention: If any writer after the time of Daniel chose to express his ideas in the form of an apocalypse, he used the standard features of apocalypses, including pseudonymity. There was nothing dishonest in this, any more than there was anything dishonest in the author of Acts putting long, detailed speeches into the mouths of persons who lived fifty years before his time. This was simply the convention of history writing in the ancient world.

**THE BOOK OF
REVELATION**

Having surveyed the book of Daniel and the characteristics of the literary genre it represents, we can turn our attention to the most notable apocalyptic work in the Bible, the book of Revelation. As a means of approaching this work, we can profitably view it as it would strike a reader today who sits down to go through it for the first time. What are its surface features, the elements that immediately come to the fore on an initial reading?

We note straight off that the book is called "a revelation [that is, apocalypse] of Jesus Christ, which God gave him" (Rev. 1:1), and that Christ has passed it on to someone named John, who in turn is passing it on to his readers because what the revelation predicts is about to happen. But then the writer surprises us by apparently starting over, for he interjects a formula of the sort Paul uses to introduce his letters, informing us that what follows is intended to be, in some sense, a letter to seven churches in the Roman province of Asia. The writer announces that while he was in exile, to which he had been sentenced for his Christian activity, he received the revelation that he is about to describe. There follows a series of messages from Christ to each of the seven churches concerning their spiritual condition (chapters 2–3). Next John tells of how he was given a vision of heaven—indeed, of the very throne of God and those who worship there (chapters 4–5). In the vision a scroll with seven seals is produced, and a lamblike being offers to open the seals so that what is in the scroll can be revealed; for making that offer, the lamb is praised in the same terms as God himself has been. As each seal is opened, potential destroyers of the earth appear and natural disasters begin to occur on the earth (chapter 6). All heaven waits in expectation for the opening of the final seal; when that happens, it brings forth—not some ultimate horror, as one might suppose, but seven angels with trumpets, thus setting up a second sequence of seven actions to parallel the first sequence. As each trumpet is sounded, John sees dreadful calamities fall upon the earth (chapters 8–11). Between the sixth and seventh trumpets, he hears a message from seven thunders but is not permitted to pass the message of that particular sequence of seven on to his readers. He sees Jerusalem undergoing a time of anguish and destruction; then the seventh trumpet sounds and great songs of praise to God are sung in the heavenly court.

The scene changes suddenly (chapters 12–13). A woman about to give birth appears in the sky, pursued by a dragon intent on devouring her child the moment it is born. But the child is carried away to God and the woman is enabled to flee to safety. The furious dragon, with forces of angels at its command, battles with the great angel Michael and his heavenly host; the dragon is defeated and thrown down to earth where it begins a campaign against humankind. The dragon creates a godlike beast and a second beast to serve as enforcer of worship for the first beast. Only those inhabitants of earth who carry the mark of the first beast are permitted to buy and sell and carry on the business of life. Several brief visions of encouragement and warning follow as a kind of interlude (chapter 14). Then, once again, a dreadful series of seven occurs: Seven bowls of wrath are emptied, one after another, pouring out terrible plagues upon the world, climaxing in a battle at Armageddon that involves all the kings of the earth (chapters 15–16).

Once again the scene shifts (chapters 17–18). John sees a woman in royal dress riding on a seven-headed scarlet beast; she is called "Babylon the Great" and is "drunk with the blood of the saints" (Rev. 17:5, 6). The woman and the beast are for a time in close league; but they fall out with one another, and the beast (which turns out to be the "first beast" mentioned earlier) destroys the woman and causes great lamentation among her former admirers.

But now it is time to begin the conclusion of things (chapters 19–20). John hears a shout go up in heaven in anticipation of victory over the forces of evil, as Christ himself goes forth on a white horse to do battle with the dragon and his beasts. The beasts are taken prisoner by Christ's army and thrown into a lake of fire, and the dragon is seized and chained up for a thousand years. During that millennium those who have been killed for their Christian faith are resurrected and live a life of bliss on earth with Christ. But then the dragon is set loose and, surprising to say, once again seduces vast numbers into following him in a war against God. Heavenly fire descends to end this final rebellion, and the dragon is thrown into the lake of fire to be tormented forever. All that remains before the inauguration of the future age is for judgment to be pronounced on all the living and dead in earth's history and for the damned to be thrown into the lake of fire. For the blessed, the new age begins when a heavenly Jerusalem is brought down to replace the earthly city (chapters 21–22). Here humans will dwell in a face-to-face relationship with God, sustained by the water of life, the tree of life, and the presence of God. The vision concludes with an angel saying to John, "Do not keep the prophecies in this book a secret" (Rev. 22:10), for the time of its fulfillment is at hand. Readers are warned that those who add anything to the book will have the plagues it describes added to them and that those who take anything from the book will have their share of future bliss taken away from them.

The foregoing summary of the main features of the book of Revelation appears complex, but it really boils down to this:

CHAPTER 1	John's greetings and the circumstances of his call
2–3	The letters to the seven churches
4–5	The heavenly court and the lamb
6–7	The seven seals

8–11   The seven trumpets

12–13   The pregnant woman, the dragon's defeat, the two beasts

14   Visions of reassurance and warning

15–16   The seven bowls of wrath

17–18   The whore of Babylon and the beast

19–20   The defeat of the dragon's forces, final judgment

21–22   The new Jerusalem, conclusion

REVELATION AS A
TYPICAL APOCALYPSE

What is one to make of all this? How are we to interpret the lamb, the dragon, the seals, the trumpets, the beasts, and all the other bizarre elements of Revelation? Certainly opinion varies widely on what the book means. But among modern scholars there is more consensus than might be imagined as to its essential meaning. Anyone who has thoughtfully read this present chapter thus far and who has also read Daniel and the sixth-century prophets will be in a good position to define that essential meaning for himself or herself. The book of Revelation is, simply put, an apocalypse. It employs most of the standard features of that literary genre and was written for the same reason that other apocalypses were written— namely, the author believed his own days to be the worst possible days and thus surely the last days; therefore the faithful were to be encouraged to persevere during this bad time, for their deliverance was soon to come.

What was the time in which and for which Revelation was written? There is general agreement that it was composed in the last decade of the first Christian century when the populace of the Roman Empire was being required to participate in at least the outward forms of emperor worship. Because many Christians (though not all) could not in conscience do this, they were suffering severe persecution, often to the point of death, for the sake of their faith. This is exactly the situation that Jews of the mid-second century B.C.E. had faced under Antiochus IV Epiphanes. To comfort those among the Christian faithful who were puzzled (to say the least) at their lot in life, the author of Revelation did for Christians just what the author of Daniel had done for Jews two centuries earlier. He wrote a book to demonstrate that because the mightiest of human agencies (the Roman Empire) was opposed to the one true faith (Christianity), God himself (in the person of Christ) must break into human history and bring it to an end. Thus the stage would be set for judgment of both the persecutors and those who had betrayed their faith under persecution, on the one hand, and for rewarding the faithful by inaugurating the new age, on the other. One must understand that an apocalypse was not the only literary form the author could have used to express his message. He could, for example, have written to the seven churches a direct and matter-of-fact letter of the sort that Paul wrote to churches for which he was concerned. But this writer chose to use the form that had worked so well in the book of Daniel—a seer's report of visions embodying symbolic things and symbolic actions—and to enrich it with the language and perceptions of the ancient prophets and mythic materials of the ancient religion of Israel.

With what characters would the author of Revelation populate his cosmic drama? Well, there would be God himself on his throne (borrowed from Daniel's representation of the "One most venerable" [Dan. 7:13]) and, of course, Christ—sometimes in human form, and at other times in the form of a slaughtered lamb (borrowed from Isaiah); there would be angels in great numbers, both bad ones and good ones, including the archangel Michael (mentioned in Daniel); there would be elders representing the twelve heads of the tribes of Israel and the twelve apostles; there would be a good woman who gives birth to a child (perhaps the woman is Mary, perhaps the nation Israel, but she is also a constellation of stars) and a bad woman (the city of Rome); there would be heavenly creatures (drawn from Ezekiel's chariot vision) and mythic beasts; there would be ghastly horsemen of destruction (drawn from the prophecy of Zechariah) and terrible locusts "like horses armoured for battle" (Rev. 9:7) (drawn from the prophecy of Joel); there would be the souls of the martyrs and throngs of saints before the throne of God; there would be the resurrected dead gathered for judgment (as in Daniel); and finally there would be the inhabitants of the new Jerusalem, blessed forever in the presence of God and the lamb.

Given that cast of characters and the conventions of the apocalyptic genre to satisfy, how would the author organize his material? He decided to begin by establishing the credentials of his speaker, John, as the chosen instrument of God's communication, and then—briefly employing the epistolary (letter) form—by delivering direct messages appropriate to each of the seven churches of Asia. That accomplished, the author chose to use the time-honored device of transporting John in a vision to the heavenly court, where he could both observe the ongoing worship of God and witness a series of scenes representing the future of heaven, earth, and underworld. It is as though John from his vantage point in heaven witnesses a celestial picture show, one that the inhabitants of heaven both appear in as performers and observe along with John as audience.

The show itself—what John sees—is in two parts: The first depicts the cosmic and earthly events at the end of the Present Age; the second depicts the eternally static situation of the Age to Come. The events that conclude the Present Age are introduced by sets of sevens that are associated with the communication of messages: seals on a scroll and heralds' trumpets. The events themselves are divided into three phases: the defeat of Satan in the heavenly sphere (by the birth of Christ), in the earthly sphere (by the destruction of Rome), and in the underworld (by the final elimination of the Devil, Death, Hades, and the sea at the very end of the age). Poised between the two ages is the thousand-year reign of Christ on earth, which shares something of both future and past: It provides a foretaste of the bliss of eternity future, but the evil of the past lies there waiting to exert itself yet once again. When this final resurgence of evil has been suppressed and ultimate judgment passed, the New Age can begin in all its perfection. The only thing the author had to add was a warning that the last days were at hand and that no one should tamper with the words of his book.

To think of Revelation in this way—as a piece of literature in which religious and mythic materials are shaped in the form of a conventional apocalypse, for the conventional apocalyptic purpose of providing comfort to the suffering faithful—is to cut through much of the mysteriousness of the book. (Not all of the mysteriousness, of course, for symbol and myth by their very nature cannot have firm walls set about them to confine their meaning.) Most of the images in Revelation had had a long previous history of religious and literary use before this particular author employed them; inevitably a multiplicity of meaning had become attached to them. Consider the rich effect of designating the personification of evil as "the great dragon, the primeval serpent, known as the devil or Satan" (12:9). This would have brought to the mind of a contemporary reader the complementary images of the serpent that misled Adam and Eve, the "adversary" who brought about the terrible suffering of Job, the sea creature Leviathan described in the book of Job, and perhaps the fabled sea monster Tiamat who (in Babylonian legend) was destroyed by the deity at the time of the Creation but must be destroyed once more at the end of time. Or again, any one of the several victories over the Devil—whether at the birth of Christ or the fall of Rome or before or after the millennium—is in a sense the same as every other one of the victories. That is why in Revelation we so often find an instance of the destruction of evil followed by rejoicing in heaven, only to be followed in short order by the return of what seems to be the same evil in different guise. Layer is placed upon similar layer, and it is not always evident just where we are in the continuing story at any given moment. But there will come an end to that story, the author of the book assures his readers, when the great instigator of evil and his entire domain will be flung into the fiery lake; and then the warfare of the saints will cease forever.

As was noted earlier, the author of Revelation departed from the conventional form of the apocalypse in one major respect: He did not assign the visionary experiences described in his book to some ancient hero of the faith like Enoch or Daniel but to one John, living at the very time the book was being written and at the very time of the suffering it describes. Because of the directness of the book and the sincere tone of its language, scholars have generally assumed that "John" and the author are one and the same. Perhaps so. (But we should not forget that this long book is a carefully crafted literary work and that, if it indeed reports genuine visionary experience, this experience was nevertheless deliberately shaped in conventional apocalyptic terms.) Or perhaps John was a Christian prophet of such great reputation that he could be chosen by another man, the author of this apocalypse, as the contemporary equivalent of an Enoch or a Daniel. Whatever the case, the decision not to assign the report to some figure from the past meant that it was not necessary to develop the sort of pious fiction found at the end of the book of Daniel, where, when Daniel complains that he does not fully understand what he has heard, an angel says to him, "These words are to remain secret and sealed until the time of the End" (12:9). The "time of the End" for the writer of Daniel was, of course, the writer's own time—the time of the persecution under Antiochus IV Epiphanes—and thus he was representing his work as being a message from the past that was presumably discovered in his own day.

JUDAISM'S REJECTION OF APOCALYPSE

In the year 66 c.e., a rebellion against Roman power began in Judea; by the time it ended in the year 73, the Temple had been destroyed and thousands of Jews had been slaughtered by the Romans. The Jews who began the strife did so in full awareness that they alone could not win a war of liberation against Rome. They were convinced, however, that what the apocalyptic writers had said was so: When the Jewish people were pushed to their final limit, the forces of God would intervene to save them and the present age would come to an end. Therefore, these patriots decided to force God's hand, so to speak, and, by starting the war, obligate God to step in and conclude it. Certain more realistic Jewish leaders, fearful of what so futile and dangerous a move could lead to (and, in fact, did lead to), were allowed by the Romans to establish a new center for their religion at Jamnia, a town near the seacoast to the west of Jerusalem. During the course of the struggle in Jerusalem, as we indicated in chapter 6, a number of Jews managed to escape the city and find their way to Jamnia, and it was there that the surviving leaders of the nation gathered to discuss the meaning of what had happened and what would be in store for Judaism in the years to come.

Out of that discussion came decisions that were of immense significance for the Bible, particularly with respect to which pieces of Jewish religious writing were eligible for inclusion in the canon of sacred scripture. One such decision, reflecting in large part a strong distaste for apocalyptic literature because it had inspired the zealous patriots to rebel against Rome, was that no piece of writing composed after the time of Ezra (in the late fifth century B.C.E.) should be admitted to the canon. Why then admit the book of Daniel, composed (as we now believe) in the second century B.C.E.? Partly because the book claimed to have been written before the time of Ezra, during the Babylonian Exile (though why that claim was believed when similar ones in other apocalypses were not is not clear), and partly because the work had always been and still remained popular. Judaism turned away from the apocalyptic fashion of thought and concentrated instead on working out the contemporary significance of the ancient law.

CHRISTIAN USE OF APOCALYPSE

But the same was not true of first-century Christians. If apocalypse had led only to bitter disappointment for Jews, for Christians its promises seemed on the verge of fulfillment. They had to admit, of course, that God's power had not yet been openly displayed before the world in behalf of the faithful: The Son of God had appeared on earth as a humble Galilean, not as the commander of a host of angels. But if the matter were understood correctly, then Christ's life, death, resurrection, and return to heaven could all be seen as constituting his *first* coming, which was necessary to achieve the salvation of those who would put their trust in him. And now all was in readiness for his *second* coming, in power, to end the present age and inaugurate the final one. His triumph over death in his first coming was an earnest of his ultimate triumph over the forces of darkness in his second coming. Some adaptation of the older apocalyptic expectations was necessary to make them fit the circumstances of first-century Christianity, but the early Christians believed that what the apocalypticists had looked forward to had already begun and would soon be completed.

Whatever suffering they had to endure as bystanders in the terrible war with Rome that the Jews had undertaken for apocalyptic reasons, the Christians did not thereafter reject apocalyptic thinking, as the Jews did, for such thinking was at the very center of their religious system. They enthusiastically quarried materials from the book of Daniel (with what result is evident in the book of Revelation), 1 Enoch (see that book cited in Jude 14–15, for example), and other such writings, and they were equally enthusiastic in composing their own apocalypses. Several passages in Paul's letters—most notably 1 Thessalonians 4:13–18 and 1 Corinthians 15:20–28—can be considered to be apocalyptic, as can Mark 13 (generally referred to as "The Little Apocalypse") and many additional passages in the gospels and later letters of the New Testament. Some time after the composition of the book of Revelation, near the end of the first Christian century, there appeared an apocalypse concerning the supposed visionary experiences of Peter and, over the several succeeding centuries, of Paul, James, Thomas, the Virgin Mary, and others. One of the most interesting of apocalypses is the book called 2 Esdras, the core of which (chapters 3–14) was probably written by a Jewish writer in about 100 B.C.E.; an introduction and conclusion were added in later centuries in order to Christianize the work. What happened to that original portion of 2 Esdras well illustrates the process by which the early Christians took over the Jewish writings that touched on matters of Christian interest and shaped them to their own ends.

CONTINUING APPEAL OF APOCALYPSE

There has been a continuing Christian interest in ancient Jewish and Christian apocalypses right up to the present day; indeed, composing visionary works about the "time of the End" or "last days" and ascribing them to ancient figures is not unknown to the twentieth century. The reason for the appeal of the apocalyptic form is as obvious as, and about the same as, the reason for the appeal of the prophets. Here are biblical (or quasi-biblical or pseudo-biblical) works that give solemn assurances of great days coming for the faithful—and those days have not yet come. In any troubled age, and ours is certainly one, such writings have their appeal. Apocalyptic writing has the advantage (if it can be called that) over prophetic writing that its proposed solutions to problems are more abrupt, more dramatic, and more violent. As we have stressed earlier, in apocalyptic writing—unlike prophecy—there is no room for repentance and thereby a change in God's intentions. There is only a sudden flash of divine power followed immediately by harsh punishment for those who have taken advantage of the misery of others and glorious rewards for those who have suffered undeservedly. In certain moods we can all probably find satisfaction in such an ordering of things.

The ancient apocalyptic writers gave much of their attention to the final times and supposed that their own days were the worst of days and therefore were the prelude to the end. We know that they were wrong, but, in our human arrogance, we may well suppose that *our* days are the worst of days and thus are the *real* prelude to the end. The habit of interpreting biblical apocalypses as though they were written to apply directly to the situation of the interpreter bears eloquent testimony to the perceptiveness of those ancient writers who attempted to provide comfort to their suffering coreligionists. It is legitimate to draw what comfort we can from those works, but we should do so with the awareness that we are only the latest in a long line of readers who have felt that the works were composed with them in mind.

SUGGESTED FURTHER READING

James H. Charlesworth, ed., *Apocalyptic Literature and Testaments*. Vol. 1 of *The Old Testament Pseudepigrapha* (Garden City, N.Y.: Doubleday, 1983).

Norman Cohn, *Cosmos, Chaos and the World to Come: The Ancient Roots of Apocalyptic Faith* (New Haven: Yale University Press, 1993).

Paul D. Hanson, *The Dawn of Apocalyptic* (Philadelphia: Fortress Press, 1975).

Edgar Hennecke, Wilhelm Schneemelcher, and R. M. Wilson, eds., "Apocalypses and Related Subjects," in *New Testament Apocrypha*, vol. 2 (Philadelphia: Westminster Press, 1964), pp. 579–803.

E. W. Nicholson, "Apocalyptic," in *Tradition and Interpretation*, ed. G. W. Anderson (Oxford: Oxford University Press, 1979), pp. 189–213.

Christopher Rowland, *The Open Heaven: A Study of Apocalyptic in Judaism and Early Christianity* (London: SPCK, 1982).

David S. Russell, *Divine Disclosure: An Introduction to Jewish Apocalyptic* (Minneapolis: Fortress Press, 1992).

The Interpreter's Dictionary of the Bible, ed. George A. Buttrick et al. (Nashville: Abingdon Press, 1962). See articles on Apocalypticism and Daniel. Supplement, 1976: See articles on Apocalypticism and Revelation, Book of.

The Anchor Bible Dictionary, ed. David Noel Freedman et al. (New York: Doubleday, 1992). See article on Apocalypses, Apocalypticism.

XI

The Intertestamental Period

In the Bible as commonly printed, going from the Old Testament to the New is just a matter of turning a page. There was Malachi, here is Matthew. Hence readers whose attitudes toward the Bible are governed by their experience with modern books are likely to assume that New Testament writers picked up immediately where Old Testament writers left off, making the two Testaments as close together in history as they are in print. But they were not. Many years intervened between these two collections of writings, years that saw profound cultural, social, and political changes in Palestine. Reading the Old Testament alone, no matter how thoroughly, leaves one unprepared for meeting and understanding the world of the New Testament. How, for example, did the Romans get into the picture? Who are these Pharisees and Sadducees? What is a synagogue? Where did belief in demons come from? Why is it that the translation we are reading is from the Greek rather than the Hebrew? The New Testament itself, which is addressed primarily to contemporaries of the writers, naturally assumes that everyone knows. But if we are to join them in this understanding, we need to make an independent study of what scholars customarily call the "intertestamental period."

The Babylonian Exile is both a convenient and a proper place to begin our story. Nothing else in the history of the chosen people since the settlement of the Promised Land had so profound an effect on them. They went into exile in Babylonia as Israelites, but they returned as Jews.

A proper description of the Babylonian Exile must begin with events in the eighth century B.C.E., as narrated in the later chapters of 2 Kings. The general outlines of the story are well known. How in 722 B.C.E., after a long siege, Samaria fell to the Assyrians, who then deported large numbers of its inhabitants to Assyria and replaced them with their own peoples, effectively ending for all time the northern kingdom of Israel. How, a little more than a century later, the new Babylonian empire that had taken over from the Assyrians began its campaigns to subdue rebellious Judah, the remaining fragment of the state originally created by David and Solomon. And how, after an initial conquest in 597 B.C.E., which resulted in the capture of the royal family and the deportation of many citizens of Jerusalem, that city was definitively captured in 587, the Temple burned to the ground, and Judah itself, as it seemed, now blotted from the pages of history.

About the conditions of the Exile, the historical books of the Old Testament say nothing, but from contemporary prophets (Jeremiah and Ezekiel) and other sources we can put together a reasonably accurate picture of the period, even though it is lacking in details. The result is rather surprising. It turns out that the Exile, in contrast to the bitterly remembered loss of Jerusalem, was not altogether a bad time for those who had to endure it. Deportees were resettled in Mesopotamian communities near rivers or irrigation canals and were allowed some degree of local freedom. They could marry and conduct normal family life. Some of them owned homes. Insofar as it was possible, they practiced their own religion. When in 538 B.C.E. Cyrus issued an edict, permitting the exiles to return to Jerusalem (an occasion joyously celebrated in the poems of Deutero-Isaiah), many of them chose instead to remain in Babylonia, preferring the security of their present situation to the perils and uncertainties of life in their impoverished former homeland—a land that most of them had never seen. In Babylonia the Jewish community grew and prospered, lasting well into the tenth century C.E. despite episodes of persecution, and supplying Judaism with some of its most authoritative scholars and commentators on the Law.

One of the occupations that must have been pursued in Babylonia had nothing to do with making a living: It was the study and copying of the documents that had escaped the burning of the Temple and been brought along into captivity. Although the Bible story is oddly silent on this fact, the Ark of the Covenant (the throne of Yahweh in the Holy of Holies) and its contents (the physical evidence of the Covenant) had disappeared forever. The only physical evidence of Israel's past was now the written word. Thus the imperative need to bring this word into order as a substitute for the other physical evidence now lost. The making of a formal scriptural canon began.

Obviously, this was a time of retrenchment and reappraisal, a time for withdrawing toward the ideological center of the cult. Yet the exiles could not help being influenced by the foreign culture surrounding them. This influence begins to show up in the adoption of Babylonian names and in the shift to a Babylonian calendar. One of the most widespread changes, accelerated if not begun by the experience of exile, was the replacement of Hebrew by Aramaic as the common spoken language of the people.

Aramaic is a member of the Semitic family of languages to which Hebrew also belongs, and resemblances between the two are quite obvious to a linguist. But, like modern French and Spanish, they are different enough so that a person brought up speaking only one of them would be able to understand very little said to him in the other. Aramaic is as old as Hebrew and was widely spoken in the Near East; from about 1000 B.C.E. onward it began to replace the native language of the Assyrian conquerors within the empire they had created. The Assyrians found it convenient to adopt a form of Aramaic for official use, as did the Babylonians and Persians after them, insuring its status as a lingua franca, that is, a mutually agreed on medium for the conduct of business among persons with differing native tongues (as English has become in international relations today). It is likely that some of the exiles to Babylonia knew Aramaic already because the Assyrian conquest of the northern kingdom would have favored its introduction there, and the evidence of 2 Kings 18:26 shows that it was not unknown in the south. The rapidity of the change is testified to by Nehemiah 8:7–8, which describes how, after Ezra had read from the Mosaic Law to the assembly of the people in Jerusalem (about a century following the return from exile), his reading had to be followed by an exposition (in Aramaic, we believe) so that the people could understand what they had just heard. Much later, in the synagogue service, it became standard practice to add a "targum"— an interpretive paraphrase in Aramaic—following the reading of the scripture in Hebrew. Hebrew did not die out entirely, however. Because of its prestige as the language of scripture and ritual, there were always educated persons who could use it, but in the end it was probably no more intelligible to the general population of Palestine than Latin is today to most Roman Catholics.

THE RETURN FROM EXILE

What further cultural assimilation might have occurred during the Babylonian Exile we cannot tell, for the picture was abruptly changed with the conquest of Babylonia by Cyrus the Persian. Cyrus, whose imperial strategy was based on allowing subject peoples to retain their own national identities within the empire, reversed the policy of his predecessors and granted permission for a band of exiles to return to Jerusalem and rebuild the Temple. (Deutero-Isaiah sees Yahweh's hand in all this: Cyrus is Yahweh's instrument, indeed his "anointed" [Isa. 45:1].) The returnees quickly erected an altar on the Temple site, but the Temple itself was not finished until 516 B.C.E., and only after much delay and opposition. About seventy years later, under Nehemiah, the walls of Jerusalem were rebuilt.

Nehemiah's memoirs testify to the hardships of life in the Persian province of Judea. His account is certainly to be believed in its general outlines, although much else in this unsettled period is obscure to us and subject to differing interpretations. Everything was meager and difficult. The returned exiles, claiming land that was now in other hands, were not exactly welcomed by those who had stayed behind. Class inequities developed. Taxes were burdensome. But at least the national religion had a center again, and this Second Temple—though a building much less grand than Solomon's before it or Herod's after it—served effectively in that role longer than either of theirs did: five hundred years. Because the Davidic monarchy had disappeared (except in the vision of prophets), the Temple was no longer a royal chapel. In the hands now of a hereditary high priesthood, the Second Temple as an institution exerted more influence over the lives of its people than its predecessor had done, ruling in civil as well as religious affairs.

HELLENISM

Information about life in Judea through the fourth and third centuries B.C.E., the years between 400 and 201, is quite scanty. But one event on the world scene is well documented and had enormous effect on the Jewish people everywhere: the conquest of the Near East under Alexander the Great. In 333 Alexander defeated the Persian armies at Issus and then sped southward through Palestine to obtain control of Egypt. It does not appear that he had to fight for Judea, as he had for Tyre; the Jews acknowledged their new ruler willingly. Alexander's sudden death in 323 B.C.E., with his conquests still incomplete and no heir in sight, caused a struggle for power among his chief generals: Eventually the portion of his empire that we are concerned with was divided between two of them, Ptolemy and Seleucis (and their descendants, the "Ptolemies" and the "Seleucids"). From 301 to 198 B.C.E. the former ruled Egypt, Palestine, and Phoenicia, while the latter ruled Asia Minor, Syria, and Mesopotamia. Judea was, as usual, caught in the middle. But the Persian policy of religious toleration was continued by the Ptolemies from Egypt; and although the Greek overlords were better organized and more enterprising than the Persians, placing heavy economic burdens on the people, they allowed other aspects of Jewish life to develop undisturbed.

The Greek conquest of the Near East had one profound effect that could hardly have been anticipated when it began, and that was the rapid hellenization of the conquered peoples. Greek culture followed Greek soldiers. "Hellenism," the name given to this complex of values, beliefs, and practices by which Greeks ordered their lives, so far from being resisted as a foreign importation, was almost everywhere welcomed and imitated. The two principal carriers of hellenism were language and institutions—the Greek language foremost, since besides being the only means of getting ahead in a world run by Greeks, it was the key to the wealth of their philosophy and literature. A somewhat simplified form of classical Greek known as "Koine" became the lingua franca of this empire and remained so even after the Roman conquests, especially in its eastern regions.

The chief institutional carrier of hellenism was the distinctively Greek city, or *polis*. The *polis* was formed on a rational model and embodied a concept hitherto unknown in the Near East, that of the citizen, or *polites*. Normally all free male residents of such a city were enfranchised citizens. They elected a council that, in turn, appointed magistrates who were responsible to it. This was a long way from modern democracy, because conquered peoples (considered barbarians by the snobbish Greeks), women, and slaves were excluded from the political life of the city; on the other hand, they did benefit from its public facilities. Among these facilities were outdoor theaters, gymnasiums, baths, stadiums, marketplaces, and temples—schools, too, for the hellenized upper classes. If a Greek city was built from the ground up, as Alexander built the city on the Nile delta that bore his name, its facilities were designed and placed according to plan—a great contrast to the usual Near Eastern city, which reflected higgledy-piggledy growth interrupted at intervals by the whim of this or that king or local strongman. If an existing settlement was taken over and turned into a *polis*, as also happened, its facilities were modified and its public life adapted to some degree in imitation of the ideal concept. To be so promoted was considered an honor, and it had practical advantages, too, because in domestic affairs the Greek cities were self-governing. Alexander's successors founded cities on the Greek plan throughout the Near East, some thirty of them in Palestine alone. One of the most important Greek cities was Antioch, at the mouth of the Orontes River on the Syrian coast, which grew to rival Rome and Alexandria and became, according to Acts, the first center of Christian missionary activity. Also important was the Decapolis, a federation of ten cities located in the Transjordan east and south of the Sea of Galilee, through which Jesus passed during his Galilean ministry, according to Mark 7:31.

During the century in which Palestine and Egypt were under the same rule, Jews found it relatively easy to move back and forth between the two. Many of the Jews who left Palestine for Egypt never came back but settled permanently there, becoming part of what is called the "Diaspora" (dispersion). It was not unprecedented for Jews to live abroad—some had done so even in the days of the Monarchy— but under these more favorable conditions the rate of emigration increased greatly. Jewish mercenary soldiers, adventurers, captured or purchased slaves, artisans, farm workers, and merchants spread through the eastern Mediterranean and Near East. That greatest of hellenistic cities, Alexandria, acted as a magnet to draw Jews, who very soon made up a large and flourishing settlement within it. They could not become full-fledged members of the Greek political community, but in other respects they were well and peaceably integrated into its life. They no longer spoke Aramaic, but Greek, and their culture was significantly penetrated and altered by that of the Greeks. It was for this colony that the Hebrew scriptures were translated into Greek, a work that began in the middle of the third century B.C.E. and continued through the second, giving us the version known as the Septuagint.

JEWISH RELIGION IN
THE DIASPORA

The Jewish faith was generally tolerated and even respected in these foreign countries—in spite of, but also to some degree because of, its oddity. And it attracted converts. Jews enjoyed special privileges, such as permission to observe the Sabbath (there was no such thing in the world outside Judaism as a regular weekly day off), to collect and administer taxes for their own use, and to settle legal disputes between members of their own group. They were excused by tacit agreement from offering sacrifices to the national or civic gods. To be a Jew under these circumstances meant that one followed the Law of Moses and acknowledged the Temple in Jerusalem as the legitimate center of its tradition, sent in the annual half-shekel Temple tax for its maintenance, and, if possible, journeyed there for one or more of the three great pilgrim feasts. But it also meant accepting an alien land as one's permanent home as well as living with the tension between the need to assimilate with the local culture and the need to preserve one's special identity. On balance, the Diaspora must have offered opportunities at least as good as those in Palestine, for we find that by the first century C.E. there were more Jews living outside Palestine than in it. To this day, the majority of Jews have continued to live in the Diaspora.

It is not surprising that Jews separated from their homeland and Temple would develop some means of celebrating their faith in a regular and public manner where they now lived: hence the synagogue (from a Greek word meaning "a bringing together"). Unfortunately we do not know when synagogues first started, or where, or in what form, or even what their purpose was originally conceived to be. They may date from as far back as the Babylonian Exile, but it is much more likely that they grew up during the Greek period. The synagogue was a very different kind of institution from the pagan temple, for here a religion was practiced without sacrifices, images, or priests, and the chief object of veneration was a scroll with writing on it. The ancient world had seen nothing like this before. The synagogue service consisted of readings from the scriptures, prayers, hymns, and commentaries and exhortations by members of the assembly. The form of organization was congregational. By New Testament times synagogues were everywhere in the Diaspora—and in Palestine, too—which indicates that we should look at them less as substitutes for Temple worship than as parallel institutions with their own raison d'être. The earliest Christian preaching, we are told, took place in synagogues. Thus at the final destruction of the Temple, the synagogue stood ready to assume the full responsibility of carrying on the Jewish faith and, incidentally, to serve as a model for the emerging Christian Church.

Jews of the Diaspora adopted the Greek language readily enough (in Palestine the popular tongue remained Aramaic) and along with it many Greek ideas and habits of thought; unlike other subject peoples, however, they could not stomach Greek religion, for it was polytheistic, idolatrous, and devoid of ethical teaching. Nor did it have anything to say about the destiny of the Jews. Hellenistic influence on Jewish religion, then, came indirectly and did not affect its basic character. Signs of hellenism are almost nonexistent in the canonical Jewish Bible; one must look for them instead in the Apocrypha and Pseudepigrapha. The reason, as we saw in chapter 6, is that the canon settled on at Jamnia was a conservative one, intended to establish the Jewish scriptures on a traditional basis and to answer once and for all any claims of authority for the Septuagint. To the extent that this was a defensive move, it was directed against Christianity, not against hellenism, but everything hellenistic got excluded in the process.

BELIEF IN RESURRECTION

One of the few new developments during this period that survived to enter the mainstream of Judaism was belief in resurrection. In the older view, seen abundantly in Jewish scriptures, each individual is an undivided living whole, a *nephesh ḥayyah*. The entire destiny of human beings is here on earth, which was made expressly to house them; upon their death they go to the shadowy underground, Sheol, where they remain forever. An important corollary to this is the principle that all rewards and punishments are given here on earth, divine justice assuring that they will be distributed according to the individual person's merits. This view held sway generally until the intertestamental period. The one unmistakable assertion of individual life after death in the Jewish canon appears very late, in Daniel 12:2, written in Palestine about 164 B.C.E.* (There never had been any doubt that God *could* revive the dead or even, as with Enoch and Elijah, bypass death entirely.) It is significant that the statement in Daniel is given in the context of an eschatological prophecy in which it is the fate of groups, not of individuals, that the writer has in mind. The dead, whether resurrected or not, are always considered as members of a class. The notion that a person's chief duty in life is to win his or her own salvation independently of anyone else's was quite foreign to Judaism.

This new idea of resurrection provided exactly the needed basis for beginning to answer Job's great question (though Job himself never wavered in his belief that death ends all): Why are the good allowed to suffer in this life, if God rules the world? Because they will be amply rewarded in the next. But just because the answer fit the question so well, we must not assume that the question is what

---

*Isaiah 26:19 speaks of the dead regaining life and bodies rising again. But in context the verse seems clearly to be a figurative reference to the return of Judah from exile in Babylonia. The freed exiles will be as the dead brought back to life.

called it into being. It is more likely that the idea of resurrection arose not as a response to an abstract philosophical question but as a response to a concrete practical one: persecution of the Jews in the second and first centuries B.C.E. This is certainly borne out by 2 Maccabees, written about 124 B.C.E. in Egypt, which covers events in Palestine to a point somewhat beyond the end of Daniel. Here resurrection and eternal life are a reward for martyrdom (see 2 Maccabees 7:9–14, 23, 29, 36; 12:43–45; 14:46). On the other hand, in more tranquil days in Palestine the wisdom teacher Jesus ben Sira could still adhere to the traditional view, presenting it in language worthy of Job's comforters (see Ecclesiasticus 14:12–19).

If eventually there was widespread belief in resurrection, there was certainly no agreement on who would be resurrected—or when, how, in what form, and to what kind of future. A popular notion was that only the righteous of Israel would be brought back, in bodily form, to enjoy a perfect second life on earth, leaving the other nations to languish in Sheol. This would be at the Final Day, the end of the present age. But merely to languish in Sheol, excluded from a second life, is a pale kind of punishment; so it gradually became common to believe that the undeserving dead would be conscious of their exclusion and thus suffer from it. From this point it was only a short distance to a concept of Hell as a place specifically designed for endless suffering, physical as well as mental. Persons living in or near Jerusalem had an excellent visible analogy at hand in that city's garbage dump in the ravine called the Valley of Hinnom, with its pall of noxious smoke and its constant flickering fires; by the Christian period, "Gehenna," the form of its name in Greek, came to mean Hell and was so used by the gospel writers.

Bodily resurrection is not the same thing as immortality. Belief that humans have an immortal soul that only temporarily inhabits their bodies came to Judaism through Greek philosophy and is reflected in various hellenistic writings like The Wisdom of Solomon. In due course it entered Christian thought. In this view the body is a mere lump of clay, unworthy of its spiritual tenant; indeed, everything material is inferior to the spiritual. Such dualism was of course quite absent in the traditional Jewish belief that we described earlier.

ETHICAL AND COSMIC DUALISM

Another basic modification of Jewish tradition came from a different source and introduced further kinds of dualism: ethical and cosmic. Ethical dualism views evil in a generalized way, as a principle balancing and opposed to the principle of good; the presence of evil is explained by attributing it to some agency operating in this world that sponsors it and aims for its increase. Cosmic dualism (discussed in chapter 10) simply projects this situation onto a universal stage and makes the agencies of good and evil supernatural antagonists who are battling not only for possession of human souls but also for final victory over all creation. In Persian Zoroastrianism, considered by some to be the source of these ideas in Judaism, the authors of good and evil are, respectively, Ahura-Mazda and Angra Mainyu. For Jews the author of evil was variously personified as Satan, Mastema, Azazel, Beliar—or in the synoptic gospels as Beelzebul. (As we pointed out earlier, in the few places in the Jewish scriptures where "the satan" makes an appearance—the Hebrew word means "the adversary"—that figure is merely one of the attendants in the heavenly court. He is humankind's adversary, not God's.) Because God had angels to do his work here below, the author of evil had demons of various kinds to do his. In some views the earth is no longer humankind's destined and proper home but rather a place of trial, where evil forces operate freely—indeed, it is under their control and actually belongs to them (see 1 Corinthians 2:8 and Galatians 4:3). At all times they are to be feared, shunned, and opposed.

Perhaps no foreign influence was needed to introduce ethical and cosmic dualism: The Jewish tradition itself contains potential sources for it. It is certainly foreshadowed in the writings of the prophets. And Genesis 6:1–4 gave rise to one explanation for the presence of evil spirits on earth who seduce humans; according to the pseudepigraphal books of Enoch and Jubilees, these spirits are the offspring of the unnatural union between the "sons of God" (or "sons of the gods") and the "daughters of men." Moreover, demonology surely had roots in popular animism, the age-old superstition that all material objects in this world are inhabited by spirits, most of them untrustworthy if not actually dangerous to humans. Animism is older than organized religion, older than creeds. Belief in demons in conflict with God may not be so much a corruption or breakdown of established Yahwism as a reversion to the real religion of the people that Yahwism never totally replaced. If we have to look hard in the Jewish scriptures for any evidence of this, it is because those scriptures were largely shaped to conform to and confirm Yahwist views.

STUDY OF THE SCRIPTURES

Even as the rebuilt Temple became once again the focus of Jewish religious life, with its priestly apparatus and regular daily sacrifice, a form of devotional activity quite independent of the Temple began to grow in importance: the study and interpretation of the scriptures. This activity eventually moved into the center of Jewish religious life and in time affected the development of Christianity. According to the gospels, Christianity from the very start acknowledged its relation to the Law of Moses and found its roots in Jewish prophecy. ("Law" translates the Greek *nomos*, which in turn is the Septuagint rendering of the Hebrew word *torah*; but *torah* properly means "instruction" and includes lawgiving as only one of its components.) The New Testament as a whole shows thorough familiarity with the Jewish scriptures, from which it often quotes and to which it often refers (more than sixteen hundred times by one count). The New Testament writers used the Greek version of the Jewish scriptures, the Septuagint, which they regarded as no less authoritative than the original Hebrew text.

The cardinal Jewish doctrine was the supremacy of the Torah. Jews believed that the Torah had been given to Moses directly by Yahweh on Mount Sinai and that, indeed, it preexisted the creation of the world. It was timeless and perfect, containing no errors or contradictions. But it did not necessarily interpret itself, and even the most reverent reading of it discovered obscure passages, vagueness, duplication, and apparent omissions. The statements of Law left much up in the air. To take merely one example, the fourth commandment prohibits work on the Sabbath, but nowhere in the Torah is there a general definition of what work is. Had the Torah not been the constitutive document of Judaism, conceivably such problems might have escaped attention; but Jews who wished to live their lives according to the Law had to have more precise guidance. Thus the tradition grew (among those we would call theological liberals) that at Mount Sinai Yahweh had given Moses two kinds of Torah, not one: a written Torah and an oral Torah. In the oral Torah a great many further instructions had been given to Moses that never got written down. It became the duty of Jewish scholars and sages (the *sopherim*— who are called *grammateis* in the Greek New Testament, a word that is traditionally rendered as "scribes") to find this oral Torah by, as it were, reading between the lines of the written Torah. They believed that what they found by doing so had been all along part of the instruction given by Yahweh to Moses, even though to an outsider it might seem that they were adding new legislation rather than bringing to light meanings implicit in the old.

In the later years of the intertestamental period, "academies" grew up around certain prominent expounders of the Torah for the purpose of studying scripture in a systematic fashion and passing their wisdom on to qualified pupils. Like the ancient prophets, these men operated outside the formal religious establishment, but unlike the prophets they did not claim to speak with the voice of God. Instead they used

human intelligence and insight; contested issues might require a meeting of *sopherim* to settle them by majority vote. They did not slavishly
follow precedent, although they respected it. In some ways they operated like modern courts of appeal; however, they did not necessarily
wait for cases to be presented before making rulings, and much of
their work was what we would call theoretical, dealing with situations
of unlikely occurrence. We must also not forget that the great bulk of
their work, which has gone mostly unrecorded, was in teaching rather
than in rendering judgments on the Law.

In the process of adapting the written Torah to practical use and
to the changing times, the scholars had to perform a good many feats
of interpretation. It was normal to remove statements from context to
scrutinize them. Individual phrases and even single words were also
isolated, if by doing so they could be made to yield a meaning that
satisfied the purpose at hand. The most prominent exponent of this
method was Rabbi Akiva ben Joseph, born in the first century C.E.,
who argued that every syllable of the Torah, even down to how it
was spelled, was independently meaningful. In the synoptic gospels,
Jesus is portrayed as an expert in scriptural interpretation, so much
so that he defeats the *sopherim* at their own game, using their tactics.
His treatment of passages in the Jewish scriptures—for example, Exodus 3:6 and Psalm 110:1, as reported in Matthew 22—depends just as
much as theirs on minute scrutiny of grammatical and semantic details. And although Jesus proclaims a radical redefinition of the Law,
his respect for it is also shown; in Matthew 5:18 and Luke 16:17 he
speaks of the Law like a true and typical Jew.

Besides the splintering of the text practiced by the *sopherim*, another technique, coming from a very different source, had a great
influence on the tradition of interpretation: allegorizing. This technique originated with the high-minded Stoic philosophers of Greece
and was first applied to the text of Homer as a means of explaining
away the naively fabulous content of these poems and the unseemly
goings-on that they recorded. In the second century B.C.E. a hellenized Jew, Aristobulus of Paneas, picked up the technique and applied it to the Torah, attempting to remove the anthropomorphisms
in its treatment of God by allegorizing them. The most determined
allegorizer of all was a contemporary of Jesus, the hellenized aristo-
cratic Jew Philo of Alexandria, whose voluminous writings were devoted in large part to interpreting the Jewish scriptures (which he
knew only in the Septuagint). Philo did not deny the literal sense of
most scripture, but his whole effort was to extract higher and more
philosophical meanings from the text than showed on its surface. In

his hands the scriptures became a kind of code book to the hidden truths of Platonic and Stoic philosophy. Moses, in other words, had gotten there first. (For more on allegorizing and the allegorizers, see chapter 17.)

ANTIOCHUS IV
EPIPHANES

The peaceful development of Jewish religious life through the third century B.C.E. was eventually interrupted by the resumption of fighting between the Seleucids, based in Syria, and the Ptolemies, based in Egypt. The initiative came from the young Seleucid king, Antiochus III (great-great-grandson of the founder of that dynasty), who attempted to wrest Palestine away from the weakened Ptolemaic monarchy and, after several setbacks, finally succeeded in 198 B.C.E. The change of overlords was welcomed by the pro-Seleucid party in Jerusalem, all the more when Antiochus issued decrees supporting the right of Jews to live by the Mosaic Law and preserving the sanctity of the Temple. But these favorable conditions soon ended. Antiochus suffered a crushing defeat in Asia Minor at the hands of the Romans in 190 B.C.E. and was murdered in 187. Twelve years later the throne passed to his younger son, Antiochus IV, self-styled Epiphanes ("God Manifest"), and with that three centuries of turmoil for the Jews began.

Antiochus IV came to power in the midst of intrigue in Jerusalem over the office of high priest. Even though Jewish law gave the king no power to appoint the high priest, Antiochus was persuaded by bribery to remove the incumbent, Onias, and appoint Onias' brother Jason in his place. Then Jason, an ardent hellenizer, got the king to grant Jerusalem the status of a Greek *polis*, with appropriate civic institutions. In the new gymnasium, for example, fashionable Jewish young men took part naked in athletic contests, according to Greek custom. This intrusion of hellenistic ways into the very heart of Judaism naturally scandalized the faithful. Matters became even worse when Jason was replaced (again through bribery) by one Menelaus, who was not even a member of the Zadokites, the family with hereditary right to the high priesthood.

In 168 B.C.E., on his way back from Egypt, where he had encountered a Roman delegation that ordered him bluntly to keep hands off that country, Antiochus IV found Jerusalem in the throes of fighting between the supporters of Menelaus and those of Jason. He took his frustrations out on the Jews by violently suppressing the struggle; then he reduced Jerusalem to the status of a garrison city, quartering Syrian troops in the "citadel," a fortress overlooking the Temple mount. In the following year he issued an edict for the complete helle-nization of Judea: He forbade circumcision, Sabbath observance, and possession of the Torah on pain of death; the Temple was rededicated to Zeus Olympius, and unclean animals were sacrificed to the pagan god on its altar. In a reversal of the time-honored custom of religious tolerance within the pagan empires, an attack had been launched against the Mosaic Law itself. Judaism was swept with horror and dismay. For the first time in history, Jews were subject to death sim-ply for attempting to practice their faith. As chapter 10 has shown, these events stimulated the writing of Daniel 7–12 (where they are presented in disguised form), which in turn became the pattern for much further apocalyptic literature.

THE MACCABEES

While the author of Daniel was dreaming of end times and divine re-tribution, there occurred an unpre-dictable and originally quite insig-nificant event that abruptly altered the course of Jewish history. In a small village northwest of Jerusalem an old priest named Mattathias defied the king's order to make public sacrifice to the pagan gods and killed the officer sent to enforce it. Then with his five sons and their families and some sympathizers he took refuge in the rugged Judean hills to make guerilla warfare on the hellenizers. Thus began the famous Maccabean uprising, so called because one of the sons was nicknamed Maccabeus ("hammer"). The story is told in two books that are not in the Jewish canon, 1 and 2 Maccabees—soberly and factually in the first, with much melodra-matic elaboration in the second.

In 166 B.C.E., shortly after the revolt began, Mattathias died and his son Judas took over the leadership. Moving his men with great skill over the familiar countryside against always superior Syrian forces under the deputies of Antiochus (who was off fighting in Per-sia), Judas Maccabeus repeatedly humiliated the enemy, and in De-cember of 164, after establishing his control over all of Judea, he suc-ceeded in taking Jerusalem. The Temple was purified and sacrifice restored, an event celebrated still in the festival of Hanukkah. Mean-while Antiochus IV had died.

THE HASMONEAN DYNASTY

The next century belonged to the Hasmoneans (as the descendants of Mattathias are called), who under one title or another ruled the Jewish state until the accession of Herod the Great. They were energetic and skilful leaders, and they built Judea up into a power to be reckoned with. By the early first century B.C.E.

the Jewish state took in virtually all of Palestine and some of the Transjordan, thus reaching its greatest extent since the days of Solomon. But it was all done under the shadow of Rome. Furthermore, the material grandeur of the Hasmonean state was purchased with the abandonment of the spiritual heritage of their ancestors. The Hasmoneans themselves took up hellenistic fashions. They politicized the office of high priest, passing it around among members of their own family. Later members of the dynasty assumed the title of king. The religious fervor with which the revolt of Mattathias had begun virtually disappeared, and when the Pharisees (whom we shall meet soon) opposed the worldliness of the ruling family, they were suppressed with great force and cruelty. At some time during the reign of John Hyrcanus, Mattathias' grandson, a group of sectarians called the Essenes decided to withdraw completely from participation in what they considered a debased and illegitimate Temple cult and to found their own community, which they did at Qumran on the shores of the Dead Sea. It was this group that produced the famous Dead Sea Scrolls, which were hidden in caves near the Dead Sea in pre-Christian times and not rediscovered until the middle of our own century.

The Hasmonean dynasty ended in power struggles among the successors of Alexander Janneus (Mattathias' great-grandson), one of whom was sponsored by Antipater, an Idumean (Edomite) strongman and friend of Rome, who had taken to meddling in Jewish affairs. The final effect of this disorder was to bring in the Roman army, led by Pompey, who entered Jerusalem, laid siege to the Temple mount, and took it in 63 B.C.E. From now on whoever ruled the Jews did so with the active help or the permission of Rome.

HEROD THE GREAT

In the end it was a client of Rome, Herod son of Antipater, who took the throne in 37 B.C.E. and reigned until 4 B.C.E. Not an ethnic Jew at all but a descendant of Idumean converts and a committed hellenist, Herod married into the Hasmonean line by taking to wife Mariamne, granddaughter of Aristobulus II, the Jewish king who lost Jerusalem to Pompey. He then set about eliminating all Hasmonean claimants to the throne by murdering, in turn, Mariamne's brother, her maternal grandfather, Mariamne herself, her mother, and finally his own two sons by Mariamne, whom he ordered to be strangled in 7 B.C.E. Had crimes such as these been his only achievement, he would never have become known as "Herod the Great"; the other side of the picture is that he was a powerful leader and administrator, determined to leave his mark on the country. He

undertook public works on an unprecedented scale, the chief orna-
ment of which was the rebuilt Temple in Jerusalem, for which he
created an enormous platform thirty-five acres in area overlooking the
Kidron valley, north of the original city of David. This platform still
exists, although it now bears later structures. A portion of its supporting wall, showing the typical massive Herodian masonry, is today
exposed along the western side and has become a place of prayer sacred to Jews. Herod's Temple would have been a magnificent building anywhere in the world, and Jews were understandably proud of
it. But it was a monument to Herod's vanity, not to any love of his
for Judaism, in which he was totally uninterested. Work on the Temple was still going on after Herod's death in the lifetime of Jesus, who
was born in the waning years of Herod's reign.

The Jewish state created by Herod—nominally independent
though in fact existing at the pleasure of Rome—did not survive his
death, and thereafter it was the Romans' task to try governing Palestine through delegates (called variously "ethnarch," "tetrarch," or
"procurator"—Pontius Pilate was Procurator of Judea, 26–36 C.E.)
and even one king, Agrippa I. The common denominator in all these
arrangements was subservience to Roman authority. But no arrangement was permanently satisfactory, and a final clash between militant
Jewish piety and Roman arrogance became inevitable. In 66 C.E.,
while Nero was emperor, civil war broke out in Herod's city of Caesarea and spread rapidly through Palestine. The Jews, handicapped by
infighting and divided leadership, put up a brave resistance to the
legions of the Roman general Vespasian, but without avail. Galilee
was conquered in 67 C.E. Vespasian, having become emperor after
Nero's death, sent his son Titus to lead the armies against the last
major stronghold, Jerusalem. The city was besieged and taken, and
the Temple burned to the ground. The year was 70 C.E.

PHARISEES AND
SADDUCEES

We must now consider some elements
in the background of these historical
events that also had an effect on the
emerging Christian movement. One
of the most striking developments
within Judaism during the last two centuries B.C.E. was the emer-
gence of sectarian parties. These were loosely organized groups of
persons holding certain views in common, somewhat like our own
political parties but without a framework of representative government to operate within. The best known of these parties, because
they are referred to frequently in the New Testament, were the Pharisees and the Sadducees.

The Pharisees probably developed out of the Hasideans (from *hasidim*, "the faithful ones") of the early Hasmonean period. Pharisees were noted for their strict piety and their zeal for the Mosaic Law. Believing that the Law should govern every moment of one's daily life, they naturally had to find ways of adapting it to new circumstances, so they became the chief sponsors of the concept of oral Law. They were, in fact, liberalizers, and accepted such novel ideas as bodily resurrection followed by reward or punishment, and a form of predestination in which human beings are responsible for their own choice of good and evil. Being interested in conduct and faith and having their hopes fixed on the world to come, they had no political aims and did not object to foreign rule per se. Their chief field of activity was the synagogue, not the Temple. After the destruction of the Temple they were the only ones in a position to begin rebuilding Judaism, which they did at Jamnia by completing the scriptural canon. Judaism today is a legacy of the Pharisees.

Most Bible-reading Christians think of Pharisees as hypocrites and hair-splitting legalists because that is how they are portrayed in the gospels. But this view is a product of the era after 70 C.E. when the Pharisees, who became the official guardians of Judaism, came into conflict with the Christians. Actually, the two sects had much in common. It is more likely that the real enemies of the Christian movement in Jesus' lifetime were the Sadducees, a smaller group of religious and political conservatives whose chief interest was the sacrificial cult in the Temple. Many of them were quite well-to-do and on comfortable terms with the authorities. They rejected the concept of oral Law and denied that there would be a resurrection and judgment of the dead. They would have been particularly hostile to a prophet from Nazareth who operated outside the Temple jurisdiction and claimed to have the power of forgiving sins, especially if he also looked like a political troublemaker. Sadducees and Pharisees both served on the Sanhedrin, which was presided over by the high priest and had religious and civil power (although the extent of the latter is disputed). With the destruction of the Temple, the Sadducees disappeared because the basis of their power was gone.

Other known parties of that time were the Zealots and the Essenes. The Zealots were political rebels who conceived of Israel as ruled by God alone and hence rejected allegiance to any foreign power, specifically Rome. They were responsible for carrying on much of the rebellion of 66 C.E., and a valiant group of Zealots were the last holdouts against the Romans in the fortress of Masada, choosing to perish by suicide rather than to surrender. We have already mentioned the Essenes as making up the Qumran community by the Dead Sea. Being separatists, they had no effect on current events, but their preserved writings (the famous Dead Sea Scrolls) are valuable testimony to the nature of eschatological thinking in the later intertestamental period.

It is in the context of these Jewish partisan movements that we should look at early Christianity, for a great many features of Christianity that we might suppose to be unique turn out, in fact, to have antecedents in various strains of Judaism. This, of course, does not mean that Christian beliefs were less authentic because they were not entirely novel; neither does it mean that they were an inevitable next step in the development of Jewish religion. What is beyond question is that they *did* develop and that they did not come into existence completely out of the blue. It should be quite clear by now that Christianity is part of the general history of its times and cannot be studied in isolation.

THE MESSIANIC TRADITION

When the gospels present Jesus of Nazareth as the Messiah, they are drawing on Jewish tradition. In Hebrew *mashiaḥ* means "anointed one"; the equivalent in Greek is *christos*, hence "Christ." The title refers to the coronation ceremony: The chosen king is anointed by having oil poured over his head (the first recorded biblical instance is the anointing of Saul by Samuel in 1 Samuel 10:1), implying that the king is God's choice and reigns with divine backing. So much for its basis. What we are considering here, though, is the idea of the Messiah cut free from its connection with any contemporary king and projected onto the future as a focus for the hope of the faithful, his ultimate subjects. The earliest mention of a Messiah in this sense comes from the prophetic age, in Isaiah 11:1–9, a classic statement that contains all the elements of the idea. (See also Jeremiah 23:5, 30:9, and 33:14–16.) A king from the line of David is expected as the savior of his people. He is to be a human king, and the salvation is to be material and national, not spiritual and individual. Why did he have to be a descendant of David? Not because of any theories of genetic inheritance but because it was highly important for this king to be legitimate in the normal human sense, the throne having been promised to David's family "for ever" (2 Sam. 7:16). Hence this was a limited concept.

Nevertheless, it was a concept broad enough for Deutero-Isaiah to apply the term "messiah" to Cyrus, the emperor of Persia, because he met an essential condition in being (supposedly) the chosen instrument of God. A little later the prophets Haggai and Zechariah gave this honor to Zerubbabel, a descendant of the royal line who was active in the rebuilding of the Temple. Outside these few places in the Old Testament (and some others that have been given messianic meanings through Christian interpretation), there is not much evidence of the Messiah in Jewish biblical tradition. Messianic thinking in Judaism is chiefly a product of the last two centuries B.C.E., and it is found in such typical noncanonical writings as 1 Enoch, the Jewish Sibylline Oracles, the Psalms of Solomon, and the Dead Sea Scrolls. During this period the concept of the Messiah was worked into eschatological thought patterns and given much specific content: He was to be a major figure in the end-of-the-age scenario. This evidence of a sudden increase in messianic thinking does not, however, mean that great numbers of Jews in these two centuries were sitting around waiting for the Messiah to come. Teaching about the Messiah formed no part of synagogue worship. We must not confuse a generalized hope of relief born out of persecution with a developed doctrine. The hope was widespread; the doctrine was the property of minority sects and eccentric writers.

In the gospels the Messiah is linked with the mysterious and equally problematical "Son of Man," a term that Jesus uses constantly in apparent reference to himself. It is first found in Ezekiel, where it means simply "man"—in other words, it specifically indicates that the bearer of the name (Ezekiel) is *not* divine. It next appears in Daniel 7:13, where (in the literal sense of the original text) "one like a son of man" is a sovereign figure who apparently is to reign over the entire world at the end of the age, although he has no role in bringing the end about. But in other intertestamental writings the picture is much clearer. In the influential book of Enoch, the role of the Son of Man is undoubtedly messianic. He preexists in Heaven until the time for revelation, when he will appear on the throne of glory and pass final judgment on the world, sending sinners to eternal punishment and accepting the righteous into his presence for their eternal reward.

In Jewish thought the term "son of God" would normally have meant the same thing as "son of man." All men are sons of God because God is their Father, having given life to the first man, Adam. But as used by Paul and in the gospels it has a very restricted meaning: This man, Jesus, was God's very son and in his flesh brought the divine down to earth. In that sense the term would have met resistance from unhellenized Jews because it smacked of paganism. In the pagan world it was common to believe that certain exceptional humans were divine. The borderline was by no means as distinct to them as it is to us. Quite ancient cults in Greece grew up around semimythological figures, like Herakles and Asclepius, and successful military leaders who were unquestionably real might receive honors befitting the gods after their death (or even while still alive). Alexander the Great inaugurated the fashion among kings of claiming divinity for themselves, tracing his own ancestry back to Zeus; among his successors, the Ptolemies and Seleucids, the claim was standard. Rome picked up the custom from the Greeks. The dead Julius Caesar was deified in 42 B.C.E. by the Roman Senate, and the first emperor, Augustus, was adorned with divine titles and had temples built in his honor and sacrifices offered on their altars during his own reign. It was a device of realpolitik, which ancient kings understood just as well as modern dictators.

Behind the selfishness of the kings, and making all this possible, was the superstition of the masses. This comes out vividly in the New Testament when Paul and Barnabas are acclaimed as gods by the citizens of Lystra, who prepare to offer sacrifices to them (Acts 14:8–18; see also 28:6). Still, however debased and however dependent on superstition, the notion of the divine man deserves some respect because it shows faith in human potential and admiration for outstanding personal qualities. Perhaps the fault was not that so many people believed in divine men but that so few men really qualified for the title.

To conclude: The independent tradition of the Son of Man as eschatological judge merged in the late intertestamental period with the older tradition of the Messiah, who was originally an earthly king of the Davidic line who would rescue Israel. The next step beyond believing that the Son of Man/Messiah was heaven-sent was to believe that he himself was divine. This we see in Christianity.

We are not saying that the claim of divine sonship for Jesus was equivalent in any sense to the claims made for hellenistic/Roman kings or other famous men but only that its terminology and concepts already existed and that a great many Gentiles would have been receptive to it because it sounded familiar to their ears and agreed with their habits of thinking. Those who knew little or nothing of the Jewish Messiah would not have needed that knowledge in order to accept the Christian gospel. The minority of receptive Jews, on the other hand, would have found the messiahship a far more congenial frame of reference than that of the divine sonship. By one means or another the eschatological disposition of this period provided a route for the new teaching that was to transform the world.

SUGGESTED FURTHER READING

Gabriele Boccaccini, *Middle Judaism: Jewish Thought 300 B.C.E.-200 C.E.* (Minneapolis: Fortress Press, 1991).

Shaye J. D. Cohen, *From the Maccabees to the Mishnah* (Philadephia: The Westminster Press, 1987).

Sean Freyne, *Galilee, Jesus, and the Gospels* (Philadelphia: Fortress Press, 1988).

Martin Hengel, *Jews, Greeks and Barbarians: Aspects of the Hellenization of Judaism in the Pre-Christian Period*, trans. John Bowden (Philadelphia: Fortress Press, 1980).

Richard A. Horsley and John S. Hanson, *Bandits, Prophets, and Messiahs: Popular Movements at the Time of Jesus* (San Francisco: Harper & Row, 1988).

H. Jagersma, *A History of Israel from Alexander the Great to Bar Kochba*, trans. John Bowden (Philadelphia: Fortress Press, 1985).

Emil Schurer, *The History of the Jewish People in the Age of Jesus Christ*, rev. and ed. Geza Vermes and Fergus Millar (Edinburgh: T & T Clark, 1973).

Alan Segal, *The Other Judaisms of Late Antiquity* (Atlanta: Scholars Press, 1987).

M. Stern, "The Period of the Second Temple." Part 3 of *A History of the Jewish People*, ed. H. H. Ben-Sasson (Cambridge, Mass.: Harvard University Press, 1976).

Michael E. Stone and David Satran, eds., *Emerging Judaism: Studies on the Fourth and Third Centuries B.C.E.* (Minneapolis: Fortress Press, 1989).

Michael E. Stone, *Scriptures, Sects and Visions: A Profile of Judaism from Ezra to the Jewish Revolts* (New York: Collins, 1980).

The Interpreter's Dictionary of the Bible, ed. George A. Buttrick et al. (Nashville: Abingdon Press, 1962). See articles on City, Greek Religion, Hasmoneans. Supplement, 1976: See articles on Literature, Early Christian; Messiah.

The Anchor Bible Dictionary, ed. David Noel Freedman et al. (New York: Doubleday, 1992). See articles on Pharisees, Sadducees, Synagogue.

XII

Apocrypha and Pseudepigrapha: The Outside Books

The subject of this chapter is those ancient Jewish and Christian writings that can be found in some Bibles but not others or—though they resemble biblical writings—in no Bible at all. The books in what is generally called the "Apocrypha" are a standard part of the Bibles of Roman Catholicism and Eastern Orthodoxy but not those of Judaism or Protestantism. The books in the categories called the "Pseudepigrapha" and the "New Testament Apocrypha" are, with one minor exception, not a part of any Bible. All of these "outside" books—so designated because they now lie outside one or another canon of scripture—are nevertheless quite important, both individually as pieces of early Jewish and Christian religious literature and collectively as material that was left by the wayside in the long process of Bible making. Of great relevance to a discussion of the outside books are two matters treated in earlier chapters: the process of canonization and the history of the intertestamental period. For the sake of convenience, we shall at various points here briefly repeat information on those two matters.

THE PROCESS OF CANONIZATION

As indicated in chapter 6, canonization is the long historical process by which a religious community sorts out for itself which of its religious books are not merely religious but *sacred*—that is, derived ultimately from God himself. The process is largely an unconscious one: Rarely in the history of canonization of the parts of the Bible did a group of religious authorities sit down and make a conscious decision about what was scripture and what was not. And when they did, they were in fact only ratifying what had been arrived at by generations of the faithful before them. An individual piece of writing was *discovered* to be sacred, in effect, as it was employed by the community in rituals, read in religious services, discussed by scholars, and preached about. Gradually a consensus developed that that piece of writing was God-given, sacred from the beginning of its existence.

There is a great deal we do not know about the canonization of individual works in the Jewish and Christian traditions. We can, nevertheless, be fairly confident that by 400 B.C.E. the canonizing process had selected out the block of five writings now called the Pentateuch (or Torah) and established that block as sacred scripture in the eyes of Judaism. By about two hundred years later, a second block of writings—referred to collectively as the Prophets because they report the words of Israel's prophets and the history of prophetic times—had come to be considered sacred. And by 100 C.E. the process of canonization had defined a third set of sacred texts, called simply the Writings, and with that the Jewish canon of scripture was closed. With respect to Christian works, we know that certain blocks gradually came to be recognized as canonical in the second and third centuries C.E. and that by the mid-to-late fourth century the twenty-seven books now in the New Testament had been settled on as sacred scripture. Our focus in this chapter will be on the Bible-like writings that were left behind when the Jewish and Christian processes of canonization had run their course.

How many such individual writings were there? It is impossible to say. Well over a hundred are still extant today in one form or another; but probably a great many others vanished without a trace in the violent events that racked both Judaism and Christianity in the first centuries C.E. We know of the existence of some outside books only because they are referred to elsewhere. We know of the existence of others because copies of them, in whole or in part, have in recent centuries been turned up by archaeologists and scholars searching in out-of-the-way places. We know about still others because, though uncanonical, they remained of interest to some religious community that continued to copy and recopy them over the centuries and in doing so gave them a home; thus such books have remained living texts down to our own time. One body of writings that comes to our attention through this third means is the Apocrypha. It is that body of the outside writings that we shall first discuss.

TRANSLATING THE
SEPTUAGINT

There is no way to consider the Apocrypha without first considering the Septuagint, the pre-Christian translation of the Hebrew scriptures into Greek. In chapter 11 we discussed how the Near East became hellenized in the centuries after its conquest by Alexander the Great, how throughout the area new Greek cities were built that attracted Jewish immigrants, and how many of these Jews of the Diaspora (those living outside Palestine) adopted the Greek language—the international language of commerce and culture—as their own. But despite the hellenizing of Jewish life in the Diaspora, interest in traditional Judaism remained strong; and it was inevitable that there would be a demand by Greek-speaking Jews for a translation of the Hebrew scriptures into Greek.

Legend has it that the translating of the first part of the Hebrew Bible, the Pentateuch, was the work of seventy-two Jewish elders, invited by the Egyptian king Ptolemy II Philadelphus to perform this task so that he might have a copy of the renowned Jewish Torah for his library. (It is from this legend that the title "Septuagint" comes: *septuaginta* is the Greek word for "seventy." The Roman numeral for seventy, LXX, is often used as a shorthand title for the Septuagint.) In actuality, this Greek translation of the Hebrew scriptures was produced informally and according to no long-range plan by Jewish scholars in Alexandria (a newly established Greek city in Egypt) during the third and second centuries B.C.E. The Septuagint grew to include not only the canonical Jewish writings (the Torah and the Prophets) but also other writings that, because of their popularity, were well on their way to becoming canonical. This latter class included older works (such as Psalms and Proverbs) long known and used by Jews in Palestine and everywhere else, as well as more recent ones that had been written initially in Greek.*

THE CHRISTIAN USE OF
THE SEPTUAGINT

The Septuagint won acceptance not merely in Alexandria but in all areas where there were Jews who spoke Greek. The infant Christian Church included among its members some individuals trained in the tradition of the Pharisees, and these may have used the Hebrew text when they searched the scriptures for proofs of

---

*Comparison of an English translation of the Septuagint with one of the Hebrew Bible will reveal frequent differences in sense because, first, the Jews who created the Septuagint worked from Hebrew texts differing in places from those that have come down to us and, second, they sometimes produced interpretive paraphrases instead of translations.

Jesus' messiahship. But the great majority of early Christians would have gone to the Greek text rather than to the Hebrew. We can generalize that, as the Church moved outward from its original base in Palestine and became increasingly gentile, it increasingly employed the Greek Bible rather than the Hebrew. All of the New Testament writers quoted the Jewish scriptures in their Greek form. The reason for the appeal of the Septuagint to Christians outside—and even inside—the Holy Land was a simple one: A great many of them read Greek, very few of them read Hebrew.

But this simple fact had a significant consequence. When these Christians went to the Septuagint to search for proof texts about the Messiah, they consulted not only the traditional Jewish writings but the newest of the works it contained—those that had been composed in Greek by Diaspora Jews and still had not established themselves in the estimation of Palestinian Jews. If it was exasperating for faithful Jews to have their traditional scriptures employed in support of the Christian movement, it was even more exasperating to have writings of no particular authority employed the same way, as though they, too, had scriptural status. Of special interest to Christians were the Septuagint books The Wisdom of Solomon, 2 Maccabees, 2 Esdras, and Tobit, which could be appealed to in arguing the Christian position on such ideas as resurrection, the Messiah, angels, and demons. The authors of the New Testament were frequently influenced by these and other late books in the Septuagint.

CLOSING THE JEWISH CANON

We have already described the Roman destruction of the Temple in Jerusalem and of the city itself in 70 C.E., as well as how certain Jewish rabbis set up a new home base for Judaism at Jamnia. Cut off from the very center of its cultic life, Judaism had now only its God and its traditions and writings. As one pious Jew of the time lamented, "Zion has been taken from us, and we have nothing now save the Almighty and his Law" (2 Bar. 85:3). But what specifically would have constituted the "Law" as the word is used here? Which of the traditional writings were indeed God-given? What about the Jewish writings that the Christians were making particular use of? And what about all the other outside books for which claims could be made of divine inspiration? One of the first tasks of Palestinian Judaism as it reorganized itself at Jamnia was to decide, once and for all, what constituted its sacred scriptures. As our chapter on canon indicated, the criteria by which this decision was made are unknown. According to Josephus, the Jewish historian of the first century C.E., the Jews of his time had come to believe that inspiration had ceased long before with Ezra (who lived in the fifth century B.C.E.) and that works written after Ezra's time were therefore not canonical. It is apparent also that books written only in Greek, or composed in Greek and then translated into Hebrew, were not considered canonical. But whatever criteria were employed, the result was that the long process of Jewish canonization came to a conclusion by about the end of the first Christian century. A third division of scriptures comprising eleven books, now known as "the Writings," was included at this time in Judaism's canon along with the Torah and the Prophets. All other candidates for inclusion were to be left outside forever. And the Greek translation of the scriptures, which included some of the chief of those outside books, was itself now to be shunned by Jews. Henceforth Jewish orthodoxy would center its attention on the Hebrew text of the scriptures, with its three canonical divisions.

THE OUTSIDE BOOKS IN
THE SEPTUAGINT

But that action did not of course mark the end of the Septuagint. For wherever Christianity went, the Greek translation of the Jewish Bible went, carrying along with it those outside books that Christians felt to be every bit as sacred as those of which Jews approved. These books were not isolated as a single collection in the Septuagint but were scattered throughout. The arrangement varies in the early manuscripts of the Greek Bible now available to us, and not all of the works in one manuscript will necessarily appear in other manuscripts. As to when these pieces were composed, the best scholarly guess for most of them is the second century B.C.E.; one of them (Ecclesiasticus) was certainly composed near the beginning of that century and several others (The Wisdom of Solomon, 1 and 2 Maccabees) near the end. Some of these works were composed originally in Greek, others in Hebrew or Aramaic and then translated into Greek. The individual "books" (or portions of them—some items are only a page or two long) we are concerned with are as follows:

1. 1 Esdras (sometimes called, confusingly enough, 2 Esdras or 3 Esdras!)—a Greek version of portions of 2 Chronicles, Ezra, and Nehemiah.*

---

*Another work, now called 2 Esdras (sometimes 3 Esdras or 4 Esdras), although not among the outside books included in the Septuagint nevertheless became linked to those books in the early Christian centuries and found its way into some copies of the Greek Bible made for the use of Christians and from there into the medieval Latin Bible. This work is a Jewish apocalypse written around 100 C.E., but it has a later Christian introduction and conclusion.

2. Tobit—a delightful short story or romance about faithful Jews who undergo hardship and testing but finally triumph with the aid of an angel.

3. Judith—a romance about a beautiful Jewish woman who courageously saves her people from destruction at the hands of the Assyrians by cutting off the head of the enemy general, Holofernes.

4. The Additions to the Book of Esther—material inserted into the canonical book of Esther to give it a religious motive rather than simply a patriotic one.

5. The Wisdom of Solomon—a defense of Judaism in the face of pressures on Jews to assimilate to hellenism.

6. Ecclesiasticus (also called The Wisdom of Jesus ben Sira)—a long collection of wisdom material much like that in the canonical book of Proverbs but more religious in tone.

7. Baruch—a brief collection of prayers and of comforting words to captive Israel.

8. A Letter of Jeremiah (sometimes included in Baruch)—an attack on idolatry.

9. The Song of the Three—the hymn in praise of God sung by the three young men in the book of Daniel when they were thrown into the blazing furnace; it is prefixed by a prayer of one of them, Azariah.

10. Susanna and the Elders—the story of how Daniel as a wise young man saved a virtuous woman from being put to death unjustly.

11. Bel and the Snake—two Daniel stories in which the hero reveals idolatry to be merely that and not true religion.

12. The Prayer of Manasseh—the prayer of repentance uttered by one of the kings of Judah when, for his sins, he was taken captive to Babylon.

13. 1 Maccabees—the history of Israel's rescue from Greek oppression in the second century B.C.E. by the efforts of the family known as the Maccabees.

14. 2 Maccabees—an abridgment of a longer history by one Jason concerning the first fifteen years of the Maccabean wars (the material parallels what is in 1 Maccabees 1–7).

Throughout the first several centuries of the existence of the Christian Church, these books were not distinguished from the canonical Jewish writings. Few of the Church fathers of those centuries had any knowledge of the Jewish background of Christianity except for what was contained in their own Greek version of the Jewish scriptures, and the Greek version included those books that had been rejected by the Jews. But by the beginning of the fourth century, an awareness began to spread in the Christian Church that a legitimate question existed about their canonicity.

JEROME AND THE
APOCRYPHA

This matter came to a head in the late fourth century when a brilliant young scholar named Eusebius Hieronymus (now universally known as Jerome) undertook a new translation of the Bible into Latin. Jerome actually set out merely to revise the existing Old Latin version of the gospels, but the task expanded as he proceeded. Gradually, he accepted the responsibility for revising the whole (or at least most) of the existing Latin Bible. As he began work on the Old Testament, Jerome became convinced that for his original he should go behind the Greek version to the Hebrew. His ability to do so was enhanced when, disillusioned with what he felt to be the worldliness at the very center of Christendom, he left Rome and moved permanently to the Holy Land to take up a monastic life. There he had the opportunity to perfect his Hebrew and to be in touch with scholarly Jews who could aid him in his work.

In the course of his study, Jerome had impressed upon him what he had already been aware of—that the Hebrew Bible did not contain some of the books that were in the Greek Bible; and he came to believe firmly that the ancient Jews had surely had the right to determine what constituted their own scriptures. To those works in the Greek version that the Jews now rejected, Jerome applied the term *apocrypha*, a Greek word meaning "hidden." This term was customarily employed to designate certain Christian writings that supposedly contained secret teachings suitable for only the wisest scholars.* In using this term Jerome probably intended only to indicate that the rejected Jewish writings were outside the usual run of Old Testament books; but by extension the term soon developed a connotation of falseness, a connotation reflected directly in the fact that the primary meaning of the English word *apocryphal* is now "spurious." In Greek the word *apocrypha* is a plural form; when used in English, it technically requires plural modifiers and verbs, as in "These apocrypha are interesting books." But the word is also used in the singular as the title of a set of writings, the Apocrypha (with a capital letter), and that is how we generally use it here.

*The idea that some divine revelations are intended for a few but not most eyes is presented dramatically in 2 Esdras 12:35–38. In 14:44–46, the seer Esdras is told that he may make public certain books but that others "are to be kept back, and given to none but the wise among your people" (NEB).

Jerome would have liked to omit the works not in the Jewish canon from his translation, but he was persuaded to include some of them on the grounds that they had long been part of the Christian tradition. To satisfy his scruples on this point, however, he attached prefaces to his Latin translations indicating that these particular books should be considered less than canonical, although useful for Christian edification. As Jerome's new versions began to be used in the Church in place of the Old Latin versions and as they began to gain broad acceptance, the Old Latin versions of the rest of the Apocryphal books, those that he had not translated, were added to them. Thus for a thousand years the official Bible of the Catholic Church (Jerome's work had replaced the Greek version in this position) contained both the Latin versions of the Apocryphal books and Jerome's comments on their uncertain status.

THE APOCRYPHA AT THE
TIME OF THE
REFORMATION

Jerome's comments assumed particular significance in the sixteenth century at the time of the Protestant Reformation. In response to the Protestant view that each believer should be his or her own interpreter of the scriptures, the Bible was translated into all the various languages of Europe (Latin being no longer the language of the people). A number of scholars and translators (some Catholics among them) published and seconded Jerome's opinion of the Apocryphal books as being less than canonical; but it was the words of Martin Luther concerning this matter that gained the most attention. Luther disliked some of the Apocryphal books considerably more than others and had some harsh things to say of them (although, to be fair, we must admit that Luther also had harsh things to say about some of the canonical books). In his famous translation of the Bible into German, Luther removed a number of the Apocryphal books from the places where they occurred in the Latin version and gathered them together at the end of the Old Testament (two of them, 1 and 2 Esdras, he omitted altogether as being unworthy). Luther's Bible was not the very first to have the Apocrypha isolated in this way; but the reputation of the man and his translation being what it was, his practice influenced what was done in many Protestant versions in other languages thereafter. (The main reason why Protestants wanted to remove the Apocrypha from the Old Testament was that they wished to get back to the fountains [*ad fontes*] of their faith. But their zeal was undoubtedly augmented by the fact that some of the Apocryphal books were considered to provide support for certain Roman Catholic ideas, such as the existence of Purgatory and the possibility of prayer for the souls of the dead.)

In 1546 at the Council of Trent, the Roman Catholic Church re-acted decisively against the spreading negative reaction to the Apocry-pha. The Council pronounced that anyone who did not accept the canonicity of twelve of the fifteen books (1 Esdras and 2 Esdras and the Prayer of Manasseh were excepted) would be anathema—ac-cursed. In subsequent times Roman Catholics became freer than they were in the sixteenth century to differ with the pronouncements of ecclesiastical councils, and in modern times Catholic opinion opposed to the judgment of the Council of Trent has often been stated. But the official position of the Roman Catholic Church is still that these dozen books are part of sacred scripture. In Catholic practice, they are called "Deuterocanonical" (meaning "of the second canon") rather than "Apocryphal." The former term takes into account the fact that the canonical situation of these books is historically different from that of the books accepted as canonical long ago in Judaism. It also avoids the unpleasant connotation of spuriousness that hangs about the word "Apocryphal."

THE EASTERN ORTHODOX
CANON

The Eastern Orthodox Church de-rives its Old Testament directly from the Septuagint. In consequence it ac-cepts as canonical all of the books listed above that were not included in the Jewish canon, including 1 Esdras, 2 Esdras, and the Prayer of Manasseh, which were withdrawn from the Roman Catholic canon in the sixteenth century. In addition to these fifteen, the Orthodox canon also includes two pieces found in some manuscripts of the Sep-tuagint: Psalm 151 (there are only 150 psalms in the other canons) and 3 Maccabees (a story of how Egyptian Jews of the second century B.C.E. were threatened with destruction on account of their religion but were saved by divine intervention). Thus the Bibles of the three major Christian churches vary notably from one another in what their Old Testaments contain. And, of course, all of these Bibles differ from the Jewish Bible by virtue of the newer Testament that they add to the older one of Judaism.

THE PSEUDEPIGRAPHA

The Apocryphal books are, to Bible readers in general, the best known of the Jewish religious writings left out-side when Judaism closed its canon of sacred scriptures. Although these books were not the only such writings or necessarily the most inter-esting or important of them, they had the great advantage of being a fixed part of the Septuagint and thus were preserved anew each time copies of the Septuagint were made by ancient scribes, first Jewish, then Christian. Other Jewish works composed at approximately the same time as the Apocryphal writings were not so fortunate in their means of preservation. Some ceased to be copied and so disappeared completely, while others left behind only scattered references to their ever having existed. But some managed to hold on to a slim existence, either by having been stored safely away in antiquity and being recovered in modern times or by having caught the interest of religious groups that copied and protected them down to modern times.

One notable characteristic of a majority of noncanonical Jewish works is that they claim to have been written by persons who obviously never could have written them. They are, in short, pseudonymous and to that extent can be considered "false" writings. The Greek term for false writings is *pseudepigrapha*, and that term was applied as a designation, first, to the pseudonymous books among the body of nonapocryphal outside works and then to the whole body. Used in this expanded way, it is not really an appropriate term, for some of the works in this group are in fact not pseudonymous; on the other hand, there are pseudonymous works both in the Jewish canon and in the Apocrypha. But the term "Pseudepigrapha" has become the traditional way of referring to the whole class of Jewish writings that were not included in the Jewish canon and were also not in the Apocrypha; for simplicity's sake we shall use this term in our discussion. (As is true of the Greek word *apocrypha*, the Greek word *pseudepigrapha* is a plural form and in English is used as a plural; but the word, when capitalized, designates a body of literature and can then be considered a singular form.)

There is no agreement among scholars on how many books constitute the Pseudepigrapha. Is it right, some ask, to separate the Apocrypha from the Pseudepigrapha simply on the grounds that the Apocrypha happened to have been included in the Septuagint and thus was taken over by Christians when they adopted the Greek Old Testament? One scholar has urged, indeed, that the Pseudepigrapha and Apocrypha be considered a single body of noncanonical Jewish writings and that they be called simply the Apocrypha. Some scholars would include in the Pseudepigrapha the works composed by the Qumran community (which produced the Dead Sea Scrolls), others would not. By one count there are nearly a hundred works or fragments of works that should be included in the Pseudepigrapha; by another there are about a dozen. But despite the scholarly uncertainty as to what the category comprises, there is a central block of writings that most scholars agree belong in it. Following is a list of a representative dozen of these. Because these works are not generally known, the descriptions below are a bit more detailed than those we gave for the Apocryphal works.

1. The Psalms of Solomon—eighteen psalms closely modeled on the canonical ones, probably composed in the first century B.C.E. in Palestine. These psalms denounce a wealthy, hypocritically religious ruling class in Israel and plead the cause of the poor and pious (perhaps the Pharisees).

2. Jubilees—a retelling of the history of the world from Creation to the time when Moses was commanded to climb Mount Sinai to receive the tablets of the Law. The book takes the form of a divine revelation to Moses, although it seems actually to have been composed about the end of the second century B.C.E. It adds much traditional lore to the canonical story in Genesis and Exodus, and it smooths out the rough spots. Written from a Pharasaic point of view, Jubilees argues for the supremacy and universal validity of the Jewish Law.

3. The Letter of Aristeas—an account of how the Greek translation of the Torah came to be composed by seventy-two elders at the request of an Egyptian king. The work takes the form of a letter from someone involved in making the translation in the mid-third century B.C.E., but that date is at least a century too early for the composition of this piece.

4. The Life of Adam and Eve—an account of what happened to Adam and Eve after they were driven out of the Garden of Eden and of the last days, death, and burial of Adam and the death of Eve. A shorter, alternative version of this work, called the Apocalypse of Moses, professes to have been delivered to Moses by the archangel Michael. The original work was probably written in the first century C.E.

5. The Martyrdom of Isaiah—a story from the second or first century B.C.E. that tells how the prophet Isaiah attempts to stay out of the clutches of the wicked king of Judah, Manasseh; nevertheless, Manasseh finds Isaiah living in the desert and puts him to death by sawing him in half. (This form of execution is mentioned in the New Testament in Hebrews 11:37, which may be a direct allusion to the Martyrdom of Isaiah.)

6. 1 Enoch (also called Ethiopic Enoch)—a long compilation of lore about events in the Jewish scriptures and elements in Jewish theology. Actually a collection of works composed by different authors in the first several centuries B.C.E., the book is represented as

having been written by Enoch, the patriarch who never died because he "walked with God, then was no more, because God took him" (Gen. 5:24). Enoch is thus the one who has entered the presence of God, learned the mysteries of the universe, of nature, and of times past and future, and now reports them back to the inhabitants of this world.

7. 2 Enoch (also called the Secrets of Enoch or Slavonic Enoch)—a first-person account, composed in the first century C.E., in which Enoch describes his journey through the ten heavens to the abode of God. There the Almighty reveals to Enoch the nature of "heaven, earth, and sea" and how the universe was created. Enoch returns to earth and teaches his children what he has learned.

8. Testaments of the Twelve Patriarchs—the final words of the sons of Jacob as they near the time of their deaths. Each patriarch confesses his sins (in particular, what he did in connection with selling Joseph into slavery in Egypt) and gives edifying advice to his survivors. The individual testaments (as in "last will and testament") were perhaps composed by Jews in pre-Christian times and were then reworked by Christians to show that the patriarchs looked forward to the coming of Jesus as Messiah; or perhaps only some of the testaments were written originally by Jews and the rest by Christians.

9. The Assumption of Moses (also called the Testament of Moses)—the last words of Moses to Joshua, in which Moses gives his young successor a preview of the future of Israel from the entry into the Promised Land down to the period of persecution under Antiochus IV Epiphanes in the second century B.C.E. Apparently written not long after that period of persecution, the author (like the author of the book of Daniel) believed he was living in the last days. He composed this apocalyptic work to encourage his readers to maintain their traditional faith.

10. 2 Baruch (given that number because there is a work entitled Baruch in the Apocrypha; also called the Syriac Apocalypse of Baruch)—a first-person account in which Baruch, secretary of the prophet Jeremiah, describes a series of divine revelations granted to him concerning future history from the Babylonian destruction of Jerusalem through the reign of the Messiah and on to the end of the age. It was composed in perhaps the late first century C.E.

11. 4 Maccabees—a sermon from the first century C.E. on the use of religious reason to calm the passions, particularly the passion of fear in the face of threatened death. The point is illustrated by a long description of the calm deaths endured, despite torture in-

flicted at the order of the tyrant Antiochus IV Epiphanes, by an old Jewish man, seven young Jewish brothers, and the aged mother of the brothers.

12. The Sibylline Oracles—a direct imitation, written over a number of centuries by both Jewish and Christian writers, of the famous Greek and Roman prophecies delivered by the gloomy sibyls (female seers) concerning the end times.

THE NEW TESTAMENT
APOCRYPHA

Just as with the Old Testament, there were many writings closely resembling the canonical New Testament books that were not included in the canon. By the time the process of canonization of Christian writings was complete, in the middle to late fourth century C.E., the process had selected out the twenty-seven books with which we are familiar and had left outside many scores of similar writings, to which scholars now apply the collective title "New Testament Apocrypha" (or "Apocryphal New Testament").

In general we can say that the twenty-seven canonical books were written before most of the still-surviving outside books (although scholars suppose that a few of the outside pieces, while composed relatively late, reflect early oral tradition or actual writings contemporary with the canonical books). The letters of Paul were composed in approximately 50–60 C.E.; most of the remainder of the New Testament, including the gospels and Acts, was composed between 70 and 100. The latest canonical piece is apparently 2 Peter, which may have been composed as late as 150. The earliest surviving noncanonical works were composed in the first third of the second century (thus earlier than 2 Peter); but the bulk of the writings that we think of as being Bible-like date between 150 and 350 C.E.

From the dates of these works it is apparent that they were written in imitation of the works that had been or were being accepted as canonical. It is true of even some of the later canonical works that they were written in imitation of earlier canonical works and were assigned "false" authors—that is, they were pseudonymous in the same way that most of the Old Testament pseudepigraphal works were. It is not at all certain why 2 Peter was ultimately accorded a place in the canon while the Epistle of Barnabas, written perhaps thirty years earlier, was not. We simply do not have enough historical knowledge about the early Church to be able to answer that question, any more than we know enough to be able to say why those of Paul's letters that we call 1 Corinthians and 2 Corinthians were accepted into the canon, while two other letters by Paul, which are referred to in 1 and 2 Corinthians, were not. Canonization moved in mysterious ways, quite outside of conscious control. The early Church could always rationalize *after* the fact about canonization; but the Church could not have predicted before the fact just which works would ultimately be accepted as scripture and which would not.

The New Testament Apocryphal works can roughly be catalogued according to the kind of biblical writing they imitate: There are apocryphal gospels, acts, letters, and apocalypses. Of the two score extant apocryphal gospels, three will be of special interest to the general student of the Bible. The Protoevangelium of James (written in about mid-second century) purports to describe the life of Mary from her own birth to the birth of her child Jesus. Similarly, the Infancy Gospel of Thomas (written at about the same time) tells of the young Jesus from the age of five to twelve.* And the Gospel of Nicodemus, also called the Acts of Pilate (written in the fourth or fifth century) recounts Jesus' trial before Pilate; his crucifixion, resurrection, and postresurrection appearances; and (as a sort of appendix) his descent into Hell between the time of his death and his resurrection. All three of these apocryphal gospels and some others like them were well known during the Middle Ages and exercised significant influence on the minds of medieval writers and artists.

The canonical Acts of the Apostles served as a point of departure for about two dozen extant works or fragments of works in the New Testament Apocrypha. These tell largely fanciful tales of the travels, miracles, and preaching of such biblical personages as Andrew, Barnabas, John, and Peter. In comparison to the large numbers of gospels and acts, surprisingly few of the Christian outside writings take the form of letters; beyond a collection of letters that purportedly passed between Paul and Seneca, the first-century Roman literary figure, there are no more than a half-dozen extant apocryphal letters. With respect to the final class of Bible-like works, apocalypses, we spoke in chapter 10 of the Christian adaptation of this Jewish form and of some of the early Christian apocalypses other than the canonical book of Revelation. Several dozen apocalypses are extant that date from the early centuries of the Christian era to well on into the Middle Ages.

*Another Gospel of Thomas (sometimes called the Coptic or Gnostic Gospel of Thomas), previously known only through fragmentary references in other writings, was found among the buried collection of documents discovered in 1945 at Nag Hammadi, in Egypt. Composed in perhaps the second century, this work is a collection of what are purported to be sayings of Jesus, some of them nearly identical to those in the canonical gospels, others without parallel there.

THE PROBLEM OF
PSEUDONYMITY

No reader of this chapter can have failed to notice what a large part pseudonymity—a writer's ascribing his work to some other writer—plays in all three categories of the outside books. We have said that one of those categories, the Pseudepigrapha, takes its name from the fact that most works included in it are "false" in the sense of having been written by someone other than the claimed author. But some pieces in the Jewish Apocrypha and most of those in the New Testament Apocrypha are also pseudonymous, as indeed are some canonical books in the Old Testament (Daniel and The Song of Songs, for example) and the New Testament (1 and 2 Peter, for example). It is quite natural for us to wonder why this element appears so often in Jewish and Christian writing of the late centuries before Christ and the early centuries of the Christian era. If it were merely a matter of anonymity—of authors' not revealing their identity—we would have two ready and satisfying explanations: either the authors were simply capturing what had existed in the oral tradition and thus could not claim truly to be the authors of what they wrote down, or they felt that the spirit of God was speaking through them and thus God deserved the credit and not they, the human tools. But why did the writers so often go the next step beyond anonymity—merely not revealing their identity—and claim that some other individual was really the author of their work? The question is a difficult one, and scholars have spent much time considering it.

Some have supposed that it was because the authority of the Law had become absolute in postexilic Israel; anyone who had new religious insight to contribute would thus have had to write it as though it were a revelation given to some great person before the time of Ezra (or to Ezra himself, he being so much involved in establishing the authority of the Law in Judaism). Other scholars make much of a related idea in ancient Judaism, that prophecy had ceased with Ezra and therefore no new voice could be considered to be inspired by God; thus the need for pseudonymity in presenting a new revelation. The morality of this is questionable in modern eyes (although we seem to be less offended by it if a given work contains ideas we feel to be orthodox than if it contains ideas that were or would become heretical). But we must be careful about projecting our modern ideas of literary morality backward two millennia. Just as the borrowing of material from other writers was not frowned on in earlier ages as it is today (we call it plagiarism and consider it a grave fault), so may have been the device of claiming some notable person from the past as author of one's own work. Perhaps the best way to conceive of pseudonymity in the late biblical and postbiblical era, however, is (as we remarked in chapter 10) as a literary convention. In the absence of charismatic figures who could proclaim with certainty, "Thus says the Lord," the vacuum was filled by written documents that set forth truth as the authors saw it but that claimed as authority the names of appropriate religious figures from the great days of the past. Pseudonymity of this kind was simply an established means of communication on spiritual matters—a way you did it when you had ideas to get across.

The presentation of religious material in the form of rediscovered ancient writing has been practiced from the late biblical era right down to our own time. In 1956 the scholar Edgar Goodspeed published a book entitled *Modern Apocrypha*. In it he discussed fifteen individual works, written in the nineteenth and twentieth centuries, as well as a further volume called *The Lost Books of the Bible*, a modern collection of several dozen ancient Apocryphal New Testament works. This volume of "lost books" about which Goodspeed wrote is still in print, sometimes published in combination with another volume called *The Forgotten Books of Eden*, which is a collection of twenty pseudepigraphal books. Advertisements for the first of these volumes, often placed in tabloid newspapers and in catalogues for novelty items, insist in bold-faced type that these are "the books of the Bible banned for over 1500 years" and that they contain "the truth about the childhood and early life of Jesus" and about "the girlhood and betrothal of Mary." Nothing so sensational is really involved, as readers of this chapter now know. The process of canonization, in doing its work in the early centuries of the Christian Church, simply left some writings out of the canon and included others. The four gospels now in the New Testament were well on their way to universal approval by the time the stories of Mary's girlhood (in the Protoevangelium of James) and of Jesus' youth (in the Infancy Gospel of Thomas) were composed in the mid-second century. Both of these apocryphal gospels had their appeal in their own time and continued to be copied through the centuries and occasionally published after the invention of printing. They were thus not "suppressed by the church for over 1500 years," as the advertisements for *The Lost Books of the Bible* proclaim. They simply never gained a place in the canon of Christian scripture and so are little known in our time.

The outside books certainly deserve to be better known than they now are by general readers of the Bible. Professional biblical scholars must, of course, be thoroughly familiar with this material, both for what it reveals about Judaism and Christianity at crucial times in their history and for the light it casts on the canonical scriptures and the process of canonization itself. But nonprofessional students of the Bible should also be informed about the outside literature, if only to prevent their supposing that the canonical writings are utterly unique in form or content.

SUGGESTED FURTHER READING

James H. Charlesworth, ed., *The Old Testament Pseudepigrapha*, 2 vols. (Garden City, N.Y.: Doubleday, 1983, 1985).

John J. Collins, *The Apocalyptic Imagination: An Introduction to the Jewish Matrix of Christianity* (New York: Crossroad Publishing Co., 1984).

John Dominic Crossan, *Four Other Gospels: Shadows on the Contours of Canon* (Minneapolis: Fortress Press, 1985).

J. K. Elliott, *The Apocryphal New Testament: A Collection of Apocryphal Christian Literature in an English Translation* (New York: Oxford University Press, 1993).

Edgar Hennecke, Wilhelm Schneemelcher, and R. M. Wilson, eds., *New Testament Apocrypha*, 2 vols. (Philadelphia: Westminster Press, 1963, 1964).

Martin Jan Mulder, ed., *Mikra: Text, Translation, Reading and Interpretation of the Hebrew Bible in Ancient Judaism and Early Christianity* (Philadelphia: Fortress Press, 1988).

George W. E. Nickelsburg, *Jewish Literature Between the Bible and the Mishnah* (Philadelphia: Fortress Press, 1981).

Leonhard Rost, *Judaism Outside the Hebrew Canon* (Nashville: Abingdon Press, 1971).

David S. Russell, *The Old Testament Pseudepigrapha: Patriarchs and Prophets in Early Judaism* (Philadelphia: Fortress Press, 1987).

Michael E. Stone, ed., *Jewish Writings of the Second Temple Period* (Philadelphia: Fortress Press, 1984).

The Interpreter's Dictionary of the Bible, ed. George A. Buttrick et al. (Nashville: Abingdon Press, 1962). See articles on Apocrypha; Apocrypha, NT; Pseudepigrapha. Supplement, 1976: See articles on Apocrypha, NT; Pseudepigrapha; Pseudonymous Writing.

The Anchor Bible Dictionary, ed. David Noel Freedman et al. (New York: Doubleday, 1992). See articles on Pseudepigrapha, OT; Septuagint; Vulgate.

XIII

The Gospels

The New Testament, like the Old Testament, is an anthology of writings from numerous sources, not a single book. It lacks the variety of literary content that we saw in the older collection, and for the most part it is focused narrowly on a single historical event: the giving of a "new covenant" to all the people of the world, replacing the "old covenant" between Yahweh and the descendants of Abraham. This, of course, is the Christian understanding of the event. Jews do not acknowledge it, and thus for them scripture stops with the end of the Old Testament canon.

The term "testament" applied to these two collections comes from the Latin *testamentum*—a document that testifies or bears witness to something (usually the will of a person anticipating death). This Latin word corresponds to the Greek *diathēkē*, which was the word chosen by the translators of the Septuagint for rendering the Hebrew *berith* ("covenant"). A covenant is a solemn oral agreement between two parties, and it may or may not be committed to writing. The binding force is in the agreement itself, not in any document. (According to the Old Testament, the original covenant with the Israelites was not recorded in writing until the time of Moses.) The translators' choice that led to our "testament" is of some consequence, because it tends to shift emphasis away from the agreement to the document, making the *words* the primary vehicle of the will of God rather than the *events* they record. But this is a problem only in the religious use of the Bible, which does not concern us here.*

The New Testament canon begins with four documents labeled with the names of their supposed authors: Matthew, Mark, Luke, and John. No one would deny that the beginning is exactly where they belong, for the career of Jesus of Nazareth, which they bear witness to, is the foundation of all that follows. They are called "gospels," from the Anglo-Saxon *godspell*. This word translates the Latin *evangelium*, from the Greek *euangelion*—all of which mean the same thing: "good news." The Latin word is still in current use: "Evangelism" is the process of spreading the good news, and traveling preachers or revivalists are called "evangelists," that is, bringers of good news. Complicating the picture somewhat, the four gospel writers are also referred to as "the Evangelists" (always with the definite article), and the term "gospel" itself has two applications: either to the basic message about Jesus Christ, "the gospel," which can be communicated by word of mouth without recourse to documents, or to the four documents, "the gospels," which carry that message.

That there are four canonical gospels and not five or six or seven— or only one—is a matter of fact. It is not known who brought the four together, or why, but the earliest records show them associated as a group, and they seem to have been the first part of the New Testament to become canonical, some time before the end of the second century C.E. Then as now, all four were believed to be equally authoritative and worth preserving as distinct witnesses. This canonical verdict, which the Church has always vigorously supported, has not gone unchallenged. In that same second century, a man named Marcion tried to establish his own brand of Christianity by discarding all the gospels except a shortened form of Luke; another man, Tatian, tried to combine the four gospels into one. Neither attempt succeeded. Even today, the existence of gospel "harmonies," which print the texts in parallel columns arranged according to similarities in content, shows the deeply felt need to bring the four gospels within a single overall structure. The result is paradoxical, because what gospel harmonies unintentionally demonstrate is gospel disharmony. The

---

*Paul, writing before any New Testament existed, emphatically asserted that the new covenant exists in men's hearts, not in written records of any kind (2 Cor. 3:3–6). One wonders what Paul would think if he could be brought back to life in our own century to see the veneration of the written word of the canonical New Testament.

four works are fundamentally different in content, arrangement, emphasis, and purpose; although they all bring the "good news" of Jesus Christ, each does so in its own way. Thus we have four different perspectives from which to view the subject. Yet the disparity is not complete. Three of the gospels—Matthew, Mark, and Luke—have much in common in the way they present the story and are related to one another as texts. They are therefore called the "synoptic" gospels (those that can be "viewed together") and are usefully studied as a group. The gospel of John is obviously in a class by itself.

AUTHORSHIP

As far as the history of the gospels as documents is concerned, it is clear that each of the four was written by a single author and was meant to stand on its own. Who were these authors—"Matthew," "Mark," "Luke," and "John"? No one knows. Church tradition attempted to connect them directly or indirectly with the original twelve disciples, but there is no real evidence to support this view. The names themselves were common at the time. Moreover, none of the gospels mentions an author's name in the text itself: All four are essentially anonymous. We might stop to ask what difference it would make if we did know the names of the actual authors. Would this in itself make their work more authentic or easier to understand or somehow more personal?

PURPOSE

It is much more productive to ask what the gospels were written *for*. Is there a common motive or purpose, in spite of their individual differences?

Two popular theories are that (1) the gospels were written down because the contemporaries of Jesus were dying off and some means had to be found to preserve their witness to his career, and (2) the infant Christian Church needed propaganda documents for circulation among would-be converts to help bring them over to the faith. (The two theories are not mutually exclusive.) The first theory seems to be supported by the preface to Luke, which refers to accounts "handed down to us by those who from the outset were eyewitnesses and ministers of the word" (1:2), but scholars are now inclined to regard Luke's preface as a conventional literary device, not as a statement meant to be taken at face value. The obligation we feel to record events of great historical importance was not felt by first-century Christians, who were still close to the events themselves. The Church in Paul's time had no gospels, nor is there any hint in Paul's letters that he felt a written record of Jesus' ministry was needed or would have been helpful had it existed. It is a fact that even after the gospels were written, some persons in the Church preferred the oral tradition and could not see the need for anything in writing. As for propaganda (here understood in its literal sense, as the act of "sowing" or "propagating"), there is no evidence in the gospels themselves that their authors had in mind an audience of the unconverted. The gospels may have been designed to *confirm* faith more than to create it, which indeed is still one of their major functions. It does not take away from their importance to note that no church, Catholic or otherwise, has ever made knowledge of the written gospels, as apart from the gospel message, a condition for accepting someone into the community of faith.

A third hypothesis about the purpose of the gospels—one now abandoned altogether—is that they were written as biographies of Jesus; they were not, nor can a biography be extracted from them. For a long time it was assumed that the "historical Jesus" existed within and behind the four gospels in such a way that, by following clues given in them and combining information that they offered separately, it would be possible to construct a reliable general account of his life—sidestepping the question of his messiahship or his divinity—and thus to finish the job that the gospel writers had done only partially. Many accounts of this kind have been written. But they are acts of imaginative piety, not of history, and belong in the same class with such religious objects as paintings of the Madonna and Child. Scholars now agree that we simply do not have the data for constructing a biography. There are too many gaps in the gospel record, there is too much conflicting data, too much vagueness. Facts per se do not exist in the gospels, where even the minor details are theologically flavored and were chosen to advance a particular author's bias.

We do not mean to suggest that a biography of Jesus could not have been written had anyone at the time wanted to do so or, still less, that the man Jesus is imaginary. There is no reason whatever to doubt his existence. His historicity in a general sense is not an issue. But it has to be realized that our only sources of information about him are the four gospels and that they are *gospels*—a very special kind of composition that has no real parallel anywhere else in literature. It is doing the gospels no favor to try to make something out of them that they were never intended to be.

Perhaps the best answer to the whole vexed question of purpose—if none of the above alternatives is satisfactory—is to think of each gospel as a particular author's attempt to give permanent shape to his own conception of the career of Jesus and its meaning. Such a hypothesis at least explains why all four are different. Each author's conception was no doubt shared within the circle where he lived and worshiped. To that extent the gospels were not wholly personal documents; but the initiative to write, as well as much else, must have come from the authors themselves. In a sense they succeeded too well, for the Church, seeing that in the gospels it had testimony of great age and authority, kept all four and made them canonical, thus denying to the authors as individuals the right they had claimed to make independent assessments of the subject.

CONTEXT

The context out of which the gospels came, as we have implied, was first-century Christianity. Because the gospels stand at the beginning of the New Testament and carry the "news" of Jesus Christ, many readers assume that they brought this Christianity into being—that converts were made by people reading the gospels and that these people then grouped themselves into a primitive Christian Church. In fact, it was just the other way around (as Acts and the letters of Paul clearly show). A Christian Church with branches across much of the Mediterranean world was already in existence before the earliest gospel was written. The burden of the message was carried then, as now, by word-of-mouth preaching. This oral tradition undoubtedly included not only narratives of important events in the ministry of Jesus, a few of them already transmuted into liturgy (as we see in Paul's references to the Lord's Supper), but also rehearsals of his sayings. To the extent that a writer used this material, he would be reflecting an unofficial consensus and not speaking merely for himself. The particular consensus might have been a fairly local one, perhaps that of the Christian community at Antioch. But as the hard-won product of a generation of conscious reflection, it would not have been lightly ignored by anyone with the opportunity to experience it, and it would have been immediately recognizable as Christian anywhere within the larger community of faith.

DATING AND SOURCES

According to the nearly unanimous opinion of modern scholars, the first gospel written was Mark's, at about 70 C.E. For a long time it was believed that Matthew, which precedes it in the canonical order, was the first written and that Mark was simply a shortened version of Matthew; but this view cannot be upheld. The next two after Mark would have been Luke and Matthew, somewhere between 80 and 90 C.E. (There is a great deal of uncertainty about these dates and no real way of knowing which gospel was composed first.) John was written about 100 c.e. As we pointed out in chapter 12, the writing of New Testament books, including some romantic and farfetched "gospels," did not stop at the end of the century; but nothing in that mass of material (the present New Testament Apocrypha) ever came close to rivaling in status the four canonical gospels in the Christian Church.

The synoptic gospels—Matthew, Mark, and Luke—are related to one another and to their sources in a rather complex way that scholars are still arguing about, but the most widely accepted view is the one diagramed in figure 6. The source called "Q" (for the German *Quelle*, "source") is now lost as such, but portions of it can be seen in Luke and Matthew. It was a collection of the sayings of Jesus and probably came into being very much as the oracles of the ancient Hebrew prophets did, memorized and written down by disciples and then copied for circulation among the faithful. The Sermon on the Mount is Matthew's arrangement of sayings from Q. None of the Q material was made use of by Mark, who either did not know of its existence or (more plausibly) found it irrelevant to his purpose. Both Luke and Matthew used the gospel of Mark as a source, but neither of the two seems to have known the other man's gospel. In addition, Luke and Matthew each had a source peculiar to himself (L and M), or so the evidence suggests, for large portions of each of their gospels are found neither in Mark nor in Q.

The fact that Luke and Matthew, acting independently, each wrote with a copy of Mark open beside him has only one meaning: that each intended to supplant Mark with his own gospel. Mark was certainly not canonical to them! Their respect for Mark clearly shows in the extent to which they took over his story, but they felt free in doing so to change it to suit themselves. John, who must have known some gospel, chose to go his own way in an even more decisive fashion. Thus the gospel writers began as rivals, not as collaborators.

If Mark had no other gospel to follow and did not use Q, did he make up everything in his own gospel out of whole cloth? Was his truly a work of fiction? The answer, of course, is no. Mark's great source was Church tradition, that is, what believers heard and remembered and passed on to others, and acted on, within the Christian community—the consensus that we mentioned earlier. This traditional material was anything but a mass of disorganized information. By the time the gospel writers arrived on the scene, much of the material had been shaped, defined, and given functions in the life of the Church. These functions, about which almost nothing is known,

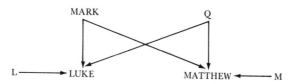

FIGURE 6. SOURCES OF THE SYNOPTIC GOSPELS

were no doubt at that stage informal; but let us not forget that Christianity arose out of a religion that was accustomed to affirming its special character through historical recitals, and there is no reason to believe that the need to do so was felt any less in the Christian than in the Jewish community. Christianity had every reason to rehearse the events of its recent past. Without a written canon to fix this material, which had passed through so many hands, some of it was probably lost and the rest underwent changes that at least tended to smooth off its rough edges. Thus the gospel writers did not deal with the raw stuff of history—they did not have it. What they had was halfway to being literature already.

LITERARY GENRES

Literature is always a *kind* of something. Works belong to genres, to general categories. The genres of the literary materials used in the gospels are fairly easy to see because inside the gospels they retain the distinctive forms that they brought with them. They include the following:

1. *Sayings and sayings stories.* Sayings are abundant in all four gospels, reflecting the importance of Jesus as a teacher. Q was largely a collection of such sayings, some with associated narrative material. Matthew and Luke drew on Q for their respective "sermons" of Jesus, providing a context for the sayings that they did not have in the original document.

2. *Pronouncement stories.* Pronouncement stories are actually a subclass of sayings stories. We do not need to draw too fine a line between them, but the typical pronouncement story is distinctive enough to call for separate treatment. It is defined as a brief episode or event that leads to and ends with a saying of Jesus. The story exists for the saying, rather than the other way round. For example, the saying, "I came to call not the upright, but sinners" (Mark 2:16–17, Matt. 9:10–13, Luke 5:29–32) belongs to an occasion when Jesus was supposedly eating in the company of taxgatherers and other outcasts.

The circumstances of the meal are not specified, and it is clear that the writers had no more idea than we do when and where it took place—or even if it did. However, that is not the point: The meal anchors the saying within a stream of events and gives it a plausible reason for existing. Sometimes the attention given to the story can actually distort its meaning, as in a common interpretation of the famous pronouncement on little children (Mark 10:13–16). It is not an expression of Jesus' love for little children; rather, the children are simply a pretext for his saying something about the nature of the Kingdom of God. One might call the passage as a whole a kind of acted parable.

3. *Prophetic citations of the Jewish scriptures.* The gospel writers were certainly not the first Christians to search the Jewish scriptures for passages that seemed to foretell Jesus and help in the understanding of God's will under the new covenant. It is likely that collections of such proof texts circulated in the early Church and thus found their way into the gospels. An excellent example would be Isaiah 6:9–10, which was used in all four gospels and in Acts.

4. *Passion narratives.* Passion narratives cover the events between the Last Supper and the Crucifixion; they may have been the first units of material to achieve independent status. Certainly an understanding of Christianity would have been very difficult without a coherent account of this event toward which the entire mission of Jesus was pointed. These are the most complex units of gospel material, and they can be broken down into subunits if one wishes to pursue the analysis further.

5. *Miracle stories.* Miracle stories, too, can be broken down into subunits, such as miracles of healing, exorcisms, demonstrations of power, and so on. The only miracle that appears in all four gospels is the feeding of the five thousand, which strongly suggests that this story had a fully realized independent form before these writers took it up and was, no doubt, often cited and recited by members of the early Church.

6. *Parables.* The parable originally was a brief story that used details from ordinary life to illustrate a moral point, like the parable of the two housebuilders in Matthew 7:24–27. It was an effective teaching device because it put things in terms that people could understand and easily remember. But parables seem to have suffered more from outside intervention than most other materials. Somewhere early in the history of their transmission they began to be regarded as mysteries, with hidden meanings, or at least as allegories with two levels of meaning, and their use in the gospels reflects this change. They are

sometimes provided with interpretations meant for the disciples alone, and in one case (Matt. 13:24–30) scholars suspect that a gospel writer made up his own parable and attributed it to Jesus. (For more on parables, see the discussion in chapter 2.)

7. *Events in the public career of Jesus.* Events in Jesus' public career would include such things as his baptism by John, the choosing of the disciples, the Transfiguration, the entry into Jerusalem, and the cleansing of the Temple.

THE PERICOPE

The literary units just described are the major building blocks out of which the gospels were constructed. The technical term for such a unit of composition is "pericope" (from the Greek, meaning "cut around"): In its original sense the term referred to something separated from or extracted from a larger context (for example, in our day, a newspaper clipping); but as now used in biblical criticism, "pericope" refers to the independent units that when assembled constitute a total work. Every student of the Bible should learn the use of this term (it applies in the Old Testament, too), for it is enormously useful, not only in providing a label for units of biblical writing when they are found but also in finding them to begin with.

Much of what the gospel writers did, then, was to put pericopes together. The narrative cement that bound them was often of the flimsiest kind—partly because the writers had no specific data about times and places and occasions, partly because they did not care much about matters so dear to modern journalists and writers of realistic fiction. For example, as Mark's narrative picks up momentum in his second chapter, we find a string of pericopes introduced by such phrases as "He went out again" (2:13), "When Jesus was at dinner" (2:15), "It happened that one Sabbath day" (2:23), and "Another time he went into the synagogue" (3:1). This kind of vagueness is typical in the gospels.

Although much of their basic material was provided in traditional pericopes, the gospel writers were still authors in the full sense of the term. They did not mechanically paste pericopes together but exercised considerable freedom in arranging them, and they habitually modified them to suit their own purposes. An edition of the New Testament with cross-references is handy for tracing this process at work. A good starting point would be to follow the set of pericopes that introduce John the Baptist and his baptism of Jesus through Mark 1:2–11, Luke 3:1–22, Matthew 3:1–17, and John 1:15–34, noting both similarities and differences. To see how freely pericopes could be moved around within the story, one could compare the placement of the "master even of the Sabbath" pericope (a pronouncement story): In Mark 2:23–28 and Luke 6:1–5 it follows the "old law and the new" pericope, whereas in Matthew two full chapters of other material intervene. In Matthew it comes after the Sermon on the Mount; in Luke it precedes the counterpart of that sermon. The "Jerusalem, Jerusalem" pericope, in which Jesus apparently foretells the destruction of the Temple, occurs in Luke in 13:34–35 *before* Jesus reaches Jerusalem; in Matthew 23:37–39 it occurs *during* his final visit to the city. The gospels, as we see, both do and do not tell the same story.

We are prepared now to look briefly at the gospels individually. They will be taken up, not in canonical order but in the presumed order of composition.

MARK

We said earlier that the majority of scholars believe that Mark's was the first of the four gospels written. It is also likely that Mark's was the first gospel *ever* written. Mark seems to have invented the gospel as a literary form. There are no real precedents for it in ancient literature. It might be said that the revolutionary nature of Mark's message needed—and was given—a revolutionary literary form for its conveyance to the world. Whether someone else would eventually have invented the gospel form if Mark had not done so we shall never know, but in the absence of such knowledge let us give all credit to Mark for his originality.

Whoever Mark was, he was apparently a gentile Christian, writing for other gentile Christians in a time when the turmoil visible on earth (especially the Jewish war of 66–70 C.E.) suggested strongly the imminence of the end, which would be accompanied by the *parousia*—the return of the Son of God—and a conclusive battle between the forces of good and evil. This view is reflected in the apocalyptic discourse in Mark's chapter 13 and in his persistent ethical dualism, that is, the belief that the world is an arena for struggle between the forces of good and evil. It is not surprising, then, that Mark's tone should be so urgent, his language so strong, his narrative style so abrupt.

Mark's sixteen chapters make his gospel considerably the shortest of the four. He cannot be bothered with a birth story, such as Luke and Matthew later produced. Instead, he begins with the blunt and straightforward announcement of Jesus Christ as the Son of God, followed by a messianic prophecy involving John the Baptist and a brief

account of Jesus' baptism; and then in 1:14 he launches into the ministry of Jesus in Galilee.* From there Mark hastens his narrative toward
the Crucifixion (which cannot be much more than a year away) because for him this was the single event that gave meaning to all the
others. Jesus was born to be crucified.

That being the case, it is no wonder that the details of Jesus' ministry as recorded by Mark are so sketchy. The traditional view of Jesus
as a teacher finds little support here. There is no extended discourse,
like the Sermon on the Mount, and the ethical rules that Mark does
give are scattered and uncoordinated, although they do have two
things in common: They are all quite absolute in their strictness, and
none of them is presented as a ticket to Heaven. Even the "first of all
commandments" pericope of 12:28–34 is better understood as a response of Jesus called forth in debate with his Jewish opponents than
as a piece of ethical teaching offered for its own sake. If good conduct
can take a person only so far (as implied in 10:27 and 12:34), and
salvation comes through repentance and belief in the gospel message
(1:15), one would like to know more specifically about this message
and about the nature of belief in it, but the answers will not be found
in Mark. This author is completely uninterested in abstract questions
of the sort that Christianity has had to wrestle with in the centuries
down to our own era. His eye is all on the fact: This man was the
Son of God, these things happened.

In Mark's version of the story, Jesus passes through the world
largely unrecognized except by demonic forces—his secret opponents
who unfailingly identify him—and at his death by a lone Roman centurion. His disciples are no better than the others at understanding
what he is about; at the moment of his arrest, they cap their failure
by deserting him. No humans in this story come out very well, not
even the people of Jerusalem, who hail him erroneously as the restorer
of the kingdom of David—that is, as a human king and not as the Son
of God. When at the end Jesus answers "I am" (Mark's is the only
gospel that records this as an unequivocal affirmative) to the High
Priest's question, "Are you the Christ, the Son of the Blessed One?"
(14:61–62), his secret remains intact, for obviously the High Priest
and all the rest believe that he is lying.

From the perspective of Mark, Jesus was a thaumaturge, a wonder-
worker, and his career was like that of a particularly brilliant meteor
that flashed briefly across the darkness of the night, illuminating this

---

*A birth story would also—for Mark—have had the disadvantage of calling atten-
tion to the human side of Jesus, which Mark wished to deemphasize.

world only long enough to indicate that something extraordinary had happened. Essential to this concept is the blackness of the surrounding sky, which can be understood to symbolize the ignorance of the people who, according to Mark, observed this phenomenon during its momentary passage. This ignorance and this incomprehension, however regrettable they may seem, are essential, for they serve to define, by contrast, the brilliance of the meteor.

More than that, as Mark presents him (and, to a lesser degree, as do Luke and Matthew), Jesus actively works to keep his true nature secret from everyone save a few intimates. His apocalyptic discourses about the end of the world are not given to the world but to his disciples. Again and again he enjoins secrecy on those whom he heals. He avoids crowds. His parables, as recorded, frequently require sophisticated explanations before they can be understood. We may assume that if the disciples needed to have them explained, they must have sailed over the heads of the other listeners.

How accurately does this picture represent Jesus' actual behavior? The question is unanswerable because the objective facts are many centuries beyond our grasp. What we have now is what we always have: a writer's conception of a subject. The "messianic secret," as it has been called, is true to a conception of Jesus' career as having been, in earthly terms, a failure. He was briefly popular within a limited area of Palestine. He attracted a small band of close associates. But he earned the notice, hence the antagonism, of the Jewish religious establishment and the Roman government and so was quickly tried for sedition, executed by crucifixion, and removed from the scene. The story was intended to end that way; therefore Jesus' failure was, in fact, his success. He was rejected because he was misunderstood. Because he was supposed to be rejected, it followed that he was supposed to be misunderstood, and so everything in his career should be brought into line with this conception of it.

Mark's gospel in the best manuscripts ends abruptly at 16:8. The so-called "long ending," verses 9–20, although canonical for Catholics and Eastern Orthodox Christians and generally accepted by Protestants, was almost certainly not written by Mark but was supplied later in order to give Mark's gospel what it lacked and the others had: a postresurrection appearance. One can only speculate as to why Mark left it out, but a very good theory is that he did so because in his view the crucified Christ rose directly from the tomb into Heaven, where he will stay until the *parousia*. A reappearance to the disciples or to the women in attendance, however brief, would for Mark have been an act of reconciliation with or a gesture of confidence in the very humans who had failed him. It was now too late for that. (The textual problem of the ending of the gospel of Mark is discussed at length in chapter 15.)

LUKE

The gospel of Luke contrasts noticeably with that of Mark both in style and conception. Yet Luke owed to Mark the idea of writing a gospel in the first place, and he adopted Mark's outline of Jesus' career for his own work, incorporating into it about half of what Mark had written. To this he added material from Q and numerous extracts from a source that has been dubbed "L" because this material appears in Luke's gospel only. More of Luke's gospel comes from L than from either Mark or Q. In addition, a certain amount of the material in Luke's gospel may be original: the famous birth and infancy narratives and the genealogy of Jesus seem likely to have been Luke's own compositions. Thus Luke's gospel presents a more complicated object for study than Mark's. Certainly it is the product of a more self-conscious literary artist, a man who writes excellent Greek and can adapt his style to the situation, who knows how to make a smoothly connected narrative, and who addresses an audience capable of appreciating these qualities.

Luke's gospel is actually only the first volume of a two-volume work, the second of which is Acts. If the two volumes were originally united—and we cannot be sure of this—they became separated at some early stage in the formation of the canon. Each is dedicated to a certain Theophilus with a short preface which, alas, tells us much less than it seems to, although the mere existence of a preface is itself significant. Theophilus might have been an actual person—perhaps a wealthy patron of Luke—but he might also have been a literary fiction, standing, as his name suggests, for all the devout persons to whom Luke's gospel is addressed. Luke himself tends to recede from us as we attempt to pin his name to an identifiable historical figure. There is no compelling reason to accept the traditional view that he was a traveling companion of Paul, the "dear friend Luke, the doctor" mentioned in Colossians 4:14 and the author of the "we" sections of Acts that begin at 16:10. If he was, then it becomes necessary to explain how his theology could differ so markedly from Paul's and how he could misrepresent (perhaps deliberately, perhaps out of ignorance) certain key aspects of Paul's career that we know about from Paul's letters. (We shall look at this problem further in chapter 14.)

In any case Luke was a gentile Christian of good education and cosmopolitan outlook, well acquainted with the Septuagint. In adopting Mark's gospel, he considerably modified its sternness and urgency. This he did in the first place by adding so much from Q and his own sources that Mark's contribution is simply less prominent. The additional material is written in a more urbane style and is responsible for a number of deservedly famous pericopes, such as the parables of the Rich Man and Lazarus, the Good Samaritan, and the Prodigal Son; Mary's hymn of gratitude (the "Magnificat"); and stories of the baby Jesus in the manger, the angels appearing to the shepherds in the fields, the boy Jesus teaching in the Temple, the repentant criminal on the cross, and the appearance of the risen Christ to the disciples on the road to Emmaus. (It is also worth noting that Luke shows a special interest in the women in his story, who are much more prominent in his gospel than in the others and are always presented sympathetically and in a lifelike way.) Luke (21:5–36) dutifully reproduces Mark's apocalypse (chapter 13), but he also softens it. In his own view the Last Day is not necessarily imminent: In the meantime there is work to be done by believers, work that he will spell out carefully in the second volume of his history. This work requires some sort of accommodation of the faithful to the world, which cannot then be entirely bad and fit only for destruction, as it is in Mark's view.

Both volumes of Luke's work were written under the conviction that the age he describes was chosen by the Holy Spirit for its decisive intervention into human history, first through the birth and ministry of Jesus Christ and second through the Church formed to complete his mission. Luke adopts the outlook of the Jewish scriptures, in which the world from its very start has been part of a divine plan, and he adopts also the prophetic anticipation of a final end to this history and a salvation of the righteous few. The Jewish scriptures as Luke and others read them pointed to the coming of a Messiah, the Son of God. These prophecies are seen as fulfilled in Jesus. But Jesus is also the Son of Man: Luke's genealogy takes his ancestry all the way back to Adam, including David but sidestepping the royal line that followed from David, as if to imply that he, Luke, is dealing here with a quite different kind of king than some expect. Indeed, the most striking characteristic of Luke's conception is his universalism. Salvation is intended from the very beginning to be available to all humankind, although the Church at first does not realize this. The rules of the game, so to speak, require that salvation be offered first to the Jews, but by Luke's day the Jews had decisively rejected it and showed no signs of ever changing their minds. Luke does not condemn the Jews unreservedly, as does Matthew, yet he cannot bring himself to treat them generously. Those who show up to best effect in Luke's stories are invariably Gentiles; often they are officials of one kind or another, who even if they themselves reject Jesus do so without fanaticism or malice. Roman sovereignty, as Luke sees it, is a guarantor of law and order, providing splendid opportunities for the growth of the Christian movement.

MATTHEW

A decade or so after the writing of Mark's gospel, the man whom tradition calls Matthew sat down with a copy of Mark to revise and expand it. His object was to produce a gospel more in line with his own conception of Jesus' earthly mission, doubtless one that was current in the Christian community where he himself lived. This is thought to have been in Antioch, on the Syrian coast north of Palestine. If Luke's gospel preceded his (the order of the two is disputed), Matthew does not seem to have known about it. But his respect for the gospel he did know, Mark's, is shown by his inclusion of about 90 percent of that gospel in his own and by his following its general order of events. To this he added material not found in Mark: sayings from Q and some material of his own (for example, a birth story of Jesus), thus filling out the gaps left by Mark and in the process making his gospel about one-third longer and quite a bit smoother to read. (The expansion of Mark does not result from a looser writing style, however; where he does follow Mark, Matthew is frequently more concise.) As in the case of Luke, we must assume that Matthew found Mark's gospel inadequate and intended to replace it with his own, not merely to supplement it.

The Jesus whom Matthew presents is a Jewish prophet–teacher with unique authority that derives from his being both the Son of God and the direct descendant of King David, thus fulfilling in his person all the messianic prophecies of the Jewish scriptures. Matthew's gospel is addressed to readers who had to be, like himself, convinced of the authority of the Jewish scriptures and practiced in the way of reading it that had become institutionalized in the postexilic Jewish world, where not only short passages, sentences, and phrases but even single words could be taken out of context and examined for their prophetic implications.* Scholars have traced more than sixty direct quotations from the Old Testament in Matthew;

---

*Matthew's Old Testament was in Greek, not Hebrew. In common with all Christians by this time, he regarded this translation (the Septuagint) as fully authoritative. It was not, of course, known then as the Old Testament.

there are also more indirect allusions to it or echoes of it. Matthew's most notable use of the Old Testament is in the "fulfillment formula," which occurs eleven times in his gospel (see, for example, the virgin birth passage in 1:22–23). He also sees typological connections between the Old Testament and the events of Jesus' life. Two of the best known of these are his placing Jesus' first major teaching discourse on the top of a mountain (which corresponds to Mount Sinai, where Yahweh gave the Law to Moses) and his addition to the Jonah pericope in 12:40, where the three days and three nights that Jonah spent in the fish's belly correspond to the period of time of Jesus' burial (contrast with Luke 11:29–30).

In light of his respect for Jewish tradition, it is remarkable that Matthew should be so bitter and unforgiving toward the bearers of this tradition. Matthew's bias almost deserves the modern term "anti-Semitism." In his eyes the Jews had missed their historic opportunity by turning against the Messiah sent to rescue them and consequently deserve no pity.

Matthew's harshness toward the Jews may be seen in his version of the Q parable of the wedding feast (22:1–14) when it is compared to Luke's (14:15–24). The point of the parable is the same in both versions: The Jews have been invited to enter the kingdom of Heaven but have refused; therefore their places will now be taken by Gentiles. They have lost their chance. But Matthew (who makes the parable, according to his custom, much more openly allegorical) has the king take reprisal on those who mistreated the servants sent out with his invitation (that is, Christian missionaries): "He despatched his troops, destroyed those murderers and burnt their town" (22:7)—perhaps a veiled reference to the destruction of Jerusalem in 70 C.E. This detail is missing in Luke. Matthew concludes with a surprising coda (22:11–14, also not in Luke) in which the king, finding a guest at his table who is not properly dressed, orders him to be bound hand and foot and turned out into the dark, the place of "weeping and grinding of teeth" (22:13). (This guest may symbolize Christians who are unfit to be members of the Church.) Another example of this harshness is Matthew's parable of the weeds (13:24–30) and its interpretation (13:36–43)—found only in Matthew—in which the place of "weeping and grinding of teeth" occurs linked with a story of selection and exclusion. The basic idea of the saved remnant, of course, is extremely old, going all the way back to the stories of the Flood and of Sodom; but its application here by Matthew is noteworthy for the attention paid to the fate of the excluded and to the way in which a Jewish notion has been turned against the Jews, for originally they were to be the ones who would be saved.

In the passion narrative, Matthew's bias produces the awkward and improbable story of the Roman guards at Jesus' tomb (not found in any other gospel), but its most dramatic—and many feel, most unfortunate—appearance is in 27:25, where Matthew (again alone of the gospel writers) has the Jewish crowd unanimously take upon itself and its descendants the guilt for the Crucifixion.

Yet Matthew's gospel is also noteworthy for the amount of fine ethical teaching it contains, with emphasis on brotherly love and forgiveness. Unlike Mark, Matthew believed that Jesus' teaching was a central aspect of his career, not an incidental one, and that this teaching could be understood to the profit of his listeners. Hence Matthew organized his gospel carefully to highlight the teaching by gathering that material into five more or less coherent discourses, of which the Sermon on the Mount is the first and most famous.

Matthew is also much concerned with the Christian Church, which in his own day was a thriving institution, although very small in comparison with what it would become. His is the only gospel to use the Greek word *ekklēsia* ("church" or, in the NJB, "community"). It is in Matthew's version of the confession of Peter at Caesarea Philippi that Jesus adds the words "on this rock I will build my church" (16:18, KJV). And in Matthew's understanding, the post-resurrection appearance of Jesus culminates in Jesus' commissioning the disciples in Galilee to go forth and "make disciples of all nations" (28:19–20). For Matthew the gospel story is over and the Church now begins its mission. For Luke, however, who has Jesus reappearing to the disciples in Jerusalem, not Galilee, the story is not over, for the Church is still an amorphous entity that must receive further guidance from the Holy Spirit before it can spread outward, and its first growth will be the conversion of many Jews—something that Matthew seems to regard as a hopeless effort.

JOHN

To turn now to the gospel of John, if one has not had this experience before, is something of a shock. What is going on here? The style and tone of gospel writing have radically changed, familiar landmarks in the story have disappeared to be replaced by others never seen before, and even the figure of Jesus looks suddenly strange. The gospel of John seems to have come from another tradition entirely—even from another universe of thought. As one reads on, it is soon apparent why John has always received independent consideration.

The basic structure of Jesus' ministry in John is a three-year period rather than the one year implied in the synoptic gospels. It includes not just one but four separate visits by Jesus to Jerusalem as a grown man, and it concentrates his activities in Judea rather than in Galilee. Some features of the story remain familiar: John the Baptist appears prominently as Jesus' forerunner; Jesus chooses disciples, puns on the name of Peter, teaches, performs miracles, confronts Jewish opposition, and is tried and crucified at the end. But here John the Baptist does not baptize Jesus, the disciples have no missionary functions and indeed no active role of any kind in the story, none of these miracles except the feeding of the five thousand is found in the synoptics, and Jesus' teaching has no generalized moral or ethical content. And even in the passion narrative—where one might expect the most agreement among the four gospels—John's version differs in many important details, not least of which is his insistence that Jesus was crucified on the day of preparation for the Passover (in the synoptics he is crucified on the Passover itself).

These are particular differences. More generally, John's portrait of Jesus shows him not as the active, engaged, and practical figure of the synoptics, mingling in the dust and confusion of everyday life, but more like a figure on a dais, always somewhat elevated and tending to be preoccupied with his own thoughts, entering the mainstream of human affairs only now and then and only for some special reason. He possesses a divine agenda governing the order of events in his life and is little, if at all, affected by pressures from the outside. Most striking of all, in John's gospel Jesus suddenly turns loquacious and begins giving long speeches explaining who he is and what he is here for. Although his audiences are still mostly uncomprehending, Jesus persists. The reserve and secrecy that characterize Jesus' communications about himself in the synoptics have completely vanished; here he talks about hardly anything else. These messages are not delivered through parables but directly, in complete expository essays. A major theme of these essays is the relationship between Jesus the Son and God the Father, something that gets little attention in the synoptics.

In the distinctive language of John, the teaching of Jesus seems to revolve around a small number of crucial terms, frequently repeated: "light," "darkness," "life," "glory," "grace and truth," "to know," "to believe." These terms, familiar in ordinary use, here carry special significance. Indeed, John's gospel might well be called "the gospel of the deeper meanings." Again and again, responding to a question or a statement from someone who uses language in its conventional sense, Jesus will take the same language and reinterpret it in a spiritual and profound way that the speaker has been unaware of. (For example, see the reinterpretation of "temple" in 2:19–22, of "water" in chapter 4, of "food" and "bread" in chapter 6, of "sleep" in 11:13.) If these responses often look rather oblique to the occasion that prompted them, it is because what matters is the saying and not the event. In John's gospel the pronouncement story, far from being an occasional feature, is the basis of the whole work.

And yet John's work *is* a gospel—of that there is no doubt even though he never uses the word *euangelion* in it. He is concerned, as the others are, with presenting a coherent and meaningful interpretation of the career of Jesus from the perspective of a Christian believer. His presentation is chronological and focuses, like theirs, on the crucial final period. Despite the differences noted in the way John presents the Crucifixion, it is still the story of a popular prophet betrayed by one of his own disciples and brought before a tribunal of his Jewish opponents, who then persuade the reluctant Roman authorities to have him executed. John clearly was in touch with contemporary traditions of the Church, and he must have had at least the gospel of Mark as precedent. There are some twenty passages in John's gospel prior to the passion narrative that parallel the synoptics—more than coincidence could allow. Where John differs from the synoptics (as in the time of the Crucifixion), one should not assume that he is necessarily wrong because he is outvoted three to one. Many scholars are now willing to believe that John's gospel has quite authentic elements in it and should be taken seriously as a historical witness.

Whatever his sources, the material was assimilated into John's own distinctive way of thinking before it emerged to form the substance of his gospel. This gospel is more like a theological meditation punctuated with significant events than the busy narrative of activity in the synoptics. The presence of Jesus on earth as the incarnated Son of God, which was nearly all-sufficient for Mark, is pointless for John without an understanding of the *meaning* of this event. Hence the constant exposition in John, which is not concerned, as it is in Matthew, with clarifying the new Law that is to govern humankind but with clarifying spiritual truth. The occasions for doing so are a series of "signs"—acts of Jesus with particular spiritual significance—which are carefully arranged throughout chapters 2–11 to culminate with the resurrection miracle that foreshadows Jesus' own resurrection. (There are at least five signs; some scholars count as many as eight in all.) That most of these events are miracles is the least important thing about them in John's view. They are acts pregnant with extraordinary meaning, which Jesus chose to perform as a means of revealing aspects of himself. Because these acts easily impress people who are looking for wonders of an earthly and transient kind, and who then believe without really understanding, the acts are ultimately unsatisfactory as a means of creating true faith. True faith is independent of physical evidence, even though the evidence be real and authentic: This is the point of the concluding episode of "doubting" Thomas, placed very deliberately at the end of the gospel. Someone else at a later time wrote chapter 21 and added it to John's work.

John's independence shows itself further in a peculiar kind of eschatology which holds that eternal life can begin now, while the believer is still on earth (5:24, 6:47); identifies Jesus the Son with the Father (8:19, 10:30, and 14:7–11); and interprets the Crucifixion in both a literal and symbolic sense: Jesus is "lifted up" on the cross (8:28; 12:32, 34) just as Moses raised the bronze serpent in the wilderness (3:14) to save his people.

To a reader of the synoptics, it is all most strange. But then it sounded strange right from the start: *En archē ēn ho logos* ("In the beginning was the Word"), wrote John, plunging us into a dizzying passage of abstractions and paradoxes that would not be at all out of place in a work of Greek philosophy but that seems foreign to the Bible. Indeed, John's contribution to the Bible is a very special one. For all of that, John was himself firmly within the Jewish tradition and was as well informed on the scriptures as any of the other gospel writers. His dependence on those scriptures is not as readily visible, but it is certainly there; one can find fifteen prophetic uses of the Jewish Bible in his gospel, four of them in the crucifixion story. Many episodes and many terms in John's gospel presuppose a fairly intimate familiarity with Jewish thought: for example, the miracle at Cana or the key phrase "lamb of God" (1:29). The beginning of the gospel, quoted earlier, is a midrash or imitative commentary on Genesis 1 and cannot really be understood apart from it. But John's attitude toward the Jewish scriptures—and this is true also of the other gospel writers—is somewhat complicated by the fact that he is both affirming and denying, both using and replacing. This ambivalence of Christians toward the Jewish scriptures came to the surface even before gospel writing began, as we shall see in chapter 14, and it has remained visible in the Christian movement even to the present day.

SUGGESTED FURTHER READING

David E. Aune, *The New Testament in Its Literary Environment*. Library of Early Christianity, vol. 8 (Philadelphia: Westminster Press, 1987).

C. K. Barrett, *The New Testament Background: Selected Documents*, rev. ed. (London: SPCK, 1987).

Harold Bloom, ed., *The Gospels* (New York: Chelsea House, 1988).

David R. Cartlidge and David L. Dungan, eds., *Documents for the Study of the Gospels*, rev. and enl. ed. (Minneapolis: Fortress Press, 1994).

John Dominic Crossan, *The Historical Jesus: The Life of a Mediterranean Jewish Peasant* (San Francisco: Harper & Row, 1991).

Robert W. Funk, Roy W. Hoover, and the Jesus Seminar, *The Five Gospels: The Search for the Authentic Words of Jesus* (New York: Macmillan, 1993).

James E. Goehring, ed., *Gospel Origins and Christian Beginnings* (Sonoma, Calif.: Polebridge Press, 1990).

John S. Kloppenborg and Leif E. Vaage, eds., *Early Christianity, Q, and Jesus* (Atlanta: Scholars Press, 1992).

Helmut Koester, *Ancient Christian Gospels: Their History and Development* (London: SCM Press, 1990).

Burton L. Mack, *The Lost Gospel: The Book of Q and Christian Origins* (San Francisco: Harper & Sons, 1993).

Stephen J. Patterson, *The Gospel of Thomas and Jesus* (Sonoma, Calif.: Polebridge Press, 1993).

Charles H. Talbert, *What Is a Gospel?* (Macon, Ga.: Mercer University Press, 1985).

The Interpreter's Dictionary of the Bible, ed. George A. Buttrick et al. (Nashville: Abingdon Press, 1962). Supplement, 1976: See articles on Baptism; Biblical Criticism, NT; Discipleship; Genealogy, Christ; Gospel, Genre; Jews, NT Attitudes Toward; New Covenant, The; Q; Synoptic Problem.

The Anchor Bible Dictionary, ed. David Noel Freedman et al. (New York: Doubleday, 1992). See articles on individual gospels, Covenant, Gospel Genre, Passion Narratives.

XIV

Acts and the Letters

In the canonical New Testament, the four gospels are followed by a book traditionally called "Acts of the Apostles" ("Acts" for short). As we have already noted, this book is the second half of a two-part work by Luke, the first half of which was his gospel, from which it is now separated by the gospel of John. The canonical position of the four gospels has some logic: Only after their story has been told in all its versions is it appropriate to look forward in time to the events following the Crucifixion and Resurrection.

A great many events followed, of course, but the one thing that made all the difference was the establishment of a church to carry out the mission of the now-departed leader. How did this happen? What trials did this embryonic group encounter? What were its early successes? Who led it? These questions and others are now addressed by the same man who wrote the gospel of Luke.

THE GENRE, PURPOSE, AND AUTHORITY OF ACTS

The four gospels belong to their own genre; except for later imitations of them, they resemble no other literary work that we know of. Acts, on the other hand, looks very much like a book of history. Although it records miracles and other supernatural events, it is chiefly given over to the doings of ordinary humans; it has a straightforward chronological order, with specific indications of time and place; it records travels and meetings and speeches; it takes us to major cosmopolitan centers of the pagan world such as Antioch, Ephesus, Athens, and Rome; it places many known historical figures on its stage; and it seems much more concerned to inform than to preach. Readers can be forgiven for thinking that now, at last, they have emerged into the light of day and can confidently expect a factual record—factual at least within the author's ability to know the facts. This looks like a game played by familiar rules.

But appearances are deceiving: Acts is no more history than the gospels are history. It is not an impartial record of events, such as we now expect history books to be, but a deliberately constructed narrative designed, even to the smallest detail, for the sake of making certain didactic points. To this end Luke has chosen which things to record and which to ignore, ordered their sequence, created settings for them, brought his characters into dramatic relationships, composed their speeches and conversations, and in general seen to it that everything in his book contributes to its overall design. The raw materials of history exist behind the book of Acts, but what the reader sees is always Luke's conception of the materials, never the materials as such (a matter discussed in chapter 1). In all this Luke was simply doing what ancient historians always did. For them the need to instruct and edify was at least as important as the need to inform, and facts, merely as such, had no particular virtue. If readers do not recognize this fundamental difference between ancient and modern historical writing and adapt themselves to it, they will inevitably misunderstand a work such as Luke's.

The twenty-eight chapters of Acts cover events between approximately 30 and 60 C.E., that is, from Jesus' last words to his disciples to Paul's final journey to Rome. If Acts was written in about 90 C.E., according to the best guess, there is almost a full human generation between its writing and the last events it records. Luke does not claim to have been an eyewitness to any of this history. Many things that he includes (for example, private conversations), he could not possibly have known at first hand. The only indications pointing to the author's personal knowledge are the "we" sections of Acts that begin at 16:11, especially the splendidly vivid account of Paul's last voyage and shipwreck. But there is nothing here that might not have been learned from documentary or oral sources, or indeed that might not have been created out of whole cloth by a skilful literary artist such as we already know Luke to have been.

Not only does this author make no claim to have been an eyewitness, he also never tells us who he is. Both the gospel of Luke and the book of Acts are anonymous. The tradition that their author was named Luke comes to the surface in Church writings of a century later. That one of Paul's co-workers was named Luke is indicated by references to such a man in the letters (Col. 4:14, 2 Tim. 4:11, and Philem. 24), but even if all three of these refer to the same Luke, there is nothing to prove that this man was the author of Acts.

The issue is not the identity of a man (who certainly did exist, whatever his name was), but the authority of a book, since more than half of Acts is given over to an account of Paul's missionary activities. On what basis did Luke—as we shall continue to call him—form his picture of these activities? If he was a traveling companion of Paul's, as many Bible dictionaries still confidently assert, we should be able to rely on what he says about Paul's missionary work. Unfortunately for that theory, Paul's own letters present at several important points a very different picture. These disparities cannot be explained away. Luke may or may not have known what Paul was doing, but it is impossible to believe that Paul himself did not know. Hence at every point where Acts and Paul's genuine letters conflict, the student of the New Testament must be prepared to believe the letters. There is no rational alternative.

It is generally assumed that Luke wrote without knowledge of—or at least access to—Paul's letters, and so went ahead and pictured Paul's mission in the way that Luke himself wished it to be known, not foreseeing that by a historical accident the very evidence needed to dispute him would crop up later in the canonical writings of his own Church. Thus the disparities, if they are acknowledged to exist, could be explained as innocent mistakes. But perhaps we should not be too quick to assume that Luke was ignorant of Paul's letters, which could hardly have been well-kept secrets, any more than Paul's reputation was. It could be that Luke decided simply to ignore the disparities, trusting that his own work, as a formal and more or less "official" composition, would carry the day against anything as casual and ephemeral as a scattering of letters. (Remember that a canon of New Testament writings was still many decades in the future.) It could also be—and we shall look at this theory again—that Luke's work was deliberately intended to counteract the influence of Paul, which Luke was well informed about and did not entirely approve of.

In saying all this we are not attempting to belittle Acts or to deny its authority but only to determine as accurately as possible what it is an authority for. Certainly the New Testament would be a much poorer collection without it. Luke's calculated artistry, although it frustrates our search for what "really happened," is itself a fascinating object of study. And however much Acts may fall short of filling our need for a history of the early Church, it is the only document in existence that even attempts such a task.

GOVERNING CONCEPTS
IN ACTS

For the proper understanding of Acts, it is helpful to have in advance some sense of the concepts that govern and inform it. The reader will then know what to be looking for and can better appreciate its author's intentions. We have gathered these concepts under the following six headings:

1. *The preeminence of the Holy Spirit.* Consistently with what he has shown in his gospel, Luke believes that the Holy Spirit is in complete charge of the development of the nascent Church. Unless inspired and directed by it, humans take no initiative. Even the original group of apostles is strangely ignorant of the divine plan: Gathered in Jerusalem after the ascension of Jesus, they seem to learn nothing from the forty-day period of instruction recorded in Acts 1:3 (just as they learned nothing from the apostolic commissions recorded in Luke 9–10 or the specific explanation in Luke 24:46–49). They have to be led, step by step, to the realization that God all along intended to offer "the repentance that leads to life" (Acts 11:18) to Gentiles as well as Jews. Even after the Pentecostal experience, they seem to have no special insight into either tactics or goals.

2. *Movement outward from Judaism.* According to Luke, the Church grew by inching outward into the world, adding categories of converts in definite stages, each stage a further removal from its Jewish origins. The first converts are all Jews in Jerusalem, won over by Peter's sermons (Acts 2–3). In the next stage (Acts 8) Samaritans are added, neighbors to the immediate north of Judea whose Jewishness is not considered authentic but who are assuredly not Gentiles. Then, in a move closely guided by the Holy Spirit, conversion is offered to an Ethiopian, whose attachment to Judaism compensates for his foreignness. The last case is the hardest, that of the Roman centurion Cornelius (Acts 10). He, too, by sympathies and actions, is on the Jewish side; but his being both a Gentile and a member of the occupying army creates a problem of such magnitude that it has to be resolved by heavenly intervention (the vision of the "big sheet" full of animals). Once he and his household are baptized, the last theoretical obstacle to taking in Gentiles of whatever kind has been eliminated. Significantly, the resistance that has to be overcome so carefully is not

the resistance of the outside world but that of the apostles themselves. In this respect Luke's story is a documentation of Jewish prejudice against Gentiles, a prejudice that Luke himself does not share.

In Luke's view, it was ordained that salvation should be offered first to the Jews and only afterwards to the Gentiles. It is undoubtedly true that the earliest Christians were all Jews, that later on their numbers included more and more Gentiles, and that finally Jewish Christianity virtually disappeared, leaving a gentile Church. Luke does not show us the full extent of this process, which was far from complete in his own time, but its direction was already clear and it was one that he approved of. Jews, he believed, had forfeited their chance for salvation, which as children of the Covenant they were better entitled to than anyone else, and so had only themselves to blame. The bitter ending of Acts 28 makes this very clear. The Church will be a universal one.

3. *Authority of the Church.* For Luke the authority of the Church is of paramount importance. This authority comes directly from Christ, who personally commissioned the nucleus of Christian believers, the apostles, before his ascension into heaven. The site of the commissioning, at Olivet in Jerusalem, ratifies the continuity between the traditional Jewish faith and that of the new covenant. These believers are presented as a body "united, heart and soul" (Acts 4:32). When a potential source of dissension is found, as that between the Aramaic-speaking and the Greek-speaking believers (Acts 6), it is settled firmly by the original apostles acting in unison. (That there are now twelve instead of eleven resulted from their first official act, the election of Matthias in Acts 1). The seriousness of disobeying the rules of the Church is shown by the fate of Ananias and Sapphira (Acts 5). When the most divisive and troublesome issue of all arises—the demand of some zealous Jewish Christians that gentile converts be circumcised and be required to follow the Mosaic Law—the apostles and elders hold a solemn conference (Acts 15) in order to debate it. The debate determines that it is the will of God that Gentiles be accepted into the Church. But because the Church has never been given any specific rules for handling this situation, rules have to be created. Here James, the brother of Jesus, asserts his authority. Everything is orderly; there are no loose ends. Although Luke manages to create the impression that the Church's organization was informal and that decisions were made by common consent of its members following free and open debate, it is clear that he himself thinks of it as authoritarian in structure. In his view, power moved downward from above, not upward from below.

4. *Christians as good citizens.* It is most important to Luke to show

that Christians are not fanatics, troublemakers, or self-righteous ex-
clusivists. They obey all laws, civil and religious. They are decent,
peaceful, cheerful, self-reliant. In attempting to make converts, they
appeal to reason, and they are moved by a genuine desire to help
others. Quite apart from religion, they set an example for all to follow. Hence, initially at least, they are well thought of by the general
public (Acts 4:33, 5:13). The Jewish hostility that begins to build
against them is presented as being totally uncalled for, and it makes a
vivid contrast (a contrast that Luke wanted to make) with the attitudes
and behavior of the Christians themselves.

5. *The inevitable advance of the Church.* The growth of the Christian
movement as presented by Luke is an almost unbroken string of successes. The Church encounters opposition and occasional failure but
no real setbacks. Even periods of persecution (as in Acts 8:1) can be
the occasion for increasing the Church's influence. This point of view
accounts, at least in part, for the ending of Acts, which is definitely
upbeat despite the fact that Paul met his death at the hands of the
Romans (as Luke knew very well) and thus had his career tragically
cut off with much still remaining to be done. But to have said this
would have given the story a gloomy ending rather than a serene and
confident one; it also would have damaged Luke's picture of the Romans as fair and reasonable people. His villains are the Jews: It is no
accident that in Acts 28:25–28 Luke introduces, once again, the prophetic oracle from Isaiah that Christians interpreted as applying to the
Jews of their own day.

6. *Paul as servant of the Church.* Without question the dominant single character in Acts is Paul. From chapter 13 onward, it is essentially
his story. But the Paul we see in Acts is Luke's conception of Paul,
and how far this conception corresponds to the real, historical Paul is
a matter that demands thoughtful consideration. There are notorious
problems with the differing versions of events in Paul's career as given
by Luke and by Paul himself. For example, Luke has Paul visit Jerusalem shortly after his conversion and mingle with the other apostles
(Acts 9:26); but Paul himself states flatly that after his conversion he
went off at once to Arabia (probably to the Nabatean kingdom), then
returned to Damascus, and not until three years later went to Jerusalem, where he saw no Christians but Cephas and James (Gal. 1:16–
20). When the circumcision problem becomes acute (see the fourth
concept above), Paul and Barnabas are assigned by the Church to lead
a delegation to Jerusalem to have the matter settled (Acts 15:2). (By
Luke's chronology this would be Paul's third visit to Jerusalem; by
Paul's, his second.) The apostolic council described in Acts 15 must
be the event described by Paul in Galatians 2, inasmuch as Paul's

account also centers on the problem of circumcision, but nothing else in the two versions corresponds—least of all Paul's fierce assertion of his independence of the Jerusalem apostolate and Luke's smooth integration of Paul into the Church machinery.

The whole range of difference between Luke's conception of Paul's mission and teaching and Paul's own conception of them is too complicated to be presented here. But we can say that Luke's Paul, as he may be termed, is one thing that Paul's Paul was not: a loyal servant of the Jerusalem Church. The Paul that we know from his letters might have undergone the ritual of purification and put in an appearance at the Temple in order to make public display of his orthodoxy (Acts 21:23–26), but it is hard to believe that he would have done so to placate the circumcision faction within the Jerusalem Church, or indeed have taken any orders from that Church. Paul's writings constantly display his belief that as an apostle he is just as good as any of the others, since he too had seen the risen Christ and had received a direct commission to spread the faith (Gal. 1:15–16, 2:7–8, and 1 Cor. 9:1, for example).

What then is Luke trying to do in his portrait of Paul? He is trying to domesticate him, to pull him firmly within the orbit of the Church, to tone down his radical theology, to contain his freedom. For Luke, freedom of the kind that Paul represented was dangerous. Among other things it had permitted Paul to repudiate the Mosaic Law. It is not that Luke himself had any great personal attachment to things Jewish, but his sense of order and his reading of God's will as revealed in human history led him to emphasize continuity and authority, and he could not imagine a Christian faith separated from its Jewish roots, even though for him the old covenant had been effectively replaced by a new one. In this, of course, Luke is very much the prototype of the modern orthodox churchman, whereas Paul is the prototype of the brilliant nonconformist—a creative force within the tradition but always a threat to its stability.

The sources that Luke drew on for his portrait of Paul are not known (as, indeed, none of his sources for early Church history are known, although he must have had many). A critical approach to the reading of this portrait does not require one to discard everything in it, but the discovery that certain key features in the portrait are contradicted by Paul himself puts under suspicion a number of other details that are usually accepted without much question. These include the claim that Paul had a Jewish name, Saul; that he came from Tarsus; that he was a Roman citizen; that Gamaliel was his teacher; that he witnessed the martyrdom of Stephen; that his usual practice was to speak in synagogues; that he preached messianism;

that only after failing to convert Jews did he decide to turn to the Gentiles; that he circumcised Timothy; and that his conversion took place on the road to Damascus in the form described in Acts. Some or all of these details may be true, but Luke is our only authority for them.

Many of the disparities between Luke's account and Paul's are no more than what one might have expected from authors writing at different periods of time (in Luke's case, long after the events), for different purposes, and to different audiences. Nor should it be surprising that the two men brought their own individual points of view to these early events in the history of the Church and that they had no hesitation in committing these views to paper. All this is normal, and it causes no problems for readers of the Bible as literature. We have seen it happen already in the four gospel accounts of the career of Jesus. In the present case, the problem of disagreement between the two versions is more than compensated for by the opportunity we have to watch this unplanned encounter of two men of powerful intelligence, each totally sincere and each convinced that he has the most important message in the world to convey.

THE CANON OF LETTERS   The remainder of the New Testament canon after Acts consists of twenty-one so-called letters and the book of Revelation. Leaving aside Revelation (which we discussed in chapter 10) and Hebrews (which is not a letter but an anonymous theological essay), these works may be classified as: (1) the genuine letters of Paul, (2) letters supposedly by Paul but whose genuineness is disputed, (3) the "pastoral" letters, and (4) the "catholic" or "general" letters.

THE GENUINE LETTERS
   Romans
   1 Corinthians
   2 Corinthians (probably a composite of more than one letter)
   Galatians
   Philippians (probably a composite of three letters)
   1 Thessalonians
   Philemon
THE DISPUTED LETTERS
   2 Thessalonians (widely accepted as genuine, but in our view
      written by an imitator of Paul)
   Colossians (probably not by Paul)
   Ephesians (almost certainly not by Paul)

THE PASTORALS
    1 Timothy
    2 Timothy
    Titus
THE GENERAL LETTERS
    James
    1 Peter
    2 Peter
    1 John
    2 John
    3 John
    Jude

The three disputed letters are often combined with the three pastorals and labeled "deutero-Pauline," meaning that all six were written after Paul's death (during the late first and early second centuries c.e.) by followers of Paul who borrowed his name to lend authority to what they wrote, following the common practice of the times. The pastorals are so called because they take the point of view of someone advising leaders of Christian congregations on what to preach to them and what rules to enforce. Because it is clear that Paul did not write the pastorals, we shall not discuss them here, although in their own right they have a certain interest and value.

The "general" letters are a miscellaneous group. The term "catholic" or "general" means that they are addressed to the Church at large, not to specific congregations as Paul's letters were. Within this group the three books of John are exceptional in that all three are anonymous and that 1 John does not even make a pretense of being a letter (it is a tract combining theological instruction and exhortation). John's name became attached to these works because of characteristics of theology and language that they share with his gospel.

The letter of James is entirely exhortation on morals and conduct, and contains so little Christian doctrine that it has been suspected of being a Jewish wisdom treatise touched up by a later Christian author. Its emphasis on works (deeds) rather than faith has not made it popular with Protestant theologians (Luther called it "a letter of straw"). It is particularly noteworthy for the way it inverts Paul's interpretation of Genesis 15:6, the key statement of the faith of Abraham, in order to prove the superiority of deeds. The letter 1 Peter urges proper Christian conduct, joyful acceptance of martyrdom, and preparation for the coming end of time. The short letter of Jude is a denunciation of those who have infiltrated the Church, bringing false doctrine with them. Its author shows a knowledge of Jewish apocalyptic writing, especially of Enoch, from which he quotes. The letter 2 Peter takes the form of a final message from the historical Simon Peter to Christian believers, although it is not written by this apostle or even by the author of 1 Peter. It is of unusual interest because the author of 2 Peter knows the gospel account of the Transfiguration, borrows extensively from the book of Jude, and specifically refers to Paul's letters—evidence that a New Testament canon is now forming. For reasons such as these, 2 Peter is thought to have been written about 150 C.E. and thus to be the latest book of the Bible.

PAUL

And now—having saved the best for the last—what shall we say about Paul? Certainly he was a man of powerful mind and dominating personality, of vision and passionate conviction. Selfless in his dedication to his cause and tireless in its pursuit, he was a man without whom it is impossible to imagine the Christian religion today. We must remember that at the time of Paul's conversion experience, some three years after the Crucifixion, the writing of the first Christian gospel was still thirty-five years in the future. There was as yet no formal Christian scripture, no developed theology. Missionaries were already in the field, but more than merely gaining converts was required. There had to be a clear and effective sense of what it meant for someone to join the Christian community of faith, of who could join and under what conditions, and of what the relationship of Christianity to Judaism, the mother religion, had to be.

It was Paul's genius to see that Christianity could not survive if it were tied to the Jewish Law (the 613 separate commandments found in the Torah, with all their ramifications through the daily lives of believers) and thus to insist on a radical separation of the new religion from the old. He saw not only the practical difficulties caused by requiring gentile converts to submit to circumcision and obey other features of the Jewish Law but also the theoretical confusion that such a requirement would institutionalize, threatening the very definition of Christianity. So he challenged the so-called Judaizers directly, going straight to the heart of the issue. And most important of all, he provided something to put in place of the Law through his doctrine of justification by faith, a doctrine that he found not in the sayings of Jesus but in the Jewish scriptures themselves! How he did so we shall see in a moment.

A word or two of caution, however. We should be careful not to give Paul undue credit for the decline and eventual disappearance of Jewish Christianity, which very likely would have occurred anyway. Far more than from anything Paul wrote or did, Jewish Christianity suffered from the disaster that struck Jerusalem in 70 C.E. and destroyed its base of power. Even had this not happened, the continuing inflow of gentile converts (Paul was not the first or the only one to convert Gentiles) would have diluted the power of the Jewish faction and brought this unstable union of the old and the new revelation to some kind of crisis point. Paul's role looks large because his writings, preserved and made canonical, are there for all to see. In addition, we should not underestimate the role of circumstances in shaping Paul's views. A great deal of what he wrote was ad hoc, in reaction to various crises in the Church, and there is no doubt that opposition to his beliefs forced him to examine them more closely and helped him to perfect his distinctive theology.

THE EARLY CHRISTIAN
CONGREGATIONS

The congregations that Paul wrote to were all in cities. Cosmopolitan and Greek-speaking, they were worlds removed from the humble Aramaic-speaking peasants and tradespeople who heard the message from Jesus himself. Organized Christianity from the start was an urban movement. Christian congregations met in private homes (see Philemon 2 and Romans 16:5)—doubtless in those of their more affluent members, where there was enough space for such gatherings. The meetings would be social as well as ceremonial, without a fixed order of proceeding. There would, however, be standard elements: the eucharistic meal (or Lord's Supper), prayer, singing of hymns, preaching by members of the congregation or by visiting evangelists (like the Jewish synagogue, the Christian congregation was run by laypeople), personal testimony, perhaps healing or exorcism, and ecstatic prophecy through glossolalia (speaking in tongues). The meeting would be in the evening, whether on Saturday or Sunday evening is not clear, but in either case not on the Sabbath.*

---

*By Jewish reckoning, the Sabbath ended at sundown Saturday; by the Roman system (which we still use), Sunday lasts until the midnight after Saturday midnight. Probably Jewish Christians continued to observe the Sabbath, but with special added ceremonies to commemorate the Resurrection on the following day (early Sunday morning). Gentile Christians would have had no reason to observe the Sabbath except out of consideration for the Jewish converts. Sunday became a day of rest only in the fourth century by an edict of the emperor Constantine. The confusion of Sunday with the Sabbath is a modern one.

By and large, Christianity in the Roman empire began as a religion of people who were not strongly attached to the status quo. Some were outsiders and misfits; others were persons with ambiguous social positions—perhaps a freedman who had accumulated a lot of money as a trader but was discriminated against and held down because of his origin, perhaps one of the "God-fearers" (gentile sympathizers with Judaism) who might have taken the plunge and changed allegiance had it not been for circumcision, perhaps an immigrant who had no loyalty to the local cult or to the official state religion, or perhaps a talented and ambitious woman who found no satisfactory outlet in existing social institutions. These were people looking for more rewarding commitments, receptive to new outlooks and to new social alignments.

In many respects this Christian population was a good cross section of society, including all but the very top and bottom; Hence existing disparities in economic status were carried over into the Church. There they caused some trouble, especially on the occasion of the eucharistic meal, which was a true meal and involved the sharing of food brought to the meeting place (see 1 Corinthians 11). The very fact that Paul complained about such abuses shows that the Christian assembly was supposed to be a place where fellowship prevailed over differences of rank. What might have seemed strange to the pagan observer was not the Lord's Supper as such—for a sacred meal held at stated times and shared by initiates was a common feature of cults and associations in the Greco-Roman period—but rather the heterogeneous group that gathered to eat it. Table fellowship was a sensitive matter (see Galatians 2:12). Sitting down and eating in harmony with someone of a different class or race or religious background meant deliberately challenging the very structures that gave life meaning.*

This pagan observer, had he pursued his investigation, would also have been amazed at the intensity of the religious tie binding members to one another and at the way this new association dominated their lives. The ancient world was full of clubs and guilds and societies of this and that, but very few were truly democratic and inclusive, and none of them had anything like the shared emotional fervor of the Christian meeting.

---

*The social composition of early Christianity explains why Paul wrote in Greek to the congregation at Rome. Although they lived in Rome, they were not Romans: They were human flotsam thrown up in Rome from the farther reaches of the Mediterranean world, which retained its hellenistic culture long after political conquest by the Roman empire.

THE LETTER

To understand Paul's letters properly, we need to know something about the practice of letterwriting in the ancient world—the context out of which they come to us. Certainly the writing of letters was a common activity in those days: A steady stream of correspondence went by land and sea between all centers of population. Much of this mail was governmental, carried by official couriers, but much was commercial and personal, between private citizens: merchants ordering goods, sending instructions, arranging delivery, paying bills; children writing home to parents; homesick soldiers writing to wives; ordinary persons inquiring after the health of friends, promising visits, requesting favors. Those who could not read and write could hire scribes. Because there were no postal systems to carry private correspondence, letters were given to friends traveling to the desired location, or to trustworthy strangers who would be paid for their trouble and given specific instructions on where to deliver the letter, there being no street addresses either. Under these circumstances, successful communication required luck and a certain amount of persistence. Letters were written on almost anything that would bear writing (for example, pieces of broken pottery); but the standard material was the sheet of papyrus, which was written on one side, folded over to make an oblong parcel, tied, and fastened with a clay seal.* Because papyrus survives indefinitely in a very dry climate such as Egypt's, a great many of these ancient letters, in whole or in part, still exist. They offer modern scholars valuable insight into the way everyday life was conducted in the hellenistic world and are of great help in understanding the New Testament.

The hellenistic letterwriting tradition called for certain stereotyped forms. There would be a salutation ("A to B, greetings") and a wish for the health of the recipient; in the body of the letter there would be much conventional language that sounds stilted to our ears; and at the end there would be a farewell formula (rarely a signature). The literary polish of these letters was normally not very high, but ambitious writers who had studied the art of letterwriting in school could draw on a considerable body of rhetoric to help them.

Paul's letters are part of this tradition. They are genuine letters, not "epistles" (an epistle is an artificial or make-believe letter written

*Letters like those of Paul, which are quite unusually long, would have needed several sheets of papyrus glued end to end to make a small scroll. The charming letter to Philemon is the only one of Paul's that much resembles, in length and in content, the usual first-century papyrus letter.

for publication rather than for mailing). They are written in the standard Koine Greek. But they are more elaborate than most other letters, and they show some characteristic modifications of the stereotypes. In particular, Paul chose his own form of salutation, dropping the standard *chairein* ("greetings") and substituting *charis humin kai eirēnē* ("grace to you and peace")—a deliberate change, with theological overtones, since it combines the Christian idea of grace with the Hebrew greeting *shalom* ("peace"). In addition, most of the letters begin with a formulaic phrase that Paul used to authenticate his role as an apostle of Christ. Most of them also include a thanksgiving formula—not for the good health of the recipients but for the saving gift of faith in Christ that they possess. In some cases the ending includes greetings from Paul and others to various specific persons known to be in the group addressed (twenty-six are named in Romans), and all the letters end with a benediction. There is no mistaking the fact that Paul had turned this common vehicle of communication in the hellenistic world to a new use and, one might also say, brought it into a new dimension.

The body of the Pauline letter frequently gives the impression of hasty, impromptu composition, as though Paul were pacing the floor and dictating as fast as his secretary could take the words down, hardly able to restrain the torrent of his thoughts. Undoubtedly, there are spontaneous elements in the letters, but detailed analysis of their structure brings to light many features suggesting premeditation and at least a certain amount of care in expression. The actual letter sent would have been in the hand of a secretary, but it may well be that Paul worked out a draft or drafts beforehand by himself. In any case we know that they were not written for publication in a book, where they now stand. Paul would no doubt be astonished if he could be brought back to see what has happened to these letters that were all written for immediate occasions and specific uses, with no thought of creating scripture for the ages. Their use at the time is clear enough: They were meant to be read aloud to the assembled congregation that Paul addressed. The letters did what Paul himself would have done had he been there in person. They are a substitute for Paul and thus are really a by-product of his missionary activity.

We may be sure that a congregation that received letters from Paul not only read them aloud in meetings but also kept and treasured them. There is no evidence to suggest that the letters circulated beyond the original recipients until someone, after Paul's death, thought of assembling them and publishing them as a collection. By that time, inevitably, a number of letters were either unavailable or missing.

(We cannot believe that in all his career Paul wrote no more than this handful.) So great was Paul's prestige that imitators quickly followed him, borrowing the form of the Pauline letter and sometimes Paul's name. Thus Paul was responsible—quite unintentionally—for the introduction of a biblical genre, the letter, just as Mark was responsible for introducing the frequently imitated genre of the gospel.

The New Testament canon does not make a distinction between the genuine and the disputed letters of Paul as we do here. Instead, it simply gathers all the letters attributed to Paul into two groups, letters to churches and letters to individuals, and arranges them within each group in order of length. A proper chronological order of the letters would be much more helpful, but the dating of the letters, either absolutely or relative to one another, is a difficult problem. Because they bear no dates, all the evidence has to be inferred from their contents. There is good reason to call Philemon the last of Paul's letters, and 1 Thessalonians may well be the first. Romans would come near the end, probably just before Philemon. They all seem to have been written within the decade of the 50s—although this point too is somewhat in dispute. In any case we are reminded, once again, how soon after the Crucifixion Christian missionary activity began and by how many years it preceded the writing of the gospels.

KEY FEATURES OF
PAUL'S THINKING

It is impossible to do justice to Paul's letters one by one in the space remaining here; on the other hand, to lump them together for some kind of summary treatment would be most unfair to them as documents with distinctive individual characteristics. The texture of Paul's writing, the immediacy and the drama of his grappling with issues, must in any case be appreciated through a firsthand reading of his text and cannot be replaced by a commentator's detached analysis. What we can do, however, is call the reader's attention to certain key features of Paul's thought and belief as they are manifested in his writing taken as a whole. The following discussion will help readers orient themselves within the Pauline universe, as it might be called, and come to a proper understanding of Paul's achievement.

1. *Expectant waiting.* No belief is more characteristic of Paul or more fundamental to everything he wrote and did than his belief that the world was very soon coming to an end. It is the absolute foundation of his thought. A reader who misses this point can understand very little else about Paul. As far as anyone can tell, Paul's eschatol-

ogy—to use the convenient technical term—derives not from any specific prophecy or saying of Jesus but from Paul's own reading of the significance of the Incarnation of God's Son in Jesus of Nazareth: Humankind was being offered a brief but sufficient chance to reconcile itself with God before a final judgment and the end of human history. (This reading depended also on the Jewish apocalyptic tradition that Paul inherited and was thoroughly familiar with.) It never occurred to Paul that the *parousia* or Second Coming ("the Day of Jesus Christ" [Phil. 1:6]) might be postponed beyond his own age into the distant, hardly imaginable future—as the Church eventually had to acknowledge. He died before the issue ever came up. One aspect of the delay, however, did show itself and require his attention: the fact that Christian believers were now beginning to die and be buried in the normal fashion. What did it mean that some of the faithful would be alive at the last trumpet call and some not? Paul addresses these questions in 1 Corinthians 15 and 1 Thessalonians 4.

Paul's sense of the coming end, which he and the Christians contemporary with him awaited expectantly, gave special urgency to his writing and his missionary efforts. For all he knew, at any moment this work might be abruptly halted by the return of Christ in triumph. This eschatology also powerfully supported his ethical teaching; Paul never viewed human conduct abstractly, divorced from the present historical situation. The burning question, given that we are on the stage of history near the end of the final act, is how should Christians behave? What is appropriate *now*? Issues that might have been important under different circumstances—such as whether to marry or whether to take a new job—fade into insignificance.

2. *Christology*. The four gospels have much to say about the ministry of Jesus but nothing to say about Christianity. They are narratives only: They present the story but do not pause to reflect on it. They seem to assume that the titles given to Jesus and the claims made on his behalf are self-explanatory, for at no point is this scattered material brought together and explored for its meaning. Some of the details in the story are surely more important than others, but which are they and why are they more important? The gospels do not say. Moreover, when the gospels reach the end of the period they cover, they stop and do not look beyond it. (The postresurrection appearances of Jesus in three of the gospels are epilogues to the story already told rather than prologues to the future.) All this accounts for the frustration that readers commonly feel when they finish reading the gospels and look for some way to connect the story with their own lives. They are standing on the outside looking in; there is no hope

for a time machine to whisk them back into the first century c.e. If
the story is meant to apply to persons living now, exactly how does
it do so?

What the gospels lack is a "Christology"—an organized, rational,
and comprehensive theory about the meaning of the Jesus story for
human beings at all times and everywhere. This need was anticipated
by Paul, working as a missionary years before the gospels were written, who found, doubtless by hard experience, that if people are to
be converted to belief in Christ, they must be told specifically what it
is that they are to believe and why they should believe it.

On their part, Paul's writings conspicuously lack what the gospels
have in abundance: an interest in the life and teaching of Jesus. In the
genuine letters there are only four references to anything that Jesus
taught (Rom. 14:14; 1 Cor. 7:10, 9:14, and 11:23–26) and the only
episode in Jesus' life that Paul pays any attention to is the Crucifixion—indeed, for Paul (who resembles Mark in this respect) it is the
one event that really matters. Of course Paul had not known Jesus; as
he tells it, his contact was with the risen Christ. But we cannot be
sure how far Paul's neglect of the life and teachings of Jesus was due
to circumstances—that is, to his ignorance about these matters—and
how far it was due to deliberate choice. It is hard to believe that Paul
was so far out of touch with Church traditions that almost none of
the material later incorporated into the gospels came to his notice.
More probably, he omitted from his own teaching what it did not
need and focused instead on what it needed.

Paul's teaching is "Christocentric" in the highest degree: centered
on the figure of Christ. In Paul's writings "Christ" becomes a proper
name instead of a title. At the same time Paul treats it so as to make it
almost a metaphysical abstraction. Believers are "in" Christ (a favorite
expression); Christ is "in" them; they are "one" with Christ and even
possess "the mind of Christ" (1 Cor. 2:16). The title that Paul chose
to add to the name "Christ" was "Lord" (Greek *kurios*). Paul did not,
of course, invent the word *kurios*, which had been used for a long time
to refer to human masters, especially kings, and which had also been
used by the Septuagint translators to render the Hebrew *yahweh*. But
Paul's use of *kurios* is quite specific and distinct; it is not at all meant
to suggest that Christ is like an earthly king (except in demanding
complete and unquestioning loyalty) nor, on the other hand, that
Christ is God. The title does not assimilate Christ into any category
other than the single one he occupies. It is so pregnant with meaning
that the simple phrase "Jesus is Lord" (Rom. 10:9, 1 Cor. 12:3), if
properly understood, conveys the essence of Christian faith. Paul also

has much to say about the "Spirit" (Greek *pneuma*), the agent through which God works on human beings. The Spirit transcends time, space, and all the boundaries separating humans from the divine and from one another; it communicates directly to the inmost center of our being; and it operates through us or its own purposes. Paul uses the concept of the Spirit in a fluid and impromptu way, not bothering to differentiate clearly between the Holy Spirit and God himself or to specify the relationship of the Spirit to Jesus Christ; thus it is wise for us not to press him for definite answers on such points.

3. *Justification by faith.* Paul's cardinal doctrine is justification by faith. It is his way of simultaneously (a) fitting the Christian revelation into the world picture that he inherited from his own Judaism and (b) resisting the effort of that Judaism to swallow up and negate what is unique in Christianity. Paul had to come to terms somehow with the Law of Moses. (The Septuagint regularly translated the Hebrew word *torah* as *nomos* [the Greek for "law"], thus emphasizing its legalistic aspect; it is this aspect that is usually uppermost in Paul's mind.) As a well-educated and zealous Jew, Paul was intimately familiar with the Law. As a Christian, however, he was convinced that the Law no longer applied. Paul could not repudiate the Jewish scriptures, believing as he did that they were in fact the true record of God's relationship with human beings. But Moses, the author of the Law, with all the immense authority that his name had acquired by Paul's time, appeared to stand in the way of the new covenant. What Paul did, then, was to skip over Moses and go back to Abraham. Abraham had received a covenant from God before Moses ever existed. Abraham was not circumcised when God chose him—hence, Paul reasons, this key requirement of the ritual Law (certainly a key requirement in Paul's day) must have been only a temporary measure and is not necessarily valid for all time. As circumcision for Paul symbolizes the whole Law, Abraham symbolizes all humankind. Even the Gentiles, who are not biologically descended from Abraham, are children of Abraham in this special sense (Rom. 4:11, Gal. 3:7). And why did God choose Abraham? Because Abraham had faith in God (Rom. 4:3 and Gal. 3:6, quoting Gen. 15:6). Was Abraham a wholly righteous man, blameless of transgression? The question cannot be answered now, says Paul, nor could it have been answered then, because in the absence of Law there was as yet no way to reckon sins. The machinery did not exist. But this did not matter, because Abraham's faith supplied whatever he needed to render him wholly fit for the covenant he was given. His faith acquitted him. He was "justified" (made or pronounced just) by his faith. Justification by faith operated outside

the Law for Abraham, Paul argues, and it will do the same for all who believe in Jesus Christ.

The break with Judaism here is absolute. In Judaism the sinner identifies his sins according to a prescriptive code, repents, and performs the ritual specified to bring about his reconciliation with God, his atonement. Not only is the process complicated, it is endless: There is no way to wipe the slate clean and keep it clean. In Pauline Christianity, on the other hand, the chance of an absolute break with one's sinful nature is offered by the free grace of God. Human action does not figure in the process at all. The initiative was God's, who, following the plan he had in mind all along (but which is only now revealed to us), sent his Son to provide the means of reconciliation. One has only to accept, to take the opportunity offered.

Given this interpretation of the story of Christ, Paul was quite correct in believing that there was no way to combine the Mosaic Law and the Christian revelation. Grace is not grace if it has to be earned; there is no point in working for something if it is offered to you free of charge. We do not know to what extent Paul reasoned all this out in advance of its application in his missionary work, but we do know that he used it in his struggles against the Judaizers. It is likely that these opponents within the Church, by forcing the issue, caused Paul to define and sharpen his own position in a way that he would not have done otherwise. (Let us remind ourselves that Paul's opponents, as far as the letters indicate, were all Christians. It was the Christians, not the Jews or the Gentiles, who obstructed his work and aroused his resentment.)

4. *Christian conduct.* Paul preached that the person who believes in Christ's sacrificial death on the cross for humankind is saved, freed from servitude to sin, guaranteed a place in the heavenly kingdom that will follow the return of Christ and the Last Judgment. Meanwhile, what? Christians still have to live in the world, mingle with others, carry on daily affairs—in other words, encounter the same temptations and opportunities to sin that they faced before. Does their salvation mean that they cannot sin any more? Does it mean that anything they now do is all right? Is God's forgiveness like an infinite bank account that one can draw upon every day as needed, without worrying that it will ever become empty? These are very real questions, and in Romans and 1 Corinthians Paul wrestled with them. Apparently some Christian converts that he knew interpreted their faith as a license to indulge themselves, and many Christians (as one might expect) behaved after their conversion pretty much as they had before it. Paul's response was to condemn such conduct, but, of

course, he could not do so any longer by appealing to the Mosaic Law. What he did, in effect, was to denounce this conduct as unchristian. Christians just don't *do* these things. It is unseemly, inappropriate to behave thus. Paul takes it for granted that fornication, idolatry, backbiting, drunkenness, swindling, and the like are wrong per se, and he makes no effort to explain why. Nor does he provide any clear answer to the question of whether Christians can unsave themselves by misconduct. Strongly coloring all Paul's thinking, as we have said, is his conviction that the world is shortly to come to an end, and this gives a special urgency to his denunciation of such practices. It is as though the people of his time were passengers aboard a sinking ship, fighting among themselves over the division of food rations and hatching petty conspiracies to take over direction of the voyage, even as the waves lap closer and closer to the gunwales and certain death prepares to swallow them all.

Paul has been accused of being an antinomian, one who opposes or rejects moral law, and apparently this criticism was already current in his own day. Obviously this is not so: Paul's distress at any kind of misconduct is eloquently displayed in his writings, and he spends a great deal of his time giving advice on behavior to his audiences. The problem is not whether to oppose misconduct but on what grounds, and it is here that Paul has much trouble. The seeds of antinomianism are present in Paul's thought, if undeveloped. It is significant that the Christian Church had to counter this tendency by instituting a series of cultic requirements, the sacraments, and working out a system of rewards and punishments based on specific evaluations of human conduct. The effort is already apparent in some of the post-Pauline writings, especially in the gospel of Matthew, which emphasizes Christ the lawgiver, the second Moses. Pauline freedom remained dormant until revived by Martin Luther fifteen hundred years later.

5. *Paul's Jewishness.* In spite of his radical break with Judaism and his continuing problems with the Judaizers, Paul remained a Jew himself and never escaped the influence of his Jewish heritage. He was proud of that heritage and his own place within it (see Philippians 3:5–7). This attachment to Judaism was personal and sentimental; but it was also a matter of intellectual conviction, for Paul believed that the Jewish scriptures were inspired and that God had indeed chosen the Jews to be his own people. At the proper moment in the working out of God's plan for them, they had been given the Law by Moses. Until the time of Christ, this law had been all-sufficient; obedience to it was required by the terms of the Covenant. But—and here Paul

deviated sharply from Jewish tradition—the Mosaic Law was meant all along to be only an interim arrangement before the next phase of God's plan unfolded with the coming of Jesus Christ. For those who believe in Christ, the Law is completely invalid.

Paul did not consider his position on the Law to be a repudiation of the Jewish past. As far as he was concerned, Jewish history continued through the lifetime of Jesus as a seamless whole, directed as always by God. More than that, the earlier stages of Jewish history carried prophetic meaning for the later ones, even to the point where some of the events in the past seem to have been deliberately arranged to provide lessons for the present. This is the sense of Paul's interpretation of the Exodus story in 1 Corinthians 10, an interpretation that does much to undermine the reality of those events and reduce the actors in them to figures in a charade. But if Paul was aware of this tendency in his view of Jewish history, he shows no signs of it.

If Paul was so Jewish, why did he abandon his people to become a missionary to the Gentiles? Luke's answer (which we have already had occasion to question) is that Paul tried first with the Jews and failed. But we find no evidence of this experience in Paul's writings, and in the famous opening chapters of Galatians he presents himself as having been intended from the very start to preach to the Gentiles. Contradicting this is Paul's statement in 1 Corinthians 9 that he went so far in his efforts to win over Jews that he even accommodated himself to the demands of the Law for their sake. There appear to have been Jewish Christians in all the congregations that Paul wrote to (except perhaps the Thessalonians), and the arguments in Paul's letters draw constantly and heavily on the Jewish scriptures in a way that shows that both he and his audiences accepted the authority of those scriptures. We have here a historical puzzle that yields no clear answer.

In any case, in forming our estimate of Paul it is important to remember that there was no such thing as orthodox, rabbinic Judaism in Paul's time. Judaism contained many parties, schools of thought, and varieties of observance. Before his conversion Paul was a Pharisee, which means that he was exceptionally devoted to observing the Law in all its aspects but also that he interpreted the Law freely and was sympathetic to such new doctrines as that of bodily resurrection. Moreover, Paul was not an Aramaic-speaking Palestinian Jew but a Greek-speaking Jew of the Diaspora—born and brought up outside Palestine and deeply influenced by hellenistic culture—for whom the Septuagint was unquestionably valid as the text of the holy scriptures. His use of these scriptures raises the eyebrows of modern students, who see him lifting passages out of context or combining passages from disparate sources, laying special significance on arbitrarily chosen words, ignoring the original author's intention, and, above all, bringing in allegorical or typological explanations when it suits his purpose to do so. Yet this practice is perhaps the most Jewish, the most traditional, aspect of Paul! To have been really revolutionary, Paul would have had to do the impossible: insist on interpreting the scriptural text according to the apparent intentions of its authors in the historical contexts when it was written. No one in Paul's time was capable of even imagining the need to do so.

Paul simply does not fit into the categories of religious faith that have been worked out since his time. He was both a radical and a traditionalist; a man of surprising narrowness and equally surprising breadth; a thoroughgoing Jew and a cosmopolite moving comfortably within the social structure of the hellenized Roman empire; a man who had ecstatic communication with the world of spirit and spoke in tongues, although he chiefly valued decorum and plain talk; a man who subordinated his whole life to the discipline of a religious vocation but who could never take orders from anyone; a colleague and friend of many exceptional women, who nevertheless endorsed keeping women to strictly defined traditional roles within marriage and the Church; a man who worked for a living and paid his bills and kept his appointments all the while he yearned for and anticipated the imminent end of the world.

If it is hard to imagine what Christianity would have been without Paul, it is not at all hard to imagine what the New Testament would have been without him: It would not have existed, at least not in a form that we would recognize. Qualitatively, Paul's writing stands with the best that we find in both the Old Testament and the New Testament: with some of the psalms, with Job, Ecclesiastes, Ruth, the history of the reign of David, the Joseph story, the oracles of Isaiah and Jeremiah, and other works that make the literary study of the Bible so rewarding. But it does not resemble any of them; it is starkly and radically original. And Paul is not a shadowy, unknown figure to be hypothesized from scattered clues within the text that he authored, or a Peter, identifiable and real but so encrusted with legend that the man himself will never be known. If Paul sometimes seems larger than life, it is because—believe him or disbelieve him, admire him or hate him—he was a genius, the kind of person whom we instinctively measure ourselves against, who draws us into the sphere of his power whether we will it or not, and who holds up before us the living evidence of what the human spirit can accomplish.

SUGGESTED FURTHER READING

Paul J. Achtemeier, *The Quest for Unity in the New Testament Church: A Study in Paul and Acts* (Philadelphia: Fortress Press, 1987).

David E. Aune, *The New Testament in Its Literary Environment*. Library of Early Christianity, vol. 8 (Philadelphia: Westminster Press, 1987).

William G. Doty, *Letters in Primitive Christianity* (Philadelphia: Fortress Press, 1973).

Martin Hengel, *Acts and the History of Earliest Christianity*, trans. John Bowden (London: SCM Press, 1979).

Martin Hengel, *The Pre-Christian Paul*, trans. John Bowden (London: SCM Press, 1990).

Craig C. Hill, *Hellenists and Hebrews: Reappraising Division Within the Earliest Church* (Minneapolis: Fortress Press, 1992).

John Clayton Lentz, *Luke's Portrait of Paul* (New York: Cambridge University Press, 1993).

Gary Ludemann, *Opposition to Paul in Jewish Christianity*, trans. M. Eugene Boring (Minneapolis: Fortress Press, 1989).

E. P. Sanders, *Paul* (New York: Oxford University Press, 1991).

Alan F. Segal, *Paul the Convert: The Apostolate and Apostasy of Saul the Pharisee* (New Haven: Yale University Press, 1990).

Bruce W. Winter and Andrew D. Clarke, eds., *The Book of Acts in Its Ancient Literary Setting* (Grand Rapids, Mich.: Eerdmans, 1993).

The Interpreter's Dictionary of the Bible, ed. George A. Buttrick et al. (Nashville: Abingdon Press, 1962). See articles on Christology in the NT; Letter; Literature, Early Christian. Supplement, 1976: See articles on Acts of the Apostles, Paul the Apostle.

The Anchor Bible Dictionary, ed. David Noel Freedman et al. (New York: Doubleday, 1992). See article on Paul.

XV

The Text of the Bible

The Bible that we read in English is a translation of works originally composed in Hebrew and Greek. In chapter 16 we will deal with the problems inherent in the translation process. But before going on to the process of translation, we must give some consideration to what is being translated. Crucial questions immediately come to mind. How, for example, have the ancient biblical texts been transmitted to our time? In what forms do they now exist? To what extent can we trust those forms to embody the actual words of the original authors?

Three points should be made at the start. (1) Although one can buy printed Hebrew Bibles and Greek New Testaments, each having the appearance of authority, those printed books are artificial entities, created in relatively modern times from the evidence in ancient manuscripts. (2) The surviving manuscripts are themselves simply copies of other manuscripts, going back in time to the *autographa*—the very texts as written by the original author's hand—all of which disappeared long ago. (3) While one hopes that the text of a given biblical manuscript will represent exactly what its author wrote, there cannot be any guarantee that this is so. The very act of repeated copying introduces alterations—the longer the chain, the greater the likelihood—and in addition to these unconscious changes, there will be those made deliberately by copyists attempting to "correct" or "improve" the original.

Given these facts, the task of biblical textual scholarship is not easy. Its theoretical goal is to weed out alterations of any kind and thus to arrive back at the original texts as the authors wrote them. But that goal will never be reached because there will never be enough evidence to accomplish it; indeed, the very notion of a *single* original text may itself be an illusion. The more realistic goals that biblical scholars settle for are (1) to determine which are the earliest among the variant readings that ancient manuscripts provide, and (2) to understand the development of biblical texts during the history of their transmission. This chapter is concerned with the evidence available to biblical textual scholars as they work toward those goals. We shall start with the later history of the New Testament text and proceed backward in time to its early centuries, and then repeat that pattern for the Hebrew Bible.

NEW TESTAMENT
MANUSCRIPTS FROM THE
FOURTH TO THE
SIXTEENTH CENTURIES

Handwritten Bibles or portions of Bibles were the commonest books produced in the centuries before the invention of printing. These, of course, were always copies of other, prior, copies. We now have more than 5,350 different manuscripts of the Greek New Testament that survive from this copying activity. These include (1) almost one hundred papyrus fragments from the third to the eighth centuries; (2) about 270 "uncials," manuscripts written in large, rounded letters (fourth century onward); (3) about 2,800 minuscules, manuscripts written in small, cursive letters (mostly late); and (4) about 2,200 lectionaries, compositions intended to be read during church services, that included words from the Bible (also mostly late).* Of all these manuscripts, only about sixty are (or at one time were) complete New Testaments. The earliest complete manuscript is from the fourth century. Remarkably, no two of these manuscripts are identical in wording! To be sure, the great majority of the differences are minor, involving such matters as spelling or grammar; but many are major, affecting the sense of entire passages. How is it possible for textual scholars to find their way, given so much diversity in the manuscripts with which they must work? Modern scholars have

*In addition to the Greek manuscripts, the scholar can also consult some ten thousand additional manuscripts that provide translations of the Greek New Testament into other languages and thousands of New Testament quotations in the works of early Christian writers.

attempted to bring some sort of order to this bewildering number of texts by dividing the manuscripts into "families."

The best known and still most widely used system is that of the great nineteenth-century scholars B. F. Westcott and F. J. A. Hort, who assigned the manuscripts to four main families, associated with centers of Christian learning: (1) the Byzantine (sometimes called the Antiochian) family, which encompasses most of the Greek manuscripts but unfortunately is now accorded the least authority because most of its manuscripts are very late; (2) the Western family, believed to issue from a third or fourth century "exemplar" (the first in a chain of copies), perhaps in Palestine; (3) the Alexandrian family, believed to be based on an early Egyptian text; and (4) the Neutral family, a term Westcott and Hort used to cover those manuscripts containing readings not associated with any of the other families. Today, scholars speak of a fifth family, the so-called Caesarean text, issuing from an exemplar from Caesarea. Other systems have been devised since the time of Westcott and Hort, such as that of Hermann von Soden or the more recent classifications of Kurt and Barbara Aland, which describe the manuscripts somewhat differently; but they all attempt to bring this mass of texts into a coherent framework.

ERASMUS' GREEK NEW TESTAMENT

The first Greek New Testament was published by the great humanist Desiderius Erasmus in 1516.* Unlike modern scholars, Erasmus had available to him only New Testament manuscripts from the twelfth century and later, and so in a sense circumstances had simplified the task for him. But he had a major problem: reconciling his Greek manuscripts with the text of the Latin Bible, which had been the only Bible known in Europe for over a thousand years. Quite often when the Latin and Greek did not agree, as in the last six verses of Revelation, he simply translated the Latin text into his own Greek. On at least one occasion, however, Erasmus ignored the pressure of tradition (which assumed the accuracy of the Latin Bible). In his Greek manuscripts there was no original for 1 John 5:7 as it stood in the Vulgate: "And there are three that bear record in heaven, the Father, the Word, and the Holy Ghost; and these three are one." This trinitarian formula was extremely important to the Church, and its omission by Erasmus caused an uproar.

*A Greek New Testament was printed in Spain in 1514, two years before that of Erasmus; but it was not "published"—that is, circulated—until 1522.

As a result, he agreed that he would include it in further editions if he could find even one Greek manuscript that contained the disputed words. He carried out his part of the bargain after having been shown a Greek manuscript that did so—one from his own time! Because of his prestige, this late reading was adopted in sixteenth- and seventeenth-century English Bibles, including the King James Version, where it remained until comparatively recent years.

Erasmus' printed Greek New Testament became known after the edition of 1633 as the Textus Receptus ("the received [or authoritative] text"). Unfortunately, the manuscripts he used were all representative of the Byzantine family described earlier and are accorded practically no status now. It is principally because this "Received Text" was based on later, inferior manuscripts that New Testament textual scholarship since the nineteenth century has had such a negative flavor. That is, modern textual critics are not only engaged in recovering the oldest readings but also in justifying the removal of words and passages that have long been a part of church tradition. There are many passages in the traditional Greek New Testament that all will agree are later additions to the text but are nonetheless a part of the canonically received literature, either because of the preeminence of the Latin Vulgate for so long in Catholic tradition or the Textus Receptus in Protestant tradition. (Conspicuous among these passages is the well-known ending of the Lord's Prayer in Matthew—"For yours is the kingdom and power and glory forever. Amen"—which is missing in Luke.) In the most recent and now standard edition of the Greek New Testament, edited by Kurt Aland and others, doubtful passages are assigned to the margins or printed within brackets. It then becomes the responsibility of the English translators to choose among the textual variants and inform their readers of the alternative possibilities.

NEW TESTAMENT UNCIAL MANUSCRIPTS

Today we assign the greatest authority to the uncial manuscripts from the fourth and fifth centuries. These are distinguished not only by their use of large letters but also by the material on which they were written, parchment rather than paper. They are all codices, that is, pages bound in book form rather than scrolls. The most valuable of these for textual purposes are Codex Vaticanus, Codex Sinaiticus, Codex Alexandrinus, and Codex Cantabrigiensis. Since these are now the indispensable tools of New Testament textual criticism, we shall describe them briefly.

Codex Vaticanus, so named because it is in the Vatican library, originally contained the entire Bible in Greek with the exception of the Prayer of Manasseh and Maccabees. It seems to have been completed some time around the middle of the fourth century C.E., possibly in response to the emperor Constantine's order that fifty manuscripts of the Bible be prepared on parchment for the new churches of Constantinople. Over the centuries, several leaves of the manuscript have disappeared, particularly at the beginning and end (the most vulnerable parts of any book). The New Testament portion is missing part of the book of Hebrews, the Pastorals, Philemon, and Revelation. Scholars had known of the existence of this codex since the fifteenth century, but very few—and certainly no Protestants—had been allowed to study it. Erasmus, for example, knew that the manuscript existed, but he had no direct access to it. Consequently, it played no role in the making of the printed Greek New Testament and the English translations that followed. Finally, in the latter part of the nineteenth century, after hundreds of years of having been furtively copied and randomly collated with existing texts, it was published in a photographic edition.

Codex Sinaiticus is the other great fourth-century uncial. It was discovered by the German scholar Constantine Tischendorf at St. Catherine's Monastery (located at the traditional site of Mount Sinai) and retrieved piecemeal during a series of visits he made there between 1844 and 1859. As recently as 1975, scholars were still finding leaves belonging to this volume in the monastery. It too originally contained the entire Greek Bible. Unfortunately, the Old Testament portion is damaged and lacking many leaves, but the New Testament is intact and includes books not found in the canon today, such as the Epistle of Barnabas and the Shepherd of Hermas. It may have contained even more books, because several leaves are missing following Barnabas. Among the many interesting features of this manuscript is the pervasive presence of a "corrector's" hand—actually several hands over many centuries. Textual scholars will now refer to "the first or original hand" or to the "corrector's hand" as a means of distinguishing the authority for various readings in this codex.

Codex Alexandrinus was the first of the great uncials to become known to Western scholars. Dating from the fifth century C.E., this manuscript was a gift from Cyril Lucar, Patriarch of Constantinople, formerly of Alexandria, to King James I of England. James died before it could be delivered, and so in 1627 it was presented to his son, Charles I. Had it arrived earlier, it might have been consulted by the translators of the King James Bible (1611), the most influential of all versions. As it happened, scholars were slow to use Alexandrinus. It apparently was not consulted for the 1633 "Textus Receptus." Though its importance in offering alternative readings to this text was shown as early as 1657 (in the "London Polyglot"), the New Testament portion was not published as a whole until the late eighteenth century. Alexandrinus originally contained the entire Greek Bible, but the centuries have taken their toll. Most of the gospel of Matthew is now missing as well as portions of the gospel of John and 2 Corinthians. The noncanonical Epistles of Clement it once contained are also missing.

Codex Cantabrigiensis, also known as Codex Bezae after its former owner, Theodore Beza, probably dates from the fifth or sixth century and contains only the four gospels and the book of Acts. The gospels are in what scholars call the Western order: Matthew, John, Luke, Mark. Matthew lacks many of the original sections, but some were restored by later hands. Cantabrigiensis is an unusual manuscript in that it contains the Greek and Latin texts on facing pages. Its Greek text differs more dramatically than any other from the Textus Receptus.

The discovery of these fourth- and fifth-century manuscripts has enabled scholars to identify a number of later additions to the New Testament text. We mentioned two of these earlier: 1 John 5:7 and the doxology at the end of the Lord's Prayer. Among others is the so-called "longer ending" to the Gospel of Mark, which seems to have ended originally with 16:8, "for they were afraid." Verses 9–20 of Mark 16, which include a postresurrection appearance of Jesus and the promise that the faithful can handle snakes and drink poison and remain unharmed, are not in Vaticanus and Sinaiticus and thus can be assumed to be additions to the original text. Modern translations of the Bible treat this problem in various ways. The Revised English Bible reports in a footnote to 16:8, "at this point some of the most ancient witnesses bring the book to a close." A footnote in the New Jerusalem Bible says that "originally Mk probably ended abruptly on this note of awe and wonder" and that verses 9–20, "missing in some MSS, are a summary of material gathered from other NT writings." The New American Bible offers a choice of endings. Its first edition provided "The Longer Ending," verses 9–20; "The Shorter Ending," which is only one verse; and "The Freer Logion Ending," which is completely unlike the other two (and has since been withdrawn in revision). The foregoing, complicated as it is, does not exhaust the manuscript evidence for the numerous and variable endings of Mark.

Also belonging to the category of later additions is the beautiful story of the adulteress whom Jesus protects with the oft-quoted words, "He that is without sin among you, let him first cast a stone at her," found in the King James Bible in John 7:53–8:11. Modern translations will usually include the passage at the traditional place but provide a footnote pointing out that many ancient manuscripts omit it.

At the same time as they represent early efforts to standardize the text of the New Testament, the great codices also represent efforts to standardize its contents by bringing together within two covers everything that belonged in the canon. They are thus important witnesses to two crucial activities, both of which would eventually have to be brought to a satisfactory conclusion if the Church was to possess a usable Bible. But what was the state of affairs earlier than this period?

NEW TESTAMENT MANUSCRIPTS FROM THE SECOND TO THE FOURTH CENTURIES

The first line of evidence for the pre-fourth century New Testament text is the papyri, almost all of which come to us from the preservative climate of Egypt. Most of these New Testament papyri were originally codices—books, that is, not rolls—though many were of course in fragmentary form when discovered. (The very earliest New Testament papyrus thus far discovered, containing several verses from the Gospel of John and dating from about 140 C.E., is a scrap measuring just 3 by 2 inches.) The papyri must be used with this geographical limitation in mind, for they do not necessarily represent the state of the New Testament text in far-off places like Antioch and Rome. Inasmuch as almost all of them are discoveries of the twentieth century, they have yet to exert any significant influence on the New Testament text.

Indeed, what to make of them is a vexing question, for instead of aiding in the search for authenticity, they make it even more difficult. One would suppose that the farther back in time one went toward the original manuscripts, the *autographa*, the more agreement one would find among copies of the "same" text. But this is not the case at all. The metaphor of a stream emerging from a single source and then being diverted into many channels as it proceeds, becoming more complex along the way, is exactly wrong in this case, for it is in the oldest manuscript witnesses to the New Testament text that we find the most muddle and variation. One study of three of the most extensive papyri (Chester Beatty I, Bodmer II, and Bodmer XIV–XV), covering passages from the Gospels and Acts, has discovered more than a thousand "singular readings"—words or groups of words not found in any other known manuscript—and this does not include different spellings! None of the papyri has a text wholly like that of the later, reconstructed families surveyed earlier. To the extent that they represent a consecutive text at all and are not fragmentary (as most of them are), they are witnesses to several textual families within the same book. For example, the Chester Beatty Papyrus II, from the third century, gives us a reading of Romans that exists nowhere else. It places the doxology that now ends chapter 16 at the end of chapter 15, lending support to those who have maintained on literary grounds that chapter 16 is really a separate letter that was later attached to Romans. Both Bodmer II and the Bodmer XIV–XV contain portions of the Gospel of John, but they often differ drastically from one another.

What can account for this wide variance among documents written so early in the history of the New Testament text? The answer, leaving aside errors of copying and writing, is that Christianity was evolving rapidly during its first several centuries and the New Testament evolved along with it to meet its needs. As the young religion spread across the ancient world, it created communities of believers in widely separated places, and these communities faced different situations and had somewhat different needs. The sacred texts were adjusted to meet local conditions. One scholar refers to such changes as "reverential alterations," while another calls them "orthodox corruptions." No one has spoken better on the matter than M. M. Parvis:

> Many thousands of the variants which are found in the MSS of the NT were put there deliberately. They are not merely the result of error or of careless handling of the text. Many were created for theological or dogmatic reasons. . . . It is because the books of the NT are religious books, sacred books, canonical books, that they were changed to conform to what the copyist believed to be the true reading. His interest was not in the "original reading" but in the "true reading."*

Another possibility as to why the early New Testament manuscripts differ so much from one another is beginning to emerge, dramatically assisted by Morton Smith's discovery in 1958 of a hitherto unknown letter of Clement of Alexandria, a well-known Christian writer who lived around the end of the second century. The letter concerns a version of the gospel of Mark that was being circulated by the Carpocratians, a second-century group of Christian Gnostics. Clement writes that the version of Mark that the Carpocratians have is a false one. He goes on then to discuss other writings of Mark. According to Clement, while Mark was in Rome he wrote "an account of the Lord's doings" for the purpose of "increasing the faith of those who were being instructed." Afterwards, Clement continues,

---

*"Text, NT," in *The Interpreter's Dictionary*, ed George A. Buttrick (Nashville, Tenn.: Abingdon Press, 1962), 4, 595.

Mark journeyed to Alexandria, where he composed "a more spiritual Gospel for the use of those who were being perfected." Clement makes it clear that Mark's second edition was built on his first: "To the stories already written he added yet others and, moreover, brought in certain sayings of which he knew the interpretation would . . . lead the hearers into the innermost sanctuary of that truth hidden by seven veils."* There is now general agreement that this letter of Clement is genuine.

Clement believed that Mark had written more than one version of his gospel. If this is indeed so, it would help explain a long-standing problem of biblical criticism. It is now axiomatic that Matthew (like Luke) used Mark in composing his own gospel, yet it is striking how little verbal correspondence there is between Matthew's book and Mark's. Is it possible, as several scholars now believe, that the version of Mark available to Matthew was not the one known to us but an earlier, different one, now no longer in existence? In that case many of the textual variations in the manuscript record could be attributed to the fact that these works were published in different versions from the very beginning.

The book of Acts also presents several textual problems that might be explained in this manner. Scholars have long known that the version of Acts in the standard printed Greek New Testaments is considerably different from and shorter than Acts as represented in the Codex Cantabrigiensis. We now have confirmation from the papyri that this "alternative" text of Acts is at least as old as the standard one. Unquestionably, two different versions of Acts were circulating in the second century. Is one of them the original? If so, which one? Or are they alternative versions designed to meet the different needs of different centers of Christian activity, as was the case with Mark's gospel, according to Clement, so that the question of priority is beside the point? These are intriguing problems, brought to the surface long after one might have supposed that there was nothing left to discover in the area of New Testament textual history.

MANUSCRIPTS OF THE HEBREW BIBLE FROM THE SECOND CENTURY C.E. ONWARD

At first glance, the textual situation of the Hebrew Bible appears far more stable than that of the Greek New Testament. The Hebrew text in use between 1000 and 1450 C.E. was a remarkably uniform one, coming as it did from a Jewish scribal tradition that enforced rigorous safeguards of accuracy, such as the prac-

* Quoted in Ron Cameron, *The Other Gospels: Non-Canonical Gospel Texts* (Philadelphia: Westminster Press, 1982), pp. 69–70.

tice of counting the number of words in a text and the copy made from it and rejecting the latter unless the two agreed exactly. In the eighteenth century, Benjamin Kennicott examined over six hundred Hebrew manuscripts of the Bible and published the variants that he discovered, which turned out to be a far smaller number than one might have expected from so large a sample. Later in that century, J. B. de Rossi published a book of variants from well over one thousand manuscripts and printed editions of the Hebrew Bible, again demonstrating a striking uniformity in its texts as compared to those of the Greek New Testament.

Unfortunately, the situation is far more complicated than it would appear to be. In the first place, all of the Hebrew manuscripts in Kennicott and de Rossi are dated after 1000 C.E.—and most are much later than that. The oldest complete manuscript of the Hebrew Bible that we have, and the one that serves as the "copy-text" for the most popular of modern printed Hebrew Bibles, the *Biblia Hebraica Stuttgartensia*, is the Leningrad Codex B19A, which is dated 1008 C.E. We realize the full significance of that date when we remember that the Pentateuch was compiled around 500 B.C.E., fifteen hundred years earlier. If the multitude of intermediate copies still existed, what complications might they not introduce into its textual history!

In the second place, all of the manuscripts mentioned above represent one particular tradition, known as the "Masoretic" from the Jewish scholars, active between the sixth and ninth centuries C.E., who developed techniques to ensure a continuity in the reading and understanding of the biblical text. They were "keepers of the Masorah," the tradition. This text was consonantal: written Hebrew basically lacks vowels, and so in a time when that language was not commonly spoken it became necessary to indicate somehow the pronunciation of words. For nonliturgical Hebrew texts the Masoretes devised systems of dots and dashes placed above and below the consonants to indicate which vowels to use and thus make sure that different readers read the same words—particularly important where the Bible was concerned. (Reading in the ancient world was always done out loud.) This standardization was helped by the fact that the consonantal text had itself been stabilized around the middle of the second century C.E.

We should not suppose, however, that the urge to "correct" texts was any less active among Masoretes than among Christian scribes. Passages that suggested irreligious sentiment could be recast—the results are known as *tiqqune sopherim* ("corrections of the scribes"). One of the more interesting of these interferences was made in Job 32:3, at the point where Job's friends, finally convinced of his innocence, according to the original Hebrew "condemned God." Second-century scribes, unsettled by this apparent blasphemy, changed the text to read "they condemned Job." This reading was accepted in the Masoretic text and has been passed down to the present.

HEBREW BIBLE
MANUSCRIPTS BEFORE
THE SECOND CENTURY C.E.

As matters now stand, the Dead Sea Scrolls provide us with the earliest texts of the Hebrew Bible (though only a small part of it, unfortunately).

The Scrolls enable us to vault backward across a thousand-year period of copying, with its inevitable mistakes. Yet it is here that we are faced with the greatest variations within the texts. The Scrolls are witness, for example, to at least three different versions of Exodus and at least two different versions of Jeremiah and Samuel, to cite only those books for which we have ample texts. All were apparently in use by the Jewish community that hid them away in the Qumran caves and thus left them to posterity. Actually, even before the discovery of the first of the Scrolls in 1947, scholars knew that Exodus, and indeed the Pentateuch as a whole, existed in different forms. Both the Septuagint (the Jewish Bible in Greek) and the Samaritan Pentateuch (a Pentateuch prepared for and used in the Samaritan community of Palestine) contain versions of the Pentateuch different from that found in the Masoretic text. Before the discovery of the Scrolls, scholars could argue that the Greek Bible mistranslated the Hebrew in many places or that the Samaritan Pentateuch deliberately misrepresented the original. But the Scrolls happen to contain Hebrew versions corresponding to the Greek and the Samaritan, suggesting that the blame for variation might better be placed on the Masoretic text itself. From Cave 4 of Qumran, for example, we have a fragment of Exodus that agrees largely with the Septuagint against the Masoretic text of 1:1–6. Similarly, the Septuagint and the Dead Sea document both report the number of patriarchs who entered Egypt as seventy-five instead of the Masoretic text's seventy. Yet from this same cave comes another fragment of Exodus that agrees with the Samaritan Pentateuch against the Masoretic text. Scholars have long been aware that the Septuagint version of Jeremiah is some 12 percent shorter than the Masoretic version and is arranged differently; they now have Hebrew texts from the Dead Sea region that represent the Hebrew tradition behind the Septuagint version.

According to Frank Cross, a leading authority on these textual matters, it is the earliest of the scrolls (from the Qumran area in the second and first centuries B.C.E.) that provide examples of the greatest fluidity of the Hebrew text, whereas the later documents (from the southern Judean wilderness in the first and second centuries c.e.) are all consistent with the Masoretic text as we know it. Once again it appears that the farther back we go in the history of a given biblical text, the more diversity we find. Yet after we have considered all the Dead Sea documents, we are still centuries removed from the time of writing of most of the biblical books. What was happening to the text during those unattested years?

THE HEBREW TEXT
BEFORE THE DEAD SEA
SCROLLS

We have no Hebrew biblical manuscripts older than those among the Dead Sea Scrolls, but there is nevertheless some evidence available of the fluid state of the biblical text in its earliest centuries. This evidence is in the Bible itself, in sets of passages that are sufficiently similar to indicate that they derive from the same source (or from one another), but also sufficiently different to demonstrate that there was great freedom in handling the text even during the time the biblical books were being written. The following are examples of duplicate texts: Psalms 14 and 53; Psalms 40:14–18 and 70:2–6; Psalms 57:7–11 and 108:1–5; Psalms 60:5–12 and 108:7–14; and 1 Samuel 22 and Psalm 18. The following are triplicates: 2 Kings 18:13ff., Isaiah 36:1ff., and 2 Chronicles 32:1ff; and Jeremiah 39:1ff., Jeremiah 52:4ff., and 2 Kings 25:1ff. For maximum effect, one should of course compare these passages as they appear in the Hebrew Bible, but even in translation the differences resulting from this freedom are evident. On a much larger scale, one could profitably compare the use made of Samuel–Kings by the authors of 1 and 2 Chronicles. Some passages from Samuel–Kings are repeated verbatim, some are subtly altered, and still others show complete overhaul.

THE TEXT OF THE
APOCRYPHA

The books of the Apocrypha, which come to us from the Septuagint, are found in the fourth- and fifth-century Greek codices discussed earlier— those at any rate that were originally complete Christian Bibles. Apocryphal books in Greek also occur among the Christian papyri. Just like the Greek New Testament texts, the texts of the Apocryphal books have their variations and disputed readings. Rather than undertaking a survey of this very complicated area, we might consider the textual situation of one particular book of the Apocrypha, known to Christians as Ecclesiasticus and to Jews as The Wisdom of Jesus ben Sira. While this book was never formally admitted to the Hebrew canon, it was sometimes cited in rabbinic literature in terms suggestive of Holy Scripture and must have been popular and respected.

The book was known in the West only in its Greek and Latin form until 1896, when five incomplete Hebrew manuscripts of it were discovered in the "genizeh" (storage room) of a Cairo synagogue. These manuscripts were eventually brought to Cambridge University, where as recently as 1960 scholars were still matching the fragments to specific chapters in the book. In all, some two-thirds of the Hebrew of The Wisdom of Jesus ben Sira is now available from the genizeh. The prologue of the book states that it was originally written in Hebrew and later translated into Greek by the grandson of the author. The genizeh fragments confirmed the existence of versions in the two languages; but were these fragments parts of the original Hebrew version, or had they themselves been translated back into Hebrew from the Greek? This question was debated until 1956, when a portion of chapter 6 in Hebrew was found among the Dead Sea Scrolls in Cave 2 at Qumran. Another small portion attached to a psalter was found in Cave 11. Later, a much larger portion (sections of chapters 39–47) was found at Masada. There is now enough evidence to affirm that, although there are differences between the Dead Sea documents and the genizeh documents, the latter appear to be based on a genuine Hebrew tradition and are not translated from the Greek. Pursuing the history of this text in detail is important only to specialists, but the findings outlined here are worth our attention because they provide a measure of confidence, first in the word of Jesus ben Sira's grandson, and second in the possibility that other such sources of textual information may come to light in the future.

SUGGESTED FURTHER READING

Kurt Aland, and Barbara Aland, *The Text of the New Testament: An Introduction to the Critical Editions and to the Theory and Practice of Modern Textual Criticism*, 2nd rev. and enl. ed., trans. Erroll F. Rhodes (Grand Rapids, Mich.: Eerdmans, 1989).

F. E. Deist, *Witnesses to the Old Testament: Introducing Old Testament Textual Criticism* (Pretoria: NG Kerkboekhandel, 1988).

Bart D. Ehrman, *The Orthodox Corruption of Scripture: The Effect of Early Christological Controversies on the Text of the New Testament* (New York: Oxford University Press, 1993).

P. Kyle McCarter, *Recovering the Text of the Hebrew Bible* (Philadelphia: Fortress Press, 1986).

Bruce M. Metzger, *The Text of the New Testament: Its Transmission, Corruption, and Restoration*, 2nd ed. (New York: Oxford University Press, 1968.

Bruce M. Metzger, *A Textual Commentary on the Greek New Testament: A Companion Volume to the United Bible Societies' Greek New Testament*, 3rd ed. (New York: Oxford University Press, 1971).

Emanuel Tov, *Textual Criticism of the Hebrew Bible* (Minneapolis: Fortress Press, 1992).

The Interpreter's Dictionary of the Bible, ed. George A. Buttrick et al. (Nashville: Abingdon Press, 1962). See articles on Text, NT; Text, OT.

The Anchor Bible Dictionary, ed. David Noel Freedman et al. (New York: Doubleday, 1992). See article on Codex.

XVI

Translating the Bible

The Bible is one of the world's great books. Throughout history, millions upon millions of people have considered it to be the very word of God and have based the eternal welfare of their souls on its contents. And yet of the vast number who have held the Bible in such high esteem, not one-half of one percent have read its *actual* words. For those words are in Hebrew and Greek, languages that only a small proportion of the world's population has ever been able to read. All others have had to depend on a translation. Thus the history of Bible translating and its state today are matters of the greatest significance for students of the Bible. In earlier chapters we have frequently dealt with some of the more notable of the ancient translations, and we shall review those here before discussing Bible translating in more recent times and the challenges faced by anyone who undertakes this essential labor.

THE SEPTUAGINT

Our review must begin with the translation called the Septuagint, which we have discussed at some length in chapters 6 and 12. This was a rendering of the Hebrew Bible into Greek, for the benefit of Greek-speaking Jews living outside of Palestine in the centuries following the reign of Alexander the Great (who died in 323 B.C.E.). The Septuagint was the Bible of the earliest Christians and thus provided the form of the Jewish scriptures that the Christian Church claimed as its own and that strongly influenced the writing of the New Testament. The Septuagint was the source from which the earliest Latin versions of the Old Testament were made, and it was a strong influence on the Latin version produced by Jerome in the fourth century C.E. The Eastern Orthodox Church still employs the Septuagint as its Old Testament.

THE VULGATE

Another notable ancient translation was the Latin Vulgate of Jerome. As we pointed out in chapter 12, Jerome actually set out simply to revise the already existing Latin versions of the Old and New Testaments. For his revision of the Old Testament, he intended to rely on the authority of the Septuagint and several later Greek translations. But as his work proceeded, he gradually felt the need to go back to the Hebrew itself for the Old Testament and so to produce a new translation straight from Hebrew into Latin. Jerome's busy life did not allow him to accomplish this completely. As a consequence, the translation associated with his name (referred to in later centuries as the "Vulgate" because it was in the language of the *vulgus*, the common people of Rome) contains within it some Old Testament books that were in the early Latin form and not revised or redone by Jerome. What Jerome accomplished was nevertheless a remarkable piece of work for one man and well represented his church's understanding of the scriptures in the late fourth century. Some time passed before this translation drove out rival versions, but having done so, it became firmly established as the official Bible of Western Christianity and remained so for the next thousand years.

The fact that the Western Church used a Latin Bible and the Eastern Church used a Greek Bible was not so important as that each of them considered its version to be the *true* Bible. Western and Eastern churchmen (or at any rate the better educated among them) were not ignorant of the fact that the Old Testament had originally been written in Hebrew; but they all believed that their particular translation had been inspired by God. When medieval Jews pointed out places in the Greek and Latin versions of the Old Testament that did not faithfully represent the meaning of the Hebrew (the Hebrew, that is, of the medieval manuscripts available to them), the Christians insisted that Jews had altered the meaning of the Hebrew text to avoid having to acknowledge its foretelling of Christ. Similarly, when Eastern Christians remarked that some passages in the Latin New Testament were not faithful to the Greek original, Western Christians retorted that the Eastern Church had altered the text of the Greek New Testament to support its own heretical views. This insistence that one's own translation is the "true" Bible can be heard today from some users of the King James Version. In their view, God inspired a body of translators working nearly four hundred years ago but no translators working before or after that time. And the continuing efforts of modern scholars to understand the Hebrew and Greek originals better is of no interest to them at all.

Throughout the long period of the Middle Ages there were no challenges to the supremacy of the Greek Bible in the East and the Latin Bible in the West. Jewish scholars, of course, continued to employ the Hebrew text of their own scriptures. And Christians made translations of individual books, such as the Psalms and the gospels, from Latin and Greek into the tongues of various Western and Eastern European peoples. The entire Bible was translated in the fourth century from Latin into Gothic, an early Germanic language; a thousand years later it was translated from Latin into English by the followers of John Wycliffe, acting on the conviction that English priests and churchgoers alike needed to have the scriptures in their native tongue. But with no official ecclesiastical support for producing new vernacular translations and making them available to individual churches, and with no means for physically reproducing them except by laborious and expensive handcopying, there was no chance that a movement could get under way to put the Bible into modern languages. In the mid-fifteenth and early-sixteenth centuries, however, certain historical developments changed the picture entirely.

THE INVENTION
OF PRINTING

The first of those developments was the invention of printing (more precisely, the invention of printing from movable type). The effects of this on human civilization are so vast that they can scarcely be overstated. Before printing, only a single copy of a book could be made at one time by one copyist; and the cost of the book amounted to the cost of materials (not inconsiderable) and what the copyist had to be paid during the period required for him to complete it (perhaps a year, in the case of a Latin Bible). But once printing came into existence, hundreds of copies could be made at about the same cost for labor as that for a single handwritten copy.

This had several major consequences for the translation of the Bible. The most obvious was that, as the cost of bookmaking dropped, the prices of books dropped and therefore people other than aristocrats and clerics could afford to own Bibles. Another consequence was that the tools requisite for making translations (grammars and dictionaries of ancient languages, concordances to the Bible, commentaries on the Bible text, and, of course, the Hebrew, Greek, and Latin texts themselves) were published for scholars—and at prices they could afford. A third consequence was a great gain in the accuracy of biblical texts, for now a scholar's careful efforts to produce an accurate version of the Bible in some given language would be reflected in hundreds of printed copies. Some errors were introduced during the printing process, of course, but not nearly so many as would have been introduced by individual handcopyists producing the same number of copies.

THE PROTESTANT
REFORMATION

Still, without official ecclesiastical support, there could be no widespread effort to put the Bible into the languages of common people. That support was not forthcoming until the time of a second historical development: the Protestant Reformation. The established churches of the East and West felt no need to have the Bible translated into vernacular tongues, for they each were confident that the means of salvation and the dispensation of divine grace were in their keeping and that lay members of the Church had no need to study the Bible privately. But with the Reformation, brought to a head by the efforts of Martin Luther in Germany, the control of the individual believer's religious life was in many places in the West torn away from the Roman Catholic Church. It was Martin Luther's overriding conviction (which had come to him, he believed, by divine inspiration) that salvation was a matter strictly between the individual and God, with no intervention necessary by a church: Human beings were to seize by faith what God made available by grace. Anyone perceiving this and accepting it, Luther thought, possessed the key to the mystery of the scriptures and had a right to study those scriptures. To that end, the Bible had to be made available in the tongues of the people. Translations were required, and they had to be as faithful to the original biblical writings as possible; thus they had to be made from the Hebrew Bible and the Greek New Testament, not from the Latin Vulgate, which was itself merely a translation.

Luther set himself to the long task of making a new translation into German (earlier ones existed, but they derived from the Vulgate). In other countries where protest against the Roman Church and a call for reform began to be raised, the "Protestants" (or "Reformers") followed Luther in pressing for translations to be made into modern languages. We shall now survey the history of biblical translation into English, but the reader should realize that there is an equivalent history for every one of the major European languages.

EARLY SIXTEENTH-
CENTURY ENGLISH
TRANSLATIONS

The first English product of the Reformers' concern for biblical translation was the New Testament published by William Tyndale, in 1525 or 1526. The English church (still loyal to Rome) opposed this work and attempted, without complete success, to seize and destroy all copies of it. (The English church had an on-again, off-again view of Bible translating, sometimes opposing it because both church and government feared what would happen when laypeople read and interpreted the Bible for themselves, at other times favoring it in order to promote Reformation ideas.) Tyndale, living and working on the Continent to avoid arrest in England, turned his attention to the Hebrew Bible as soon as he had completed work on the New Testament. He learned some Hebrew and began translating, completing the Pentateuch, Jonah, and most of the historical books. Then he was arrested by agents of the Catholic Inquisition, imprisoned, and in 1536 put to death as a heretic.

A year before his death, an assistant to Tyndale, Miles Coverdale, had published the first complete *printed* Bible in English, using Tyndale's New Testament and some of the books of the Old Testament that Tyndale had translated; the remainder of the Old Testament and the books of the Apocrypha, Coverdale himself translated. Coverdale's efforts were encouraged by the English church (which had now been removed from the Roman Catholic fold by King Henry VIII) and permitted by the government; they were not curtailed as Tyndale's had been ten years earlier. The English church allowed a reworked Tyndale–Coverdale version (called the Matthew Bible) to be published in 1537 and then published its own version of this Bible in 1539. Referred to as the Great Bible because of its large size, the 1539 work was authorized to be read in all of the parish churches of England.

THE GENEVA BIBLE

But this openness to Bible translations came to a sudden end when, following the reigns of Henry VIII and his young Protestant son Edward VI, Henry's daughter Mary took the throne in 1552. She was a strong Catholic and intended to bring the English church back into the Roman fold. Those Reformers who resisted her efforts were in danger of execution as heretics. To save themselves, many fled to cities on the Continent that were sympathetic to the Reformation; many who did not flee were burned at the stake by the government of "Bloody Mary." English Protestants living in Geneva, Switzerland, set about making their own translation of the Bible into English. This work, published in 1560, was a resounding success. For one thing, the Geneva Bible was printed in easy-to-read Roman typeface rather than in the old-fashioned black-letter (Gothic) typeface, and it was divided into numbered verses for handy reference. Then, too, it had notes and headings that were strongly Protestant in nature and that suited the public mood in the decades following the death of Mary and the ascension to the throne in 1558 of the third of Henry VIII's children to rule after him, Elizabeth.

THE BISHOPS' BIBLE

Queen Elizabeth, Protestant in sympathy, was constantly watchful against Catholic efforts to regain power in her realm. But she was unwilling to allow Protestant enthusiasts to dictate how she should rule either the nation or the church (of which she was the official head). The English church that she and her advisers shaped had within it some similarities to the Roman Catholic Church—far too many similarities, in the view of English Puritans (so called because they wished to purify the church of Catholic elements). But it was distinctly independent of the Catholic Church, and it encouraged the reading of the Bible in English. To promote this, the bishops of the English church authorized a new translation to be made and used in parish churches. This "Bishops' Bible," like the Great Bible before it, was strongly in debt to the work of Tyndale and Coverdale. Published in 1568, it remained the official Bible of the English church for many decades. About the use of the Bible during the Elizabethan era we can make the generalization that, although everyone who attended church services (and everyone was expected to do so) heard the scriptures read from the Bishops' Bible and themselves recited passages from a prayerbook derived from Coverdale's Bible, most of those who participated in daily family devotions at home used the Geneva version.

THE DOUAY-RHEIMS BIBLE

Just as during Queen Mary's reign a great many Protestants fled to the Continent to escape persecution and to be able to practice their religion freely, so numerous Catholics fled during Queen Elizabeth's reign for those same reasons. Although these Catholics did not believe that a Bible in English was necessary for the good of people's souls, they did realize that the Protestants had a distinct advantage in possessing versions in English that could be appealed to in religious disputes between the two parties. Thus, despite their misgivings about putting the scriptures into the common tongue, English Catholic scholars living on the Continent in the late 1570s set about making their own English version. They translated from the Latin Vulgate, for that was the official Bible of their church; and, in order to avoid importing alien ideas into their version, they stayed as close to the Latin in form as the English language would allow (often they stayed *too* close to the Latin and produced readings that scarcely were English at all). Their New Testament was completed and published in 1582 at Rheims, in France, and their Old Testament in 1610 at Douay (or Douai), also in France. Despite its overliteralness, the Douay-Rheims Bible (as it is usually called) achieved a place as the chief Bible of English-speaking Catholics for the next three-and-a-half centuries. A revision of this Catholic version in the mid-eighteenth century by Bishop Richard Challoner removed many of its stylistic defects.

THE KING JAMES
VERSION

Just a year after the publication of the complete Douay-Rheims version, there appeared a translation that was destined to become the greatest of the English Bibles. Pressure for a new, more accurate English version had begun to build shortly after the death of Queen Elizabeth I in 1603 and the ascension to the throne of King James I. James gave his permission (which is why the resultant work came to be called both the Authorized Version [AV] and the King James Version [KJV]), and in 1604 he appointed a large body of scholars to undertake the task. The finished work, published in 1611, initially met with considerable resistance, especially from persons with Puritan sympathies who still championed the use of the Geneva Bible. A number of decades were to pass before the KJV would become the undisputed favorite version of English-speaking Protestants. But finally its triumph was complete; and as the scores and hundreds of years passed, it appeared that nothing would dislodge the KJV from its preeminent place. Its language, of course, became increasingly old-fashioned and quaint as time went by. But that language was the language of the English Renaissance and was felt by the bulk of its readers to be the appropriately majestic means of expression for the "good news" of the Bible. Thus, ironically, the making of a new English translation became increasingly necessary with the passing of time but also—given the growing veneration of the KJV—increasingly difficult to accomplish. Various one-person translations were produced in the eighteenth and nineteenth centuries, and some of them flourished for a time; but none was ultimately of any major significance.

THE REVISED VERSION

By the middle of the nineteenth century, however, there could finally be no denying that much of the language of the KJV was so out of date as to be—for beginning readers of the Bible, certainly—a hindrance to understanding. Furthermore, better textual sources than those the 1611 translators had used were now available, and much more was known about the original languages of the Bible. Various proposals for revision and attempts at revision of the KJV were made in the mid-nineteenth century. But it was not until 1870 that a committee was formed within the Anglican Church (some non-Anglicans were included) to begin the formal work of revision. The committee set as its task both to introduce into the English text such new knowledge about the Hebrew and Greek originals as had been discovered since 1611 and to alter any KJV language that was misleading. But when it did alter the language, the committee unfortunately employed only words and constructions typical of the KJV and other early translations. And in striving for faithfulness to the originals, they imposed on themselves certain severe limitations, such as trying wherever possible to follow the word order of the originals and always translating a given word in the original by the same English word. The inevitable result of this policy was to produce a very stiff English Bible marred by "translationese." The New Testament of the Revised Version, as it was termed, was published in 1881, the Old Testament in 1885.

THE AMERICAN
STANDARD VERSION

An American advisory committee had been established shortly after the revisers began their work in England. Drafts of the revised portions were sent to this committee for comment, the idea being that, where possible, the English revisers would accept American proposals for alteration, and, where not possible, the American preferences would be listed in an appendix to the Revised Version. The Americans were requested, and agreed, not to publish their own revision in competition with the English Revised Version for a period of fourteen years. When that period expired, the American committee renewed its work on the text and, in 1901, published what is usually referred to as the American Standard Version. This incorporated into the text those preferred American readings that had been listed in the appendix to the Revised Version as well as additional readings that the American committee has settled on in the years following publication of the Revised Version.

Thus, by the beginning of the twentieth century, both England and America had produced what were intended to be successors to the King James Version. But although these new versions enjoyed some popularity, neither was able to replace the KJV, either for use in churches or for private reading. They undoubtedly were more faithful to the sense of the originals than was the KJV; but the public proved to be unwilling to sacrifice the comfortable familiarity of the KJV for any mere increase in accuracy. Thus the revisions of 1881–1901 solved nothing of the problem that the KJV presented except to point the way for further work in revising or translating afresh.

THE REVISED STANDARD VERSION

A number of new and worthy translations of the entire Bible or portions thereof appeared in England and America in the first half of the twentieth century. But the next one to achieve widespread and lasting attention was begun by a body of American Protestant churches in the 1930s and brought to conclusion in 1952. This was the Revised Standard Version (RSV), a reworking of the American Standard Version—although what the translators actually hoped to replace was not the little-used revision of 1901 but the King James Version, which was now more archaic than ever in its language and more out of date than ever in its biblical scholarship. As had happened to most of the English translations before it, all the way back to Tyndale's version, the RSV was published to a chorus of negative criticism: The beloved language of such familar passages as the Lord's Prayer and the twenty-third Psalm had been altered, certain crucial theological terms had been dropped, religious credentials of some of the translators were suspect, and so on. But time has gradually reduced the dissatisfaction with the RSV, and some religious bodies that once denounced it now embrace it as the best of the recent translations. Although its translators were careful to draw on the best biblical sources and scholarship and to avoid using in their English text utterly archaic diction and obsolete constructions, they nevertheless pointedly maintained the familiar idiom of the KJV. It is probably true that most persons listening to a public reading from the Bible cannot tell whether they are hearing the RSV or the KJV.

A revision of the RSV, the New Revised Standard Version (NRSV), was published in 1989. Evident in the text of the NRSV

are changes made on the basis of recently discovered ancient manuscripts, as well as the elimination of gender-oriented language in
places where gender plays no significant part in the original.

THE NEW ENGLISH BIBLE

While work was proceeding in
America on the RSV, planning began
in Britain for a new translation by
Protestants there (ultimately British
Catholics as well as Irish Protestants
and Catholics joined the project). This was not, however, to be
merely an effort parallel to that of the Americans, which looked back
to the biblical tradition of the sixteenth century. This would be a
completely new effort, paying allegiance solely to the Hebrew and
Greek texts, on the one hand, and to the demands of current English
language, on the other. Its break with the past would be signaled by
its title, the New English Bible (NEB). The New Testament of this
work appeared in 1961, just 350 years after publication of the KJV,
the Old Testament and Apocrypha in 1970. There were some com-
plaints about the NEB—for example, that it sometimes sacrificed traditional readings too casually and that in places it tended more toward
paraphrase than direct translation. But on the whole it was judged to
be a great success, delivering the sense of the originals accurately in
language that was both highly idiomatic and suitably dignified. Beginning readers of the Bible had in the NEB a translation that spoke in
the language they actually used. And readers accustomed to the language of the KJV had in the NEB a translation that broke through
their usual understanding (or unthinking acceptance) of familiar passages and forced them to reconsider what they had been taking for
granted.

The reception of the NEB far exceeded its sponsors' expectations.
Particularly gratifying was the widespread use of this translation in
worship services, for which it had not been primarily intended and
for which it was not entirely suitable. An effort to revise the NEB
began in the 1970s and came to fruition in 1989 with the publication
of the Revised English Bible. This is a completely new translation,
often scarcely revealing its kinship with its predecessor. Particular attention was given by the translators to altering the sentence structure
and word order of the NEB so as to make the revision appropriate for
congregational, as opposed to private, reading. But no gain in one
direction on the public/private spectrum can be achieved without loss
in the other; and alterations that make the REB satisfactory for liturgical purposes inevitably limit its fidelity to the sense of the original
languages. As was true with the NRSV, the translators of the REB

paid attention to removing male-oriented language in those places where the sense of the original text of the Bible does not require it.

RECENT MAJOR EDITIONS

Along with the RSV/NRSV and the NEB/REV—the former maintaining and the latter breaking away from the prevailing biblical tradition—a number of other important translations of the Bible into English have been made in recent years. (The twentieth century can be considered the second great period of English biblical translation, as the sixteenth century was the first.) Two recent English versions have been prepared by bodies of Catholic scholars: (1) the New Jerusalem Bible (NJB), published in 1985, a translation into English from the Hebrew and Greek but influenced by the thorough research performed in connection with an earlier French translation and the earlier Jerusalem Bible, and (2) the New American Bible (NAB), published in 1970. A fresh translation of the Hebrew Bible by a team of Jewish scholars was published under the title Tanakh in 1985. For evangelical Christians a team of one hundred scholars produced the New International Version (NIV), published in 1978; earlier, for that same audience, the American Standard Version of 1901 had been revised as the New American Standard Bible (NASB), published in 1971. For readers wishing to have a Bible in the simplest sort of language, there is Today's English Version (TEV), published in 1976. For those willing to work their way through biblical texts that are more complex than usual, there is the Anchor Bible, which in most cases devotes an entire volume to a single biblical book and supplies scholarly notes and commentaries alongside the texts.

THE PROBLEM OF
TRANSLATION

We observed at the outset that most people who have read the Bible have read it in translation. That being so, it is legitimate to wonder how well readers are served who must depend on a translation of the Bible. Can such readers be confident that translations in general, or any one in particular, will give them the whole Bible and nothing but the Bible? The answer must, alas, be no. And for that state of affairs—contrary to what one might suppose—translators are not primarily to blame. There are two reasons why translators cannot simply take "the whole Bible and nothing but the Bible" in the original languages and make a completely faithful translation into another language. First, there is no such thing as a universally agreed upon text of the Bible to translate from; second, complete faithfulness in translating anything, the Bible included, is impossible to attain. In what follows, we shall expand on these two essential truths.

When translators of the Bible sit down to do their work, what specifically will they translate—what will they have before them as they work?

Among other things, they will have before them printed books, compiled by textual experts, that contain texts of the Bible in a variety of ancient languages. The number of these books may be relatively small, perhaps a half dozen or so, but those few scholarly works will be of such a nature as to provide translators with readings drawn from hundreds of ancient biblical manuscripts. As we reminded readers in chapter 15, the original manuscripts handwritten by the Bible's authors are no longer in existence, nor are the earliest collections of individual biblical writings. The oldest manuscripts that textual scholars can draw on for the benefit of translators date from hundreds of years after the biblical writings were composed. As copies were made from copies during those hundreds of years, alterations in the texts were inevitably introduced by the copyists, sometimes intentionally, sometimes unintentionally. Also, as we pointed out in chapter 15, alternative versions of some biblical books may have circulated right from the beginning. Thus our earliest surviving biblical manuscripts differ from one another at point after point.

A single set of readings drawn from all the sources at any given point in the biblical text is called the "textual variants" or "variant readings" for that point. It is important that biblical scholars, including translators, be able to know what these variant readings are for any biblical passage they are working on. In order that the variants may be readily available, they are printed as footnotes in the ancient-language printed editions of the Bible prepared for scholars. Also in those notes will be many proposed emendations of the biblical text—informed guesses about what the original text may *really* have been at this point or that point, where error or interference with the text are suspected. The total number of variant readings and proposed emendations for the whole Bible is now in the hundreds of thousands, and the number will increase as additional ancient manuscripts are found and as textual scholars continue their work. The largest proportion of variants involves minor matters of spelling and grammar, but many thousands concern matters more significant than those.

The rich resources provided by textual scholarship are a great boon to translators, of course. But they force on translators the necessity of making decision after decision about just what it is they are going to translate: At any given point should they take this reading from one manuscript, that reading from another, or a third reading proposed as an emendation? Translators can simplify their problem by following closely the form of the text given in a scholarly edition, a text based on some single ancient manuscript or one put together from many sources by reputable textual scholars. But there will still be many places where individual translators will have their doubts about the accuracy or appropriateness of the text they are following. Often they will wish to select from the footnotes an emendation proposed by some scholar or a reading found in some manuscript other than the one their chosen edition presents. Or they will wish to insert a word or passage not found in the text before them or to omit one that is.

All of this is very complex and potentially tedious. The point of presenting it here is to demonstrate that, before scholars begin translating the Bible, they must choose which overall version of the original-language texts they are going to use as the basis of their work; then they must continue choosing among individual readings as they arrive at hard spot after hard spot. Anyone who demands that they translate "the whole Bible and nothing but the Bible" is imagining an entity that does not exist. In effect, there *is* no Bible until the translators construct one out of the myriad possibilities available in their sources.

DIFFERENCES BETWEEN LANGUAGES

We said earlier that the one reason why a completely faithful translation of the Bible cannot be made is that complete faithfulness in translating is impossible to attain. The ideal of translation is to carry over, from the original language to the "receptor" language, the whole sense and nothing but the sense of the work in hand. But that ideal can never be achieved. Any translation of literary material from one language to another will always, inevitably, leave some portion of the original sense behind and will impose some additional sense of its own. This is so for the reason that no two languages have a simple one-to-one relationship with one another. That is, all languages differ from one another in ways far more profound than just having different vocabularies. Thus, translation cannot be done simply by proceeding word by word through the original and turning each word into a corresponding one in the receptor language.

Consider the problem first on the basic level of the lexicon—that is, the words available in each of the two languages a translator is dealing with. Seldom will there be words in one language that correspond exactly with words in another. There will be a great number of *approximately* corresponding words, of course, that will overlap in one or several senses; however, each word in a pair can have additional senses that the other word does not have. To capture all the possible senses of the Hebrew word *derek*, for example, a whole set of English words has to be drawn on: "way," "distance," "road," "journey," "manner," and others. Which one of these senses will be appropriate for any given occurrence of *derek* in the Hebrew Bible? Context will of course be the major determinant, for it will make plain whether "distance" or "road" is the sense of *derek* intended by the author. But context will often not be sufficient for determining which sense a word in the original should be assigned. The word *elohim* can mean either "gods" or "God"; in several significant places (Exodus 32:4, for example) it is unclear which sense the word has. Similarly, Hebrew *ruaḥ* can mean either "wind," "breath," or "spirit"; there is no certainty which of these senses the author of Genesis 1:2 intended when he wrote that the *ruaḥ* of (or from) God moved over the primeval abyss. When context does not provide certainty, translators must nevertheless choose one or another of the possible senses to put into their text. The other sense or senses they may wish to indicate in a footnote to the text.

Just as the senses of "equivalent" words in any two languages will not correspond exactly, neither will grammatical and syntactical structures. It is difficult for persons with little experience of a second language to believe that other languages work differently from their own. But it is a fact that there are languages in the world that, unlike English, lack a passive voice or that assign gender to inanimate objects or that have no rhetorical questions or that do not employ a past/present/future tense system or that always indicate whether a named person is dead or alive, and so on and so on. The original languages of the Bible have their share of features quite unlike anything in English, requiring that translators make radical adjustments in producing a readable English equivalent. Consider the following: (1) Hebrew verbs express not the time of an action but rather whether the specified action is complete or incomplete. Tenses can usually be fairly easily assigned to such verbs, although less easily in prophetic writings, where there are rapid shifts back and forth among past, present, and future. But in assigning tenses, translators will often choose to leave unexpressed the completeness/incompleteness element because English has only awkward ways to handle that. (2) Hebrew displays much less subordination of one clause to another than does English, and in general it lacks our great variety of words that indicate logical connections between clauses and phrases. In Hebrew narrative, sentence units tend to be strung out one after another in boxcar fashion and to be hooked together by means of a single, all-purpose connective that is usually translated "and" in the KJV. In Genesis 19:1–3 that connective occurs seventeen times in the Hebrew and is translated "and" every time but once in the KJV. Note the effect this has in the following verse (Gen. 19:3): "*And* he pressed upon them greatly; *and* they turned in unto him, *and* entered into his house; *and* he made them a feast, *and* did bake unleavened bread, *and* they did eat" (emphasis added). Modern translators could not get away with such choppy construction, nor would they want to. They would represent the Hebrew connective not merely with "and" but with "when," "but," "however," and so on (depending on context) or at times would simply omit it. (3) Greek verbs indicate time of action more definitely than Hebrew verbs do, but they also indicate nature of action (whether linear, recurring, or completed). Greek verbs are thus very complex in meaning, and translators who attempt to capture *all* of the potential meaning will risk overloading their English and producing a translation so clumsy as to be unreadable.

Along with words and structures not directly translatable, there are in every language idioms—certain set expressions that have developed solely within that language and have no force outside it. When idioms occur in the Hebrew and Greek of the Bible, translators must choose among several possibilities: (1) they may translate the idiom directly when the result will make some degree of sense in the receptor language; (2) they may capture just the main point of the idiom and abandon its colorful dress; or (3) they may substitute for the idiom an approximately equivalent idiom in the receptor language. We can observe all of these things happening in the translations of 1 Kings 21:21 found in the various English versions. Here is the KJV reading: "Behold, I will bring evil upon thee, and will take away thy posterity, and will cut off from Ahab him that pisseth against the wall. . . ." The phrase "him that pisseth against the wall" is a literal rendering of the original, a Hebrew idiom that occurs six times in the Old Testament, always in the setting of a curse. It is (as the reader will realize with a little reflection) a poetic if vulgar way of referring to males as opposed to females: Only males can urinate against walls. It is possible to translate this idiom literally, as was done in the KJV for an early seventeenth-century audience. But the resulting English will nowadays be considered offensive by many (particularly when the Bible is read publicly), and not everyone will understand that the phrase denotes specifically male offspring. For these reasons, some modern translations render the Hebrew idiom simply as "every male" or "every last male." But others make an effort to capture the energy of the Hebrew with a forceful English idiom, giving us "every mother's son" in the NEB and "every manjack" in the NJB. Both of these reflect the racy, hard-hitting quality of the original without forcing its vulgarity on the sensitivities of a modern audience. Idioms always provide special difficulty for translators; the satisfactory handling of them is one of the hallmarks of a good translation.

The problem of how to represent idioms leads naturally to the related problem for the translator of how to represent literary forms. In both cases the question is how to carry over to the reader not only the *sense* of the original but the *means* by which that sense is achieved. In a poetic construction, as in an idiom, the means of expression draws attention to itself and is indeed an inherent part of the sense itself. (To leave behind the vulgarity in the idiom discussed earlier is to leave behind some part of the sense.) We can see the problem of literary form acutely, although on a small scale, in the case of puns. The Hebrew Bible abounds in puns. One notable kind involves placing together words of similar sounds but different senses. The difficulty in translating puns is to come up with a set of words in the receptor language that will represent both the sound and the sense of the set of punning words in the original. Consider Genesis 2:7: "Yahweh God shaped man [Hebrew *adam*] from the soil of the ground [Hebrew *adamah*]." There exists no combination of English words that can capture both the *different senses* and the *shared sounds* of the two Hebrew words here; so translators must represent the sense and forego the sounds. Translators have better luck in dealing with the wordplay in Genesis 2:23, where Adam says concerning the newly created Eve, "She is to be called Woman [Hebrew *ishshah*], / because she was taken from Man [Hebrew *ish*]." Happily, the English words "woman" and "man" share sounds in quite the way that *ishshah* and *ish* do, while also having the required contrasting senses. But this situation is quite rare; in most cases, translators must abandon the sound element of puns to get the sense.

And what is true of puns is true of all other literary forms: The burden of sense that a literary device carries can be translated, but not much or any at all of the form itself—and thus not the nuance of meaning and potential for delight that is inherent in artistic form. Consider the challege that Psalms 111, 112, and 119 and Lamentations 3 present to the translator. Each of these chapters is, in the Hebrew original, an alphabetical acrostic poem—that is, in each of them the initial letter of the first line (or group of lines) is the first letter of the Hebrew alphabet, the initial letter of the second line is the second letter of the Hebrew alphabet, and so on through all twenty-two letters of that alphabet. Should English translators of these poems try, while capturing the sense of the words, to capture this artistic element as well? Unfortunately, there is no way they can. There are twenty-six letters in the English alphabet, not twenty-two; and English letters are not the same as those in Hebrew. The acrostic feature simply cannot be accommodated when these Hebrew poems are translated into a language with a different sort of alphabet.

In general we can say that the approximate sense of a passage, the order of the points it makes, and its tone can be translated; but most elements of its form and sound, including rhythm and rhyme and wordplay, cannot be translated, although they can be equivalenced. This does not mean, however, that poetic form should be ignored by the translator or its existence hidden from the eyes of the reader. That a poetic passage is poetic is one of the most important things to know about it: We are prepared to understand a passage that we take to be poetry differently from the way we understand one that we take to be prose. It is important then that translations of poetry at least *look* like poetry, although many of the poetic effects of the original cannot be represented in the translation. One of the great virtues of modern translations of the Bible is that in all of them attention has been given to presenting poetry as poetry.

HOW TO CHOOSE A TRANSLATION

A reader of the Bible in English is faced with a wide range of modern translations to select from. Not only are there the half-dozen major versions mentioned at the end of our historical survey of translations, there are also scores of lesser-known twentieth-century English versions of the whole Bible or its parts. How is a reader to choose among them? What criteria should guide one's choice? Should one take into account, for example, the theological position of those who produced a translation that one is interested in?

THE QUESTION OF THEOLOGICAL BIAS

Well, there is no ignoring the fact that most Bible translating is paid for by religious organizations and performed by individuals with a specific religious commitment. One can suppose that those involved have employed a chain of reasoning something like this: "The Bible is supremely important to persons of our religious persuasion. Because few of our number can read it in the original languages, an English translation is necessary. We have our own unique view of the Bible; therefore, we would do well to make our own translation." It would be a wonder if a translation deriving from this sort of reasoning did not to some extent reflect the theology of those who produced it. Wherever (as is often the case) the Hebrew or Greek texts are unclear or ambiguous—wherever words and expressions in the original can legitimately have any one of several senses—translators with a specific religious commitment will understandably (and sometimes unconsciously) be prone to choose a sense that accords with their own views and those of their intended audience.

Isaiah 7:14 is perhaps the most notable place where this happens. In this passage the prophet Isaiah tells the king of Judah that a young woman is going to bear a child and name him Immanuel (that is, "God is with us"). The prophet goes on to say that by the time this child leaves infancy, a certain military threat to Judah will have vanished. The Hebrew term for the female here, *ha-almah*, designates some specific young woman of marriageable age. When this passage was translated into Greek in the Septuagint, *ha-almah* was represented by a Greek word meaning not only "young woman" but also "virgin" in our modern sense. And when the Greek passage was later quoted by the author of Matthew's gospel, it was set in a context requiring that the young woman be understood *specifically* as a virgin and that the birth thus be a miraculous birth. Conservative Christians who hold to the view that there can be no discrepancies in the Bible (and who also place a high value on the doctrine of the Virgin Birth), believe that if the young woman is said to be a virgin in the New Testament passage, she must be understood to be a virgin in the original Hebrew passage. Thus, translations made by and for conservative Christian groups will "conserve" the traditional view and use the word "virgin" in Isaiah 7:14, although the Hebrew term in question does not require that sense and its immediate context works against it.

FORMAL CORRESPONDENCE VERSUS DYNAMIC EQUIVALENCE

Although there is plentiful opportunity for this sort of shaping of a translation in line with theological views, it in fact occurs so rarely in the major modern translations that beginning Bible readers need not hesitate on this particular score to use any one of them. A far more important element to take into account in choosing a translation is the literalness or freedom of the versions available and what one's personal preferences are in this matter. Actually the terms "literalness" and "freedom" do not well express the point at issue here. The terms "formal correspondence" and "dynamic equivalence" have been coined by translation theorists to designate the two ends of a spectrum of possibilities in translation. Formal correspondence has been defined as the "quality of a translation in which the features of the form of the source text have been mechanically reproduced in the receptor language. Typically, formal correspondence distorts the grammatical and stylistic patterns of the receptor language, and hence distorts the message. . . ." Dynamic equivalence has been defined as the "quality of a translation in which the message of the original text has been so transported into the receptor language that the response of the receptor is essentially like that of the original receptors."* Briefly put, in the former the emphasis is on the *form of the original*, in the latter on the *reader's ability to understand readily*. Note that neither of these elements is good or bad in itself and that it is legitimate to strive for either one in translating. In beginning any translating project, those involved must decide for themselves whether to favor the demands of the form or the needs of the reader. Just where the translators place themselves on the spectrum between these two positions will determine the kind of translation they produce.

When choosing among translations, readers should use the characteristic combination of formal correspondence and dynamic equivalence in each as a major factor in their choice. If one wishes to move slowly through the Bible, studying it chapter by chapter, or if one needs an English translation as a help in beginning the study of biblical Hebrew and Greek, one should choose a translation embodying a high degree of formal correspondence: the Revised Standard Version, say, or the New American Standard Bible or the New International Version. If one wishes to read through the Bible in such a way as to become generally familiar with it, or if one intends to read the Bible aloud to others, one should select a translation embodying a high degree of dynamic equivalence: the Revised English Bible, say, or the New American Bible, the New Jerusalem Bible, or Today's English Version.

Consider the differences between the following two translations of Romans 5:12–13:

> Therefore, just as through one man sin entered into the world, and death through sin, and so death spread to all men, because all sinned—for until the Law sin was in the world; but sin is not imputed when there is no law. (NASB)

---

*Eugene Nida and Charles Taber, *The Theory and Practice of Translation* (Leiden, The Neth.: Brill, 1969), pp. 200, 201.

> Sin came into the world through one man, and his sin brought death with it. As a result, death has spread to the whole human race because everyone has sinned. There was sin in the world before the Law was given; but where there is no law, no account is kept of sins. (TEV)

The first of these embodies formal correspondence, the second dynamic equivalence. One learns from the first as much as any translation can teach about the close details of sense, form, and order of the original while still being within the limits of recognizable English. The second gives the general sense of the passage in easily accessible English at the expense of the close details. A reader will find one of these appropriate in certain circumstances, the other appropriate in other circumstances.

Can translators go too far in one direction or another? Yes, without a doubt. In the direction of formal correspondence, they may go so far as to produce a work that is more Hebrew or Greek than English. In the direction of dynamic equivalence, they may go so far as to produce a text that is smoother and easier for modern readers than it ever was for its first readers. In this latter case the translators' concern for their readers' limited abilities may prompt them to indulge in interpretation rather than translation of the text. There is a fine line that divides making the sense of the original clear and interpreting that sense, and translators must be careful not to cross it. That line was certainly crossed by the producers of one nineteenth-century revision of the KJV. These revisers believed firmly that baptizing in biblical times was done in only one way—by immersing converts in water—and they felt that the world needed this truth plainly stated. So wherever they found any form of the word "baptize" in the KJV, they substituted the appropriate form of "immerse" for it. They laid hands even on John the Baptist, renaming him "John the Immerser"! This kind of "help" for readers belongs in clearly labeled interpretive notes to the English text, not, misleadingly, in the text itself.

USING SEVERAL TRANSLATIONS

To be dependent on a translation, as most Bible readers are, is to be dependent on a number of qualities in the makers of the translations: how much they know of the Bible's original languages and of the cultures, religious systems, and historical situations that produced the original texts; the depth of their commitment to scholarly objectivity; their skill and imagination in using the receptor language—in our case, English; and their awareness of the true nature of the process of translating. Average readers of the Bible are not equipped or inclined to consider whether the translation they are using was produced by persons possessed of all these qualities; they just assume that what they have before them is the Bible—and that's that. But allow us to suggest a simple way for average readers to limit their dependence on the abilities and efforts of a single translator or set of translators. That way is merely to make use of more than one translation. By consulting two or three versions, readers can double or triple the number of expert opinions available to them about the meaning of a given passage. At the very least, this will impress on readers a fundamental truth about translations of the Bible: Each one embodies thousands upon thousands of individual decisions about (1) what constitutes the proper text in the first place, (2) what its sense is, and (3) how the sense can best be represented in the receptor language. More pointedly, of course, a comparison of translations will indicate how much confidence an interpreter can put in some particular version's rendering of a given passage. Before investing too much effort in discussing a passage, the interpreter had better be sure that in the original language it has a clear sense that *can* be interpreted.

As an illustration of what one can learn from a comparison of translations, consider the form that Proverbs 18:19 takes in two English Bibles:

> An offended brother is more unyielding than a fortified city, and disputes are like the barred gates of a citadel. (NIV)

> A brother helped is like a strong city, but quarreling is like the bars of a castle. (RSV)

Note that in the first translation the brother had been offended, in the second helped. As though that were not confusing enough, there is a widely divergent reading in another version:

> A brother is a better defense than a strong city, and a friend is like the bars of a castle. (NAB)

Here there is nothing negative whatever, only the helpful brother and friend. The explanation for these considerably different versions is that there are variant readings for several words in the Hebrew of Proverbs 18:19, giving us a text sufficiently uncertain that three bodies of scholars have assigned them three different meanings.

Another instance of profitable comparison of translations involves a passage we looked at in chapter 9 but that now deserves a second look in the context of translation. The passage is a crucial one, Job 13:15, in which the tormented Job speaks of God. Consider the difference between these two versions:

> Though he slay me, yet will I hope in him;
> I will surely defend my ways to his face. (NIV)

> Behold he will slay me; I have no hope;
> yet I will defend my ways to his face. (RSV)

If the first form represents the sense of what the author actually wrote, then we have one kind of Job: a man who, with divine complicity, is suffering terribly but who nevertheless trusts in God's good intentions toward him. If the second version represents what the author intended, then we have another kind of Job: a man who sticks his chin out and says *he* is not afraid of a bullying deity and will insist on speaking the truth. These two senses are possible because there are two different readings in the Hebrew manuscripts available to us. Actually the difference lies in only one letter in one word, but what a difference it makes, giving us a Job who is either a hero of the faith or a rebel against God! The whole context of the passage favors the negative statement. But because pious translators through the centuries (and before them pious editors of the Hebrew text) have recoiled at the thought of a rebellious Job, they have acted on the slim possibility that the positive wording was legitimate and thus have tried to save Job's reputation (although Job, in fact, says very negative things about God elsewhere in the poetic portions of the book). Here we have a case where the translators' desire to maintain a religious tradition has determined what sense they assigned a passage.

A reader interested in pursuing the comparison of translations further may wish to perform the laborious but fascinating exercise of examining each of the following passages as they are represented in a number of English versions (including the KJV and the NJB): Genesis 1:1–2, 21:14; Exodus 6:4, 32:4; Leviticus 16:8; Deuteronomy 4:19; Judges 1:14; 1 Samuel 13:1; Job 9:20; Psalm 2:2, 7; Amos 5:25–26; Jonah 3:3; Micah 4:5; Matthew 6:13; Mark 7:16; Luke 9:55–56; John 14:30–31; Acts 10:19; Romans 8:28; 1 Peter 3:18–19. In examining these passages (and any of the earlier ones cited in this chapter), readers should be sure to consult the textual footnotes in those translations that have them. In one view of things, it would be well for a translation to indicate in the notes every Hebrew and Greek variant reading *not* selected for rendering into English and every possible alternative sense of the Hebrew and Greek that *was* chosen. But that would make for such a vast number of notes that the notes would bulk as large as the text itself. What translators do (those who provide any notes at all) is to present only a selection of variant readings and alternative senses. Careful readers should always make the effort to consult the notes if their translations have them, for consulting the notes—like reading a passage in multiple translations—keeps one constantly aware of the degree of uncertainty that exists with respect to both the state of the text and its meaning.

SUGGESTED FURTHER READING

S. L. Greenslade, ed., *The West from the Reformation to the Present Day*. Vol. 3 of *The Cambridge History of the Bible* (Cambridge: Cambridge University Press, 1963).

David A. Lawton, *Faith, Text, and History: The Bible in English* (Charlottesville: University Press of Virginia, 1990).

Eugene Nida and Charles Taber, *The Theory and Practice of Translation* (Leiden, The Neth.: Brill, 1969).

Edwin H. Robertson, *Makers of the English Bible* (Cambridge: Lutterworth, 1990).

Jan de Waard and Eugene Nida, *From One Language to Another: Functional Equivalence in Bible Translating* (Nashville: Nelson, 1986).

The Interpreter's Dictionary of the Bible, ed. George A. Buttrick et al. (Nashville: Abingdon Press, 1962). See articles on Versions, Ancient; Versions, English; Versions, Medieval and Modern (Non-English). Supplement, 1976: See article on Versions, English.

The Anchor Bible Dictionary, ed. David Noel Freedman et al. (New York: Doubleday, 1992). See articles on Septuagint, Vulgate, Theories of Translation, Versions.

XVII

The Religious Use and Interpretation of the Bible

In this book we have discussed the Bible in literary and historical terms. We have considered it to be an anthology of writings, composed during a thousand-year period by scores of writers, each one addressing an individual audience about some specific concern. We have said that these writings were drawn together over a long period of time from a large body of religious literature and gradually admitted by Jewish and Christian communities to their sacred canons. *In literary and historical terms*, the Bible came into existence through human agencies and processes that are not in themselves mysterious. Consequently it can be discussed in much the same way as Homer's *Iliad* or any other ancient literary work.

But plainly the literary and historical matters that we have dealt with in earlier chapters do not exhaust the possibilities of interest in the Bible. There is, obviously, the whole religious dimension of the Bible—the religious use to which it is put, once it has come into existence—which lies outside a literary-historical concern but which is the only concern that matters for most Bible readers. The majority of priests, rabbis, and ministers today have had systematic training in the human agencies and processes that brought the Bible into existence (little that is in this book would be new to them). But most members of their congregations and their Bible classes have no knowledge of and no interest in such things. For them it is sufficient to know simply that the Bible is a revelation from God (however it was that God inspired the human authors and employed human means to express his word), and they go to the Bible uncritically to draw from it what God wishes to tell his people.

WHAT RELIGION HAS
DRAWN FROM THE BIBLE

What has religion, in fact, drawn from the Bible? What has it found there to make use of? Six categories of religious material can be isolated that have their source in the Bible.

1. *Sacred history.* As we pointed out in chapter 4, historical material constitutes a large part of the Bible. Both Judaism and Christianity value this material highly, for each is a historical religion originating from specific events that believers feel were under the direct control of God. For Jews the sacred history comprises the selection of Abraham and his descendants, the bondage of those descendants in Egypt and their rescue therefrom, the establishing of Israel as God's own nation, the placing of David and his family on the throne in Jerusalem, the destruction of Israel's northern kingdom and exile of Judah to Babylon, and the return from exile and resumption of religious life in Jerusalem. Christians claim all of the preceding as part of their own sacred history and add to it the birth, ministry, death, and resurrection of Jesus and the spread of the primitive church. Both Jews and Christians are enjoined to recall the central events of their sacred histories at regular times in their religious calendars (Passover and Hanukkah, for example, or Easter and Christmas).

2. *Theological doctrines.* What can one know about God? What is his nature, what are his wishes, how does he view humankind? What happens when we die? Is there life after death? If so, what form does it take? Theological questions like these find no answers in human experience or logic. If there are answers at all, they must come from a supernatural revelation. For Judaism and Christianity, that supernatural revelation is the Bible.

3. *Moral precepts.* How are people to act toward one another? How are we to respond to others who do us wrong? Are we doing enough for others if we simply obey the law of the land? Is it ever right to kill another human being? What are the limits of legitimate sexual relations? Do some classes within society deserve our special concern? The Bible is certainly not the only place where answers to questions like these can be found: How human beings should relate to one an-

other has been a major topic of writers from the earliest times. But for Jews and Christians, the Bible is the chief source of moral guidance; in the religious view, the Bible's precepts concerning how we should treat one another derive directly from God himself.

4. *Ecclesiastical structure and practice.* How should a body of believers organize themselves? Where does authority lie in a local congregation? Should there be human authority at a higher level than that of the local congregation? Although the Bible does not contain much material directly bearing on these questions, what it does contain has had considerable effect on the organizational structure of religious bodies at both the local and supralocal levels. Once such a body comes into existence, the question arises as to what should happen when its members gather together. In substantial part, what is said and done in Jewish and Christian services—the detail of their liturgy—derives from the Bible. Liturgy includes not only the words and actions of a service but also the pattern into which they fall—the shape the service takes—and this pattern, too, usually has a biblical basis. Within the service, individual blocks of the liturgy, such as creeds, responsive readings, hymns, and prayers, also owe much to the Bible.

5. *Ideas about the end times.* Future history is properly a division of theology—specifically, of eschatology, the study of the "last things." But for some religious bodies within the Judaeo-Christian tradition, eschatology is far more than merely one theological matter among many; it is their self-defining and all-consuming interest. Such bodies tend to employ the Bible as a guidebook to the end times and as a device for interpreting events of the present, which is understood to be a preliminary phase of the end times. Even groups not so deeply concerned with eschatology may conceive of history as heading toward a climactic point, and they too will use scriptural categories in discussing the ultimate fate of humankind.

6. *Personal guidance.* Quite apart from its use in organized religion and on public occasions, the Bible is employed by individuals for their own private purposes. Some read straight through the Bible, chapter by chapter (perhaps one a day), as a kind of religious exercise; others follow a plan of reading worked out by a religious organization or a Bible teacher. In general such an exercise serves to reinforce the reader's system of religious belief; in particular, it may serve to provide clues that the reader can interpret as divine guidance for the conduct of daily life or the making of difficult decisions. One extreme form of employing the Bible to secure personal guidance, a form probably used much more in the past than in the twentieth century,

is to flip the Bible open at random and to read the first verse the eye happens on. That verse is taken as a revelation of God's will for the given day or situation.

One further use of the Bible deserves to be mentioned, although it does not, in contrast with the previously mentioned uses, involve something drawn from the Bible. This is the use of the Bible as an *object* with special symbolic significance. In Western society, the physical Bible itself serves on formal occasions as a representative of the deity or, for the nonreligious, of whatever absolute authority or principle they believe in. When employed this way, the Bible can be referred to as an "icon." It is the Bible as icon that the witnesses in law courts place their hand upon while swearing that they will tell the truth, and that persons being inaugurated into important offices place their hand upon while swearing to undertake faithfully the duties expected of them. It is the Bible as icon (as opposed to the Bible as source of texts) that fiery evangelists wave in the air or thump on the pulpit to underline the points of their message. It is the Bible (inevitably a large-sized one) as icon that families pass on from one generation to the next and in which they inscribe the facts of family births and deaths, thus indicating the continuing commitment of the family to religion. An extension of the use of the Bible as icon is its use as talisman, that is, as a charm—as something with magical powers. In this sense the Bible functions just as the crucifix or the cross does in horror films when brandished in the face of a werewolf or demon. The soldier who carries a small Bible into battle for the sense of security it provides is employing it as a talisman.

LIMITATIONS ON THE USEFULNESS OF THE BIBLE

Religious Jews and Christians, then, draw from the Bible certain general categories of material. Does this mean that it is possible to go to the scriptures and find these categories systematically presented? And are they available for immediate application to one's life? By and large, the answer to both of those questions is no. The Bible's own categories of organization—established by its authors, editors, and copyists over many hundreds of years—correspond only approximately in some cases, and not at all in others, to the categories of modern use of the Bible. Materials that could be assigned to the six categories defined earlier do exist in the Bible but are not neatly isolated for our use. And those materials are rarely *directly* applicable to life.

One limitation on their applicability is that the scriptural passages concerned with any single topic are not necessarily consistent; their varying emphases and occasional contradictions require that careful interpretation be done before they can be applied. Another limitation is that the bulk of the Bible is not written in forms that are immediately usable for religious purposes. Sacred history can, of course, be drawn directly from the scriptures. But much that relates to theology and morality occurs in the narrative portions of the Bible; thus principles and rules must be inferred from stories. A third limitation on the applicability of biblical materials is that they were not *in the first instance* addressed to us. In the narratives about Moses, for instance, Moses is portrayed as speaking to Israelites living near the end of the second millennium B.C.E., and the persons who composed and edited those stories intended them for Israelite audiences of their own times (the tenth to sixth centuries B.C.E.). Likewise, in the narratives about Jesus in the New Testament, Jesus is portrayed as speaking to Jews of the early first century C.E., and the persons who wrote those narratives did so for specific audiences in the late first century. How then do the words of Moses or the words of Jesus apply to persons of later times? Are Jews today remiss if they do not make animal sacrifices as Moses required? Are Christians at fault if they buy fire insurance or contribute to a pension fund and thus violate Jesus' command to take no thought for the morrow?

Plainly, those who use the Bible in a religious way must first select from the Bible what seems to them to be significant and then must interpret it in such a way as to make it consistent within itself and relevant to some overall religious system. It will be instructive to see how this process works in relation to several of our six categories.

SELECTING DOCTRINAL MATERIAL

Consider first the category of theological doctrine and, for purposes of illustration, the doctrine of baptism in particular. Baptism is an extremely important matter for Christian churches, for in the view of most of them it is one of the crucial steps toward becoming a Christian. Yet, important as all will agree it to be, there is a notoriously wide range of views among churches about even the simplest matters connected with baptism. How is the act to be done? By immersing the new believers? By sprinkling water on them? By pouring it? Should it be done to infants or only to those who can understand its meaning? Does salvation occur simultaneously with baptism, or does it precede or follow it? Can there be salvation without baptism? Must baptism with water be accompanied by "baptism with the Spirit," as evidenced by speaking in mysterious tongues? Disagreement over these questions has split many a church through the centuries, and whole new denominations have been established by those who happen to agree on the answers.

But how is such a marked disagreement possible? What, after all, does the Bible *say* about baptism? Well, it "says" a great deal—or more accurately, baptism is touched on at many points in the separate writings that constitute the New Testament. Unfortunately, at none of these points is there a systematic discussion of the act: how and when and to whom it should be done or precisely how it relates to other religious acts. What we have instead are (1) accounts in the four gospels and in Acts of persons baptizing and being baptized, and, in some of those accounts, brief remarks made by individuals concerning the matter; and (2) unsystematic references to baptism here and there in the New Testament letters. Just as unfortunately, some of these accounts and scattered references are ambiguous, and not all can be made to square with one another. So each individual Christian denomination has selected certain biblical passages as significant to its view of baptism and has shaped from these passages a consistent doctrine for its members.

What is true of baptism is true of a great many other doctrinal matters—the nature of the Godhead (in Christian terms, the Trinity), the relationship between Jesus' human and divine natures, the problem of evil, the problem of suffering of the innocent, the meaning of the Christian communion service, and so on. About these matters there has always been a wide variety of opinion among believers. And that is so, basically, because the Judaeo-Christian scriptures contain a wide variety of materials that can be interpreted in a wide variety of ways.

SELECTING MORAL MATERIAL

Consider next, in relation to this same situation, the moral codes that religion derives from the scripture. The Bible does contain several blocks of moral prescriptions, directly stated. But are these necessarily applicable outside the situations for which they were originally intended? The Decalogue (Ten Commandments) is generally held to be so—although similar material in passages immediately following the Decalogue would not be considered moral today because it advocates taking "an eye for an eye" and speaks approvingly of slavery. The moral prescriptions in the book of Proverbs would seem to be still generally applicable, although many individual proverbs tend to be more prudential—that is, self-serving and self-protective—than altruistic. And some are morally very questionable. What is one to make, for example, of the direction in Proverbs to feed your enemy when he is hungry and give him water when he is thirsty so that "you will be heaping red-hot coals on his head, / and Yahweh will reward you" (25:21–22)? What sort of a motive for a good action is *that?* The words of Jesus in the gospels can be considered as setting a high level of morality for his followers through the ages—so high, indeed, that some of Jesus' principles are looked on as not really practicable. What nation, even if it considers itself Christian, is willing to have no army and to turn the other cheek so that its enemies might attack it not once but twice? And what can be made today of Jesus' warning, "Everyone who divorces his wife and marries another is guilty of adultery" (Luke 16:18)? To what extent is *that* hard teaching still being taken seriously, even by the churches?

Religion, then, must do some very careful selecting and suppressing when it goes to the scriptures for direct statements about morality. It must exercise equal caution when it attempts to derive moral principles indirectly from biblical stories, for some actions and persons considered laudatory in the biblical setting would be considered reprehensible in a modern setting. For example, according to the historical books Israel was doing God's will in annihilating the people of Canaan ("He left not one survivor and put every living thing under the curse of destruction, as Yahweh, God of Israel, had commanded" [Joshua 10:40]). But can we infer from this that genocide is sometimes legitimate? The Israelites were allowed to make slaves of non-Israelites. But does that justify slavery? David committed adultery with another man's wife and then had the man murdered to cover his act. Yet his son Solomon was exhorted by Yahweh to "walk before me in innocence of heart and in honesty, like your father David" (1 Kings 9:4). And more than one later king is said to have done "what Yahweh regards as right, as his ancestor David had done" (1 Kings 15:11; see also 2 Kings 18:3, 22:2). Are adultery and murder not such terrible crimes when committed by certain special people?

SELECTING
ESCHATOLOGICAL
MATERIAL

Even more than is the case with theological doctrines and moral precepts, views of the last days can be constructed only by carefully selecting some parts of the Bible's abundant eschatological materials and suppressing others and then carefully shaping the chosen parts into a consistent whole. What is selected is mainly those passages from the prophetic and apocalyptic writings that refer to the future—the future, that is, from the point of view of the biblical authors. What is suppressed is everything in the prophetic and apocalyptic writings that ties them to the long distant past. The books of Ezekiel and Revelation are much analyzed and picked through by those who construct views of the last days. Yet both these books were thoroughly concerned with the times in which they were written. The nations and territories referred to by the prophet Ezekiel constituted the world as it was known to an inhabitant of the Near East in the sixth century B.C.E. The situations of which he spoke were those of his own time and (he believed) of the near future. Similarly, the visions in Revelation referred to a state of affairs existing at the author's time in the late first century C.E. The future of which the author spoke was, from his point of view, just around the corner; at both the beginning and end of his book he said that "the Time is near" (Rev. 1:3 and 22:10). But such indications that prophetic and apocryphal works were oriented to their authors' times are inconvenient to persons who construct elaborate scenarios of the last days. Those passages are thus ignored, and others are chosen for emphasis that can be interpreted as referring to Russia and the United States and NATO and thermonuclear war, and so on.

We have been making the point that the categories of material traditionally drawn by religion from the Bible are mainly constructed categories—constructed by making a selection among the biblical texts and then shaping the pieces into consistent wholes. (Only sacred history is excepted from this judgment, since it is a category that the biblical texts do, in fact, present in a systematic and unified way.) The same categories exist for all bodies* within Judaism and Christianity, but the content of the categories varies from body to body according to the characteristic emphases and needs of each body. Naturally each body, of which there are many thousands worldwide, holds that its particular formulation—the total package that it shapes from the scriptures—is the right one. (There is no room here for modesty: Any religious body that did *not* hold its formulation to be the right one would have no reason for existing.) There are thus as many formulations of biblical material—as many packages—as there are bodies within Judaism and Christianity.

But what makes such a range of formulations possible? Does each body's characteristic package differ from all the others simply by virtue of what is selected from the Bible to put into the package? Certainly selection is important in this respect; but equally important is the difference of opinion among interpreters as to what any given passage means.

---

*We are using "bodies" here as a shorthand term for the divisions within Judaism and for the churches, denominations, and sects within Christianity.

REASONS FOR VARYING
INTERPRETATIONS

There are some obvious reasons why a passage is capable of being variously interpreted. The passage may occur in more than one form in the earliest biblical manuscripts. Or information essential to an understanding of the passage may have been lost through time (we may no longer know what some individual words mean, for example). Or the author of the passage may simply have failed to state his meaning clearly. Or he may have been purposely ambiguous, perhaps through employing figurative language (metaphors, personifications, symbols, and so on) with no certain reference. These potential explanations for the existence of varying interpretations apply to all works of literature, including the Bible. But there is another explanation that applies to the Bible alone, reflecting its unique status in the estimation of its interpreters. According to the view of traditional Judaism and Christianity, every biblical book had not merely a human author but a divine one as well. The human author employed all of his own ability and personality to compose his work, but God was so thoroughly involved in the process that the completed work says precisely what God wished it to say. And often, according to the traditional view, what God wished the Bible to say was something more profound or more far reaching than what the human author was apparently saying.

Consider the following passage in this connection:

> Get you behind your walls, you people of a walled city;
> the siege is pressed home against you:
> Israel's ruler shall be struck on the cheek with a rod.
> But you, Bethlehem in Ephrathah,
> small as you are to be among Judah's clans,
> out of you shall come forth a governor for Israel,
> one whose roots are far back in the past, in days gone by. (Mic. 5:1–2, NEB)

The human author here, writing in the mid-eighth century B.C.E., was predicting that in the not very distant future Israel would suffer a defeat, but then one of David's descendants would take the throne and presumably set things right. Eight centuries later, however, Christian interpretation (as we have it in Matthew 2:6) took this as a reference to Jesus. Micah, as author, was speaking of an event soon to happen in his own time; God, as author, was speaking of both that event *and* of the birth of Jesus. This passage could be said thus to have a literal sense and a sense that goes beyond the literal. (Whether the human author was ever aware of the more-than-literal sense is something not agreed on by practitioners of this sort of interpretation. For simplicity's sake we shall speak of the human author as not being aware—or at least not completely aware—of the full meaning that God intended.)

The literal sense of the above passage can be worked out by anyone who has a knowledge of the meaning of its words and a knowledge of events in the Near East in the eighth century B.C.E. But the more-than-literal sense can be worked out only by those to whom God has granted special insight, which is what Christians believe the New Testament writers, like Matthew, possessed. It is also what is claimed by Jews for the ancient rabbis and by Christians for the church fathers, for official church councils, and (by Roman Catholic Christians) for the Pope. And it is what has been claimed, right up to our own time, for and by a host of Bible scholars, religious leaders, and preachers. With so many and such widely differing commentators through the centuries interpreting the Bible not merely literally (a task difficult enough in itself) but more than literally, it is small wonder that for any given scriptural passage a number of diverging interpretations now exist.

Biblical interpretation is a tremendously complex and difficult field of study, made so in part by problems with the terminology employed in discussing it. Before proceeding further, we must give some attention to this matter.

LITERAL SENSE

We spoke confidently of the "literal" sense and the "more-than-literal" sense of a passage in Micah, as though those were sure and unambiguous labels. But consider the problems involved with the term "literal." It comes from the Latin *littera* (meaning "letter" and by extension "word"). So the literal sense of a passage is presumably the basic meaning of its words taken as a unit—the commonsense meaning that is there before one begins interpreting. But what is the basic meaning of the statement "Yahweh is my shepherd" (Ps. 23:1)? Is the deity really a shepherd tending sheep? Obviously not. Here, and in all figures of speech, basic meaning lies not on the literal but on the figurative level, and "literal" is thus not a logical term to apply to the basic, commonsense meaning. Despite illogicality, most interpreters do nevertheless use the term "literal" for the basic sense, expecting their readers to understand that the literal sense also at times includes meaning that is communicated by figures of speech. Some avoid the problem by using such terms for the basic meaning as "plain sense," "historic sense," "sense intended by the author," or "carnal sense" (as opposed to "spiritual sense"). Each of these has its value. But the term "literal" is so well established in this context that we shall go ahead and use it here.

MORE-THAN-LITERAL SENSE

The situation with the more-than-literal sense is, understandably, even more complex than that with the literal sense. Like the literal it is frequently designated by alternative terms: "fuller," "inner," "higher," "deeper," "figurative," "symbolic," "spiritual," "mystical," and others. Beyond that, this sense is usually conceived of as being multiple—or to put it another way, interpreters have found a number of more-than-literal senses of the scriptures. One of the most obvious of these has to do with the words of the Hebrew prophets. As we saw in connection with the passage from Micah, on the literal level prophets spoke to and about their own times; but on a higher level, according to interpreters, they spoke to and about future times—and specifically, according to the Christian view, about the time of Christ and the end times. Another of the more-than-literal senses derives from what is called "typological" interpretation, according to which certain historical events, persons, and things in the Old Testament are held to be precursors ("types") of corresponding elements ("antitypes") in the life of Christ and in the lives of Christian believers. Thus the Israelites' passage through the Red Sea (at the time of the exodus from Egypt) is a type of Christ's baptism and of baptism in general, Moses' raising the bronze serpent to heal the victims of snakebite (Num. 21:4–9) is a type of Christ's being raised on the cross for the salvation of humankind, and so on.

Yet another of these senses is the allegorical. This is an elevated spiritual or moral sense derived by projecting upwards—onto a high-level screen, as it were—some apparently unassuming scriptural passage. Thus, for an allegorist, the passage in Genesis 1:27 "male and female he created them" is allegorically a reference to Christ and the Church. The law against eating swine's flesh in Deuteronomy 14:8 is an allegorical warning against associating with swinish persons, and (in one of the largest efforts of this kind) the erotic poetry of The Song of Songs is an allegory of the relationship between God and Israel or God and the individual soul or Christ and the Church or Christ and the individual believer, depending on who is doing the interpreting. Even the dull genealogical sections of the book of Numbers will be edifying, it is sometimes claimed, if only one understands their allegorical import. Now, as we said in chapter 2 in discussing the literary devices of the Bible, there are certain scriptural passages that everyone will agree to be allegorical, for so their authors intended them. When, for example, the author of Ephesians 6:13–17 urged believers to clothe themselves in "all God's armour"—which includes a breastplate of righteousness and shield of faith, and so on—he was plainly writing allegorically; we cannot fail to understand that pieces of armor stand for spiritual qualities. But through the centuries a great many less likely passages in the Bible have also been assigned a spiritual or moral sense.

There are other more-than-literal senses that interpreters have discovered in the scriptures, including those derived from etymologizing the Bible's words (that is, going back to their source meaning), from paying close attention to certain numbers understood to be symbolic (three, four, seven, ten, twelve, and so on), and from the practice known as "gematria" (which assigns numerical values to Hebrew and Greek letters, takes some significant biblical name or phrase and totals the values of its letters, and then determines who or what in modern times has the same number-value and thus is "really" being referred to; the mysterious number 666 of Revelation 13:18 involves this sort of calculation.) One well-known medieval system of interpretation presupposed the existence in the Bible of four senses: the historical (that is, literal), the tropological (moral), the allegorical (doctrinal), and the anagogical (concerned with the final events of an individual's existence). Just what each of these senses was supposed to include varied with the interpreter, and the terms attached to them had a disconcerting way of sliding about, expanding and contracting, now applying to one thing and then to something else. The word "allegory" itself has always been troublesome in this respect. At times it has been used, as we used it earlier, to designate a standard literary form. At times its meaning has been extended to include typology. And at times the meaning of "allegory" has been so thoroughly extended as to cover all of the more-than-literal senses. Thus the very practice of reading the Bible for more-than-literal meaning is often called simply "allegorizing" and those who do it "allegorizers."

One important qualification must be made before we proceed. When we speak of interpreting for more-than-literal senses, we are not speaking of what is involved in drawing useful examples or principles or parallel cases from the Bible. When ministers say in their sermons that "just as such and such happened in biblical days, so in our time we see such and such happening," they are not allegorizing the scriptures; they are simply explaining the significance of contemporary situations in terms of familiar biblical ones. Allegorizing, as well as reading with an eye for prophetic and typological meaning, requires the prior assumption that the literal level of biblical meaning will not do, in and of itself, and that the full meaning of the scriptures reaches beyond the literal.

MOTIVES FOR
NONLITERAL
INTERPRETING

What is the reason for conceiving of the scriptures in this way? Why should interpreters insist that the Bible contains meanings that any fair-minded reader may well doubt are really there? Why not simply take the Bible's words at face value? There are two basic motives for more-than-literal interpretations, but the two are so intertwined that in practice they can scarcely be distinguished. One is backward looking, concerned with affirming the truth and relevance of the ancient scriptures. The other is forward looking, concerned with extending the authority of the scriptures up to the present in order to serve as a justification for current beliefs and practices. We might identify the first of these motives with Judaism in its early centuries and the second with Christianity in its early centuries; but in time the two motives blended and operated with equal force in both faiths.

IN EARLY JUDAISM

Consider how the urge to maintain the relevance of the scriptures demonstrated itself, near the beginning of the existence of Judaism, when the first section of what was to become the Jewish Bible was being compiled. Even as the Torah was assuming the form in which we now have it, during the sixth and fifth centuries B.C.E., much of its content was no longer capable of direct application to contemporary Jewish life. Its material, derived from the ancient past, was revered as being the word of God, but it needed some sort of adaptation if it was to be relevant to the real life of Jewish communities. One means of adapting it was to extract from the particulars of the Torah a set of religious and moral principles and instructive parallels to situations in contemporary life. But beyond this the Torah required to be read figuratively so that senses other than the timebound literal one could be drawn from it.

The same is true of the second division of the Jewish Bible, the Prophets. The prophetic works were valued as a record of the words that the great ancient spokesmen for God, such as Isaiah and Jeremiah, had directed to their own times. But there was much that the prophets had said would happen that had not in fact happened. How could this be if the prophets' words were God's words? Might it be that the prophets were actually speaking not of their own time but of future times and that their prophecies were yet to be fulfilled? We can see the book of Jeremiah being interpreted this way by the author of Daniel in the second century B.C.E., and perhaps it was the very possibility of such interpretation that brought the Prophets into the canon of sacred scripture not long before Daniel was written.

It was not only the passage of time and changing conditions that raised doubts about the continuing value of the scriptures. After the Greek conquest of the ancient world and the spread of hellenism, the question inevitably arose for Jews about the relevance of their ancient scriptures in the light of the claims of Greek philosophy. That question was addressed by Philo of Alexandria (discussed briefly in chapter 11), an Egyptian Jew of the first century C.E. who wrote a large body of work designed to prove (more to pagans than to Jews, perhaps) that there was no conflict between the Jewish Bible and Greek philosophy. To achieve this end, Philo had to resort heavily to allegorical interpretation of the scriptures. Anything there unworthy of God (or the Greek conception of God), anything inelegant or insufficiently elevated, anything apparently too simple-minded—all had to be seen as having a higher, philosophical or ethical sense. This might strike us as doing violence to the Bible, but it is better thought of as honoring the Bible. The interpreter's bedrock assumption is that the scriptures are eternally true and that, if only we make the effort, we can discover how the very greatest of philosophical principles are contained within it. Some of Philo's interpretive methods had already been employed by Jewish rabbis, who were concerned not only (like Philo) to maintain the reputation of the scriptures but to extend their authority to new situations. The second of these motives was also particularly at work among the Christians of the first century C.E.

IN EARLY CHRISTIANITY

Reading the Jewish scriptures for their more-than-literal sense began very early in the history of Christianity. We know from Paul's letters (the earliest part of the New Testament to be composed) that this kind of interpretation was well developed in Christianity by the middle of the first century C.E. According to the gospels, Jesus himself practiced it: for example, "Then, starting with Moses and going through all the prophets, he explained to them the passages throughout the scriptures that were about himself" (Luke 24:27). The scriptures of the Jews were the scriptures of Jewish-Christians as well—indeed, those were the only scriptures recognized by Christians during the first century. If Christians were going to find their movement authorized in God's word, it would have to be within the Jewish scriptures. Jews who were not Christians objected to this, quite understandably, arguing that their Bible had Jewish meanings exclusively. To counter this, Christians insisted that the Bible had meanings that transcended the "carnal" Jewish ones. Very often in Paul's letters we can observe him assigning more-than-literal senses to passages from the Jewish scriptures. He does so in 1 Corinthians 9:9–10, where he is trying to prove that he has a perfect right, although he won't act on it, to take some personal advantage of his ministry. Note how he contemptuously dismisses the literal sense: "In the Law of Moses we read, 'You shall not muzzle a threshing ox.' Do you suppose God's concern is with oxen? Or is the reference clearly to ourselves? Of course it refers to us, in the sense that the ploughman should plough and the thresher thresh in the hope of getting some of the produce" (NEB).

In another place, Paul is arguing that the followers of Christ, not the Jews, now constitute the true people of God. To make the point, he employs the account in Genesis of how Abraham had fathered sons by two women, one the slave Hagar and the other his wife Sarah. "There is an allegory here: these women stand for the two covenants" (Gal. 4:24).* Here he does not dismiss the literal or historical sense, as he does in the passage about the ox. But he has no interest in it—only in what he can make of it. Yet Paul can be quite literal-minded when it serves his turn. In a passage that occurs in Galatians just a chapter earlier than the one cited, he is arguing that God had in mind the salvation of the Gentiles through Christ even as he was appointing Abraham to father his special people. For biblical evidence, Paul turns to Genesis and insists that God's promises made to Abraham's "progeny" were really made to Christ, for the Hebrew word used in Genesis 12:7 (and other similar passages) is singular, not plural. "The words were not *and to his progenies* in the plural, but in the singular; *and to your progeny*, which means Christ" (Gal. 3:16). Paul manages here to use a highly literal reading of the Genesis passage for a considerably more-than-literal purpose: "progeny" *must* refer to Christ and not—as common sense seems to require—to Abraham's descendants.

The gospels and Acts, as we reminded the reader earlier, were written two to four decades after the first of Paul's letters, at a time of increasing animosity between Jews and Christians. Their writers frequently cited the Jewish scriptures, most often the prophets, for

---

*In modern critical terminology this would be an instance of typology rather than allegory.

evidence that events in the life of Christ and in the early years of the Christian church had been referred to in scripture long before and thus were part of God's plan—indeed, were the great end toward which God had been directing all of history. In working out a sense applicable to their own situations, Christians ignored the sense intended by the authors of the prophecies. All four gospel writers, for instance, report that John the Baptist's activity as the forerunner of Jesus had been foretold by Isaiah: John was "a voice of one that cries in the desert: / Prepare a way for the Lord, / make his paths straight" (Mark 1:3). But these words, as they occur in Isaiah 40:3, refer to the time in the sixth century B.C.E. when the Jews living in exile in Babylonia were about to journey to faraway Judea. The voice is a heavenly one, instructing that a highway be prepared in the wilderness for the exiles to travel upon in comfort. Note that in the New Testament form of the passage, it is the *voice* that is located in the wilderness, not the highway. By that alteration, the passage was made particularly applicable to John the Baptist, a rough figure of a man who did his preaching and baptizing in the wilderness along the Jordan River.

The gospels contain relatively little typological and allegorical, as compared to prophetic, interpretation. Mark and the other synoptic gospels do present one very significant piece of allegorical interpretation, that of Jesus' parable of the sower. The parable is quite able to stand on its own as a parable, which seems to be all that Jesus intended it to be. But, as we indicated in chapter 2, the first gospel writer, Mark, probably capturing a reading of it that was current in the early Church, provides it with a full-dress allegorical interpretation (4:13–20). Elsewhere in the New Testament, the epistle to the Hebrews displays all the kinds of more-than-literal as well as literal interpretation of the Old Testament. Its author treats Old Testament material literally when he uses great characters from the Bible as illustrations of godly behavior or when he describes Jewish religious rituals. He interprets prophetically when he says that Jeremiah's promise of a new covenant has now been fulfilled through the death of Christ. He interprets both allegorically and typologically when he cites as a "type" of Christ, who is God's eternal priest, the Old Testament figure Melchizedek (a priest whose parentage and death happen not to be mentioned in the Genesis passage where he is treated, and who can therefore be thought of allegorically as never having been born and never having died!). Throughout, the author of Hebrews speaks in terms of shadows and symbols—of old covenant elements which, when understood aright, prefigure the elements of the new covenant secured for humankind through Christ's mediation with God. The book of Hebrews provides a dazzling display of how to interpret scripture for senses beyond the ones intended by the authors of scripture.

The tendency toward more-than-literal interpretation that is evident in the New Testament was confirmed and reinforced for Christians when the individual writings that now constitute the New Testament achieved canonical status. If Paul and the gospel writers and other New Testament authors could go beyond the literal to find more profound senses, then it was certainly legitimate for other Christian interpreters to do so. The long reign of more-than-literal interpreting of the Bible was now well inaugurated.

IN LATE CLASSICAL AND
MEDIEVAL TIMES

We have said that one motive for this sort of interpreting was to affirm the eternal truth and relevance of the scriptures. This was the motive of Philo, and it was the motive that energized the Jewish and Christian defense of the sacred writings against pagan criticism in the late classical and early medieval centuries. When pagan philosophers made light of the Judaeo-Christian scriptures for containing impossibilities (for example, the sun standing still in Joshua's time, the miracles of Christ), indecencies (Abraham's marriage to his half sister, Lot's incest with his daughters), discrepancies (as to whether David or Elhanan killed Goliath), or blasphemies (God's walking in the Garden with Adam), then an interpreter could respond that the accounts of these unlikely events were actually allegories containing hidden truths.

The other motive for nonliteral interpreting, to permit the extending of biblical authority forward to cover new forms and new practices, came into play in late classical and early medieval times when both Judaism and Christianity found themselves in drastically different situations from anything the authors of the Bible could have envisioned. Judaism lost its Temple (as a consequence of the Jewish-Roman war that commenced in 66 C.E.) and then its access to the holy city, Jerusalem (as a consequence of Bar Kochba's revolt in 135 C.E.). The very survival of Judaism depended on its finding new forms of worship and religious expression—but only such forms as could somehow be justified from the Bible. For this purpose, the freedom to go far beyond the literal sense of the text was absolutely essential. Christianity experienced an equally drastic but quite different sort of change, developing as it did from a decidedly fringe religion appealing to minority groups in the first century to the official religion of the Roman Empire in the fourth. How could the teachings of Jesus, directed to his simple Galilean followers, or the advice of Paul, directed to tiny Christian communities scattered here and there in the ancient world, ever be applied to a highly organized, wealthy, and powerful Church centered in Rome? Again, the Bible had to be very freely interpreted if this ecclesiastical structure—so unlike the primitive Church described in Acts—was to find its warrant in God's word. In addition, Christians were faced with the continuing problem mentioned earlier of appropriating the Jewish Bible to their own use. To accomplish this, and to justify doing so, interpreting the "Old" Testament (a term that shows a plain Christian bias) for senses beyond the literal was a necessary tool.

REACTIONS AGAINST
SUCH INTERPRETATION

Once the habit of reading for hidden senses becomes established, whether in individual or institutional interpreting, it is difficult to limit it or ever again to give primary attention to the plain sense of the text. A few of the best-known interpreters of the Bible through the ages, such as Jerome and Luther, did indeed draw back from their early enthusiasm for allegorizing (in the strict sense of that word); but as Christians they never ceased interpreting the Old Testament prophetically and typologically. In reaction to the Christian use of the Jewish scriptures to support Christian positions, medieval Jews grew increasingly cautious about employing the more-than-literal senses. Some of the greatest medieval Jewish interpreters—among them Saadia Gaon, Solomon ben Isaac (known as Rashi), Abraham Ibn Ezra, and David Kimchi—insisted on the primacy of the literal, historical sense of the scriptures. But even they did not abandon the traditional Jewish way of allegorizing. And other medieval Jews, with a strong bent toward mysticism, pushed the art of allegorizing to extremes. The Protestant Reformation brought with it both a new appreciation for the literal sense and a new emphasis on the accurate texts and knowledge of languages requisite to the study of literal sense (the reformers having been considerably influenced in this by Jewish biblical scholarship).

But it was not until the flourishing of the "Higher Criticism"—the study of the Bible in literary-historical terms, which began about two centuries ago—that appreciation of the literal sense came into its own. The rigorous efforts of European scholars, particularly in Germany, gradually established the probable dates of composition of biblical writings, the identities and situations of their authors and first audiences, the relation of the writings to Jewish and Christian religious traditions, and the history of their editing, canonizing, and copying.

Better and better texts of the Bible in the ancient languages and faithful translations into the modern languages were produced. The means were available, and have become increasingly so in the twentieth century, for determining the literal sense of the scriptures with some degree of certainty.

These advances in knowledge of the Bible have curbed the worst of the traditional excesses in biblical interpretation. Jesus' parables can now be read simply as parables and not be forced to bear the burden of allegorization. Noah's Flood or the crossing of the Red Sea or other Old Testament references to water need not automatically make one think of baptism. Every reference to a lamb or sheep in the Old Testament need not be considered a type of Christ, nor every serpent a symbol of the Devil. Biblical names can nearly always be conceived of simply as that—names—not as cryptic terms whose hidden meanings must be determined by arithmetic or etymology. Some of the standard ideas about, and ways of interpreting, the scriptures in the past would now strike even the humblest of Bible students as absurd.

MODERN NONLITERAL INTERPRETATION

But though today, to a greater degree than in the past, the literal sense of biblical writings is permitted to stand on its own, we certainly cannot say that more-than-literal interpretation is no longer practiced. There are, to be sure, religious groups that take the Bible very literally: for example, those that handle snakes and drink poison in response to the words of Jesus in Mark 16:18. But those same groups will read the Old Testament prophetically and typologically and will not hesitate to take any part of the Bible figuratively when literal meaning does not accord with their faith and practice. And even among the mainline religious bodies there is still the historic need to defend the Bible's worth and to argue for its contemporary relevance, still the need to employ its authority to justify religious ideas and institutions that its authors could never have envisioned and did not address themselves to. Where drawing principles and examples from the literal level will not serve, then appeal can be made to the more-than-literal. The ancient conception of the Bible as a collection of oracles—individual statements that can be plucked out and used as proof texts for one's views, as clues concerning the future, and as advice for this day's business or this moment's crisis—that conception is far from extinct in our own time. To make the Bible work in any of these ways demands considerable latitude in interpreting its words, the sort of latitude that more-than-literal interpreting makes possible.

RELIGION AND THE BIBLE

In the popular conception, the Bible is the source of religion. But the facts of history argue the reverse: Religion creates scriptures, not the other way around. There was a religion of Israel before there were sacred writings in Israel; there was Judaism before there was a Jewish Bible; there was a Christian Church before there were Christian scriptures. True, many individuals in history have read and pondered the Bible and then gone out and established their own religious bodies—churches, denominations, congregations, sects. But that is never done in a vacuum. Such individuals always operate within or in response to a tradition, reacting to this or that element in some already established religion, redefining this or that doctrine or ecclesiastical pattern. The Bible is never the sole or even the chief formative element in the founding of new religious bodies.

Just as religion creates scriptures, religion interprets scriptures—or, more precisely put, it is a function of each religious body to mediate the Bible to its members. The Bible is not self-explanatory. It is a large and complex anthology of works drawn from the writings of ancient Judaism and Christianity. To function in religion, the Bible must be picked and chosen from and its materials considerably interpreted. Such a process of selection and interpretation implies a prior system of religious faith and practice. In a sense, the Bible has no religious meaning until we see it through religious eyes. Our religious eyesight has been developed through the lenses of our catechism and creed; we have learned how to see at Sunday School, Hebrew School, Daily Vacation Bible School, parochial school, youth group, Bible conference, synagogue, church. Not surprisingly, what we see when we look at the Bible through our religious eyes is what we expect to see—the customary, the familiar. That is not the least of the miracles associated with this remarkable book.

SUGGESTED FURTHER READING

James Barr, *The Scope and Authority of the Bible* (Philadelphia: Westminster Press, 1980).

Stephen D. Benin, *The Footprints of God: Divine Accommodation in Jewish and Christian Thought* (Albany: State University of New York Press, 1993).

The Cambridge History of the Bible, 3 vols. (Cambridge: Cambridge University Press, 1963–1970).

R. J. Coggins and J. L. Houlden, eds., *A Dictionary of Biblical Interpretation* (London: SCM Press, 1990).

Robert M. Grant, *A Short History of the Interpretation of the Bible* (New York: Macmillan, 1963).

Frederick E. Greenspahn, ed., *Scripture in the Jewish and Christian Traditions: Authority, Interpretation, Relevance* (Nashville: Abingdon Press, 1982).

James L. Kugel and Rowan A. Greer, *Early Biblical Interpretation*, Library of Early Christianity (Philadelphia: Westminster Press, 1986).

Donald K. McKim, *What Christians Believe About the Bible* (Nashville: Thomas Nelson, 1985).

Bertrand de Margerie, *An Introduction to the History of Exegesis*, trans. Pierre de Fontnouvelle (Petersham, Mass.: St. Bede's Publications, 1993).

David Norton, *A History of the Bible as Literature* (Cambridge: Cambridge University Press, 1993).

Stephen Prickett, ed., *Reading the Text: Biblical Criticism and Literary Theory* (Oxford: Basil Blackwell, 1991).

The Interpreter's Dictionary of the Bible, ed. George A. Buttrick et al. (Nashville: Abingdon Press, 1962). See articles on Allegory; Interpretation, History and Principles of. Supplement, 1976: See article on Interpretation, History of.

Appendix I:
The Name of Israel's God

In the Hebrew Bible the deity of Israel is designated in a number of different ways. Some of these ways are clearly metaphorical, as when he is called a "rock" or a "shepherd" or a "father" or a "king," but we are concerned here only with the literal ones. The most common literal word is *elohim*, the plural of the ancient Semitic word for a deity, *el*. In a few contexts the plural form is obviously meant to be understood simply as "gods," not as a reference to the single deity of Israel; in a few other contexts it is ambiguous. Otherwise it is intended and should be understood as singular, even if accompanied by plural pronouns and verbs (for example, Genesis 1:26, "Let us make man in our own image, in the likeness of ourselves . . ."). This peculiarity—that the plural *elohim* should be read as singular—can be explained as a case of the so-called plural of majesty, an enhancing or magnifying device traditionally used by kings and queens in reference to themselves (and nowadays by not a few politicians). The biblical usage may also have been influenced by the concept of a heavenly court or council that assists in the divine rule (see, for example, 1 Kings 22:19–22). The Hebrew *elohim* and the English "God" are really titles, not names, indicating the role or position of deity, not the deity's identity. The only thing that makes the word "God" at all exclusive is the capital "G"; but even with the capital in place, much depends on the religious frame of reference within which the word is found, or who is using it.

There is, however, a word in the Hebrew Bible that has no ambiguity about it at all, because it is the personal name of the deity of Israel. That is the word we represent in this book as "Yahweh." The name is said to have been revealed to Moses at Mount Sinai when he received a commission to lead his people out of slavery (Exod. 3:13–18). (The story encourages us to suppose that until that moment the divine name had been kept a secret from the very people who worshipped its bearer—otherwise the revelation of the name to Moses has no point. Why the secrecy? The text offers no explanation.) The evidence of the Mosaic story is contradicted by Genesis 4:26, which asserts that the use of "Yahweh" began in the early patriarchal period, before the Flood. In any case the power of the divine name was thought to be great when it was used in prayers, curses, and blessings, in conformity with the universal belief that the essential nature of anything was concentrated in its name. Divine help is requested by invocation, literally by "calling forth," using the name of the deity—and this remains true in our own day. So important did the concept of the divine name become in Israel that the Deuteronomic writers customarily referred to the Temple in Jerusalem as the dwelling of Yahweh's "name." Here the name is virtually identical with the deity himself.

The personal name of the deity appears in the Hebrew Bible as *yhwh* (and is hence called "the tetragrammaton," the four-letter word). That it was written in consonants made it like any other Hebrew word; the original Hebrew writing system did not provide letters to represent most vowel sounds. But this name was sharply distinguished from all other words in the language because it became subject to a taboo: It could be written but it could not be pronounced. There is no real evidence that this had originally been the case; the older Yahwist faith seems to have permitted the use of the divine name by the ordinary believer as a matter of course,* and the name itself is extremely common in the Hebrew Bible (by one count appearing some 6,800 times). During or shortly after the Babylonian Exile, however, when the ancient faith was reformed into what could

*In Deuteronomy 6:13 Moses is reported as saying, "Yahweh your God is the one you must fear, him alone you must serve, his is the name by which you must swear." Joshua 9:18–20 and Jeremiah 16:14–15 both speak in positive terms of using the divine name in oaths. The story in Leviticus 24:10–16 is anachronistic, a late addition to the text.

be called Judaism, the divine name was set off as too sacred to be pronounced and strict rules were laid down to prevent its use under any circumstances.

What was a reader of the Bible to do, then, upon encountering the deity's personal name in the text? (We must remember that in those days reading, even to oneself, was always done out loud.) The solution was to substitute a neutral title, *adonai* ("my lord") whenever *yhwh* occurred. This tradition became firmly fixed and exists in Judaism to this day. Although non-Jewish readers have no reason to observe the taboo, it is respected in nearly all modern translations, following the precedent of the Septuagint and the Vulgate, which used *kurios* and *dominus*, respectively, the Greek and Latin words for "lord." In today's Bible versions, such as the New Revised Standard Version, the Revised English Bible, and the New International Version, the word is printed as "Lord," using a combination of large and small capitals. Every instance of "Lord" in an English text has *yhwh* behind it.

Most of the biblical quotations in the book you are reading have been drawn from the New Jerusalem Bible. The reason is that, among the modern translations of the entire Bible, the NJB is the only one that sets tradition aside and gives the deity of Israel the name the Hebrew texts give him: Yahweh. The great majority of readers take for granted that some word equivalent to "Lord" is in the Hebrew text, but it is not. The word "lord" is a title, not a name; and putting it in capital letters does nothing to change this fact. Of course, there is nothing wrong with designating the deity with a title. But where the Bible specifically has the personal name, translators should not take it upon themselves to make a substitution. The use of "Lord" instead of "Yahweh" effectively depersonalizes the deity, turns him into a kind of vague abstraction, and implicitly rejects the repeated emphasis in the Bible on his unique personal relationship with Israel. It also disguises the fact that Yahweh is a *character* in the biblical drama, with entrances and exits and a role to play, all assigned by the writers.

The objection to designating the deity by a title does not apply to the New Testament, in which *kurios* is the original word used in the text, without any sense of substitution for another word—that is, without any sense of the history we have just outlined. Along with *theos* ("God") it is a proper way of referring to the deity in the New Testament and should be rendered in English as "Lord."

The situation described above is unfortunately complicated by the persistence in English of a wholly mistaken word, "Jehovah." To understand how this came about we must return to a consideration of the Hebrew writing system and what had to be done to prevent *yhwh* from being pronounced. When vowel points—small dots and dashes placed above and below the Hebrew consonants—were developed in early medieval times to indicate which vowel sounds a reader was to use in pronouncing words, every word except *yhwh* was so marked. In the case of *yhwh*, care had to be taken to see that it was *not* pronounced. The device adopted was to add the vowels of *adonai*, already in use as the standard substitute, to the consonants of *yhwh*. (The effect is more apparent in the original Hebrew than in the English transliteration.) Thus the reader coming upon the sacred name would be reminded to say *adonai*. The danger of pronouncing *yhwh* was surely more imaginary than real, for how could people have known how to sound a word they had never heard? But scribal traditions often have very little connection with reality, and in any case the divine name was not a word that could be left open to even the theoretical possibility of misuse. So through thousands of recopyings the hybrid form persisted: the consonants of one word with the vowels of another.

In comparatively modern times (the early sixteenth century is a date often given) someone unacquainted with this tradition took the hybrid form at face value and transliterated it as "Jehovah." Thus "Jehovah" appeared in several English versions, including (in a few places) the King James Version, and became acceptable in English usage. But however *yhwh* is to be pronounced (there is a certain amount of guesswork in rendering it as "Yahweh"), it is definitely not "Jehovah." This fact is now well known, and modern translations of the Bible have virtually dropped the incorrect "Jehovah."

SUGGESTED FURTHER READING

The Anchor Bible Dictionary, ed. David Noel Freedman et al. (New York: Doubleday, 1992). See article on Names of God in the OT.

Appendix II:
Writing in Biblical Times

In societies where Christianity and Judaism are principal religions, the Bible is a common and easily obtainable book. If we happen not to have a copy, the nearest library or bookstore will supply one—indeed, will doubtless offer a choice among versions—and we need not travel with one because there will be a Bible waiting for us in our hotel or motel room when we arrive. Thus, it is hard for us to realize that during the greater part of the Bible's span of existence, copies were scarce and hard to come by, even in nations where there was no political or religious authority anxious to keep it out of the hands of the people. The major reason for its limited availability was simply that the Bible, like all written documents, had to be reproduced by the laborious and expensive process of manual copying. Until the invention of printing from movable type around 1450 C.E., most ordinary persons could never have dreamed of owning—or even handling—a Bible.

The Bible was a handwritten book for a much longer time than it has been a printed book. But what was required to create a formal written document during those many centuries before the printing press took over? As a preliminary to learning the answer to this question, we must free ourselves of preconceptions about writing derived from our own modern culture, with its typewriters and word processors and ballpoint pens and—especially—its abundance of cheap paper. This appendix will survey the technology of writing in order to shed light on the conditions that gave us the Bible as a physical object.

Writing is a very old human invention. But this is not to say that literacy, the ability to read and write, was common anywhere in the ancient world. Though today more than half of the world's adults are literate (in some countries the figure approaches 100 percent), three thousand years ago fewer than 1 percent would have been. Writing in those days was the tool of state and religious authorities—to record their laws, tax receipts, legal decisions, and historical chronicles—and to some extent of merchants and landowners. On the one hand, there was the great mass of illiterate common people who neither produced nor consumed anything written; on the other, the comparative few whose occupations required writing skills and who (as the surviving evidence shows) were responsible for the accumulation of huge archives of written documents dedicated to their own narrow interests.

Given the need for writing in ancient societies, by what physical means was it carried out? Writing can be done on any relatively smooth surface with any tool that can make a mark on that surface. Kings who wished to set before their people a code of laws could command that it be inscribed on a stone pillar or on the face of a cliff—but the people would have to come to the inscription; it could not be taken to them. Merchants with a tally to keep or a letter to write could do so on a piece of bark or broken pottery; but obviously what could be inscribed on such a medium was necessarily quite brief. For messages of any considerable length, there was not a wide choice of media in antiquity, and in most areas there was only one obvious material upon which to write. In lands lying along rivers, clay was readily available at little cost for processing; in lands in which the grazing of flocks provided meat and wool or hair, animal skins were available as a by-product; and in lands with marshy lowlands, the papyrus sedge was available for the taking. Clay, animal skin, and papyrus were the three chief materials employed for writing in the West and in the Near East until the introduction of rag paper from the Far East long after the biblical period.

WRITING ON CLAY

Clay seems to us today as the least likely of these writing materials. Our modern process of writing by hand, in which a pointed marker is moved across a relatively unyielding surface, does not work well on clay. Those who first began to use clay as a writing medium would quickly have discovered that it is easier to press or poke a writing implement into it than to scratch lines across its surface. Thus, the device that came to be used was a stylus with a tip that made a little wedge, an elongated triangular impression, in the clay. The writing system (called "cuneiform," meaning wedge-shaped) required that the thin tip of the wedge be pointing sometimes downward, sometimes directly or obliquely to the right, and that words be constructed from standardized combinations of these marks. The clay itself was prepared for writing by being shaped into tablets of a size that could be held in one hand while the other worked the stylus. The finished tablets could be dried in the sun, fixing the inscription, and thus be suitable for most kinds of communication and record keeping. If the tablet later got wet but was not otherwise mishandled, it could be dried out slowly and again be as good as new. Tablets baked in a furnace were impervious to any damage except being smashed. Hundreds of thousands of whole and partial tablets, baked and unbaked, bearing records related to every aspect of life, have been recovered from the ruins of ancient civilizations. Clay tablets cannot contain nearly so much information as books of the same size or weight; but such tablets are inherently a far more permanent medium of written communication than anything but stone surfaces.

WRITING ON PAPYRUS

Egypt had its river and could have developed the use of clay as its major writing medium. But close to its centers of government and religion, in the marshes of the Nile River delta, Egypt also had papyrus growing in great abundance; and from this it obtained a remarkable fibrous material for making not only food, medicine, rope, sandals, clothing, boats, and sails but an excellent equivalent of our modern paper. The triangular stalks of the plant, measuring anywhere from seven feet to twice that in length and one to three inches in thickness, were cut and carried to nearby factories while still fresh. There the stalks were reduced to lengths and the tough rind removed, revealing the fibrous pith, thin strips of which were torn or cut off along the length of the piece, every effort being made to get as wide a strip as possible. Sheets could then be produced by laying out a number of strips side by side, covering those with strips running in the other direction, and then hammering the two layers of pith into a single sheet. The resultant piece could be rubbed smooth, trimmed to size, and glued to other sheets to form one long roll, ready for a variety of uses, most notably writing. The writing was done with a reed brush, the end of which had been rendered fibrous, perhaps by chewing. The ink, if black, was made from carbon in the form of lampblack or soot, or later from oak galls and iron sulfate; red ink was made from iron oxide.

Papyrus was a remarkable writing material, in use for at least four thousand years. (Papyrus documents still exist that were written as early as the thirtieth century B.C.E. and as late as the eleventh century C.E.) When carefully made from the choicest part of the stalk (nearest the center), papyrus had the virtues of the best modern paper. Shipped from Egypt all over the ancient world, it was the writing medium employed by both Greece and Rome in the thousand years of the classical period. We know something of its manufacture and use from what writers of the time said about it, and even more from studying papyrus documents that have survived. But survivability was the weak point of papyrus. Unlike clay (and like our paper), it could not retain its integrity in the presence of moisture. The documents that have lasted through the centuries have done so largely in the dry sands of Egypt, where they were thrown out as trash or tossed away by conquerors interested in spoils of a more substantial kind.

WRITING ON ANIMAL SKIN

The third of the major materials upon which writing was done during the biblical period was animal skin. There were actually two categories, determined by the way the skin was prepared. One was leather, which was produced by first soaking the hide in a lime or salt solution to loosen the hair so that it could be removed and then tanning the skin with substances that would preserve it and give it the qualities appropriate for whatever use the leather was intended for. (Between these two essential steps, unhairing and tanning, the skin was sometimes split into several layers, making it thin enough to be especially suitable for document use.) The other category of animal skin used for writing was parchment, produced by soaking and unhairing the hide, as was done in the case of leather, but then drying the dehaired pelt under tension and (generally) not tanning it; the piece could then be scraped smooth and whitened or otherwise tinted. The best quality of parchment is called "vellum," a word originally applied to parchment made from the hide of calves.

Animal skin in the form of leather was probably used as a writing medium from the very outset. But it was not until about the third century B.C.E. that parchment was recognized as a particularly desirable form of writing material; and another five or six centuries went by before the use of parchment surpassed the use of papyrus. Parchment could take rougher handling than papyrus and could more satisfactorily be written upon on both sides. But it did not survive exposure to moisture much better than papyrus; although it would not disintegrate so quickly when damp, it would lose its shape and have to be restretched while being dried if it was to regain its original dimensions.

Once the individual skin was processed into leather or parchment suitable for writing, it—like the individual sheet of freshly made papyrus—would be trimmed into pieces of an appropriate size and sewn together into a roll (or scroll—the words are interchangeable). An author or scribe who intended to produce a book in roll form would simply have unrolled a short length of the material and begun to write. If the writing were of the kind that is read from left to right, as are Greek and Latin (and English), the writer would put the fresh roll to his right side, pull a portion toward him and write a column, then pull a further portion toward him and write a second column to the right of the first one, and so proceed to the end. If the writing were of the kind that is read from right to left, as are Hebrew and Aramaic, the process would be reversed and the resulting book would be a mirror image of one in Greek or Latin.

ANCIENT ISRAELITE
WRITING

All three of the writing materials we have been discussing would have been known to the ancient Israelites. They probably would not themselves have employed clay tablets to any great extent, but while under the rule of Assyrian and Babylonian overlords (from the mid-eighth to the mid-sixth centuries) they would have had more opportunity than they wished to receive missives on clay from those Mesopotamian powers. The Israelites themselves would have used papyrus or animal skin when writing anything more substantial than would fit on a piece of broken pottery (a "potsherd") or a wooden tablet.

The Bible contains a number of stories concerned with the writing process and the media of writing. Moses, of course, is said to have written on stone tablets that he then carried down the mountain (Exod. 34:28–29), just one of the remarkable deeds that are attributed to that great man. Joshua, a junior version of Moses in a number of respects, performs a similar feat when, as the Israelites stand before him, he engraves on stone "a copy of the Law of Moses" (Josh. 8:32). Isaiah is directed by Yahweh to take a "large tablet" (Isa. 8:1) and write a symbolic name upon it "in common writing" (or perhaps "with an ordinary stylus"—the meaning of the Hebrew is uncertain). Ezekiel is given a scroll to eat by Yahweh, which is perhaps to be understood as a small roll of papyrus (Ezek. 3:1–2). Jeremiah dictates an oracle concerning Judah's coming destruction to his secretary Baruch, who takes it down in ink on a scroll. Baruch reads the scroll first to the people gathered at the Temple and then to the chief men of Judah, who are so dismayed at what they hear that they report it to the king, Jehoiakim. The king orders that the scroll be brought and read to him; but so little impressed is he with what he hears that he takes a penknife and cuts pieces off the scroll even while the reading is in progress and drops them into the fire (Jer. 36). (Some commentators think that this scroll would have been leather, but others say that leather burning in an indoor fire would have smelled so bad that the scroll must have been papyrus.) At the end of 2 Timothy the author requests that "the scrolls" be brought to him, "especially the parchment ones" (2 Tim. 4:13), these probably being notebooks. And in the book of Revelation, John is required, Ezekiel-like, to eat a little scroll, which is sweet in his mouth but sour in his stomach (Rev. 10:9–10).

WRITING THE BIBLE   The individual books of the Jewish and Christian scriptures were initially written down during a span of a little more than a thousand years, beginning sometime before 900 B.C.E. (parts of the Pentateuch) and ending by about 150 C.E. (the letter called 2 Peter). On what materials would they have been written? We can suppose that the first versions of all the older books of the Jewish Bible—the Pentateuch, the history books, Job, Proverbs, and some of Psalms—would have been written on leather (locally available in Israel) or papyrus (from Egypt). As for the prophetic books, it was the nature of prophets to be speakers, not writers, and their oracles probably circulated orally (as Jesus' sayings did) among their disciples and then might have been written down in the form of notes on potsherds or "notebooks" of papyrus or wood. But as time passed and the traditions of the individual prophets became established, the books as we now know them—embodying both the oracles and stories about the prophets—would have taken shape and been written down in a more substantial form on leather or papyrus. During the centuries when the last books of the Jewish scriptures were being composed and the Pentateuch and then the Prophets were beginning to be thought of as canonical (i.e., not merely religious but sacred) and were copied and recopied, parchment came into general use and could have been chosen by a writer as an alternative to the other two materials. In the

Roman Empire, during the time those documents were being composed that would someday be gathered and canonized as the Christian New Testament, writing was done primarily on papyrus, with the use of parchment gradually increasing.

FROM ROLL TO CODEX

Until near the end of the first century C.E., books on papyrus and animal skin would have taken the form of a roll. But late in that century a new form of book was developed, the codex, which is the form all modern books take. In a codex the sheets of papyrus or skin material were not fastened end to end, as in a roll, but stacked one on top of the other. To make a book, the stack was folded in half (creating right- and left-hand leaves) and then sewn together at the fold. Because a thick stack is hard to handle, the practice developed of folding and sewing just a few sheets at a time—four, say—and then stacking the resulting units in proper order and binding them together into a whole.

People tend to be conservative in their tastes with respect to books. The new codex form—despite the fact that it was easier to handle and could hold more writing than a roll (because its leaves could be written upon on both sides)—was not at first widely adopted in the Roman world. But it apparently appealed to one group within the Roman Empire, the Christian Church. If we examine the still-surviving Christian books from the second through the fourth centuries, we find that the great majority of them are codexes (or codices, the classical plural form). Why the young Church should so readily have adopted the new book form when non-Christians were slow to do so is not certain. One reason sometimes given is that Christianity was a missionary religion and that Christians needed a handy book of texts—the Jewish scriptures, Paul's letters, the gospels—to use in the effort to convert others to their cause. According to this view, in the course of heated discussions with Jews or pagans, Christians would have been able to turn to pertinent passages far more quickly in a codex than in a roll. But this argument assumes that biblical texts of the early centuries of the Church were, like our own, divided into chapters and verses and neatly numbered for ease of reference. Those texts, in fact, were lacking most of the aids that we today feel are essential for readers, even spacing and punctuation; and the individual biblical books not only lacked chapter and verse numbering but were not even divided into chapters and verses. There was nothing called "John 3:16" at the time and no easy way to locate the passage that now bears that label. Our present system of chapter division was not devised until the early thirteenth century and our system of verse division until the mid-sixteenth century.*

FROM PAPYRUS TO
PARCHMENT

Another significant change took place in the making of books during the period from the second to the fourth centuries, namely, the substitution of parchment for papyrus as the standard material of which books were made. This must not be tied too closely to the shift from roll to codex, for throughout the period there were codices and rolls made from both papyrus and parchment. But by the end of the fourth century the standard book form in the West had become the parchment codex—and so it remained for the next eleven hundred years. The fifty Greek Bibles that Emperor Constantine ordered to be made and placed in churches in the mid-fourth century were parchment codices. The copies of Jerome's Latin translation that spread across Europe in the Middle Ages were parchment codices. And even after the invention of printing, when rag paper was plentiful for use in the new bookmaking process, the printing of particularly fine copies of books was sometimes done on parchment.

What we hope has impressed the reader from the foregoing survey is the vast gulf that separates our world, with its inexpensive printed materials and widespread literacy, from the world within which the Bible developed—and within which it was disseminated for so many hundreds of years. We get our knowledge of the Bible from reading it. Before the spread of printed books (that is, before the sixteenth century), what the vast majority of people knew about the Bible was obtained at second hand, from hearing portions of it read in synagogue or church, or from looking at artistic representations of biblical stories in church windows or carvings. A handwritten and a printed book are read in exactly the same way—they are both real books—

*The system of chapter division now in use is the work of Stephen Langton, a thirteenth-century scholar who rose to the position of archbishop of Canterbury. Working in Paris earlier in his career, Langton oversaw the production of a Latin Bible embodying his chapter-division system. This "Paris" Bible, prepared by professional copyists and issued in 1231, became the model for many others, and Langton's system passed into the copies of the Bible that came off the printing presses some two centuries later.

The present system of verse division was developed by a sixteenth-century French scholar and printer, Robert Estienne (or Stephanus). In 1551, while residing in Geneva as a Protestant exile, he published a French-language New Testament with numbered verses; in 1553 he published a complete French Bible and in 1555 a Latin Vulgate, both employing his system.

but read by whom, at what expense, and in what numbers? Printing radically altered the status of all literature, but most of all that of the Bible. There are now more copies of the Bible in existence than any other book ever written.

SUGGESTED FURTHER READING

Edward Chiera, *They Wrote on Clay: The Babylonian Tablets Speak Today*, ed. George G. Cameron (Chicago: University of Chicago Press, 1966).
Naphtali Lewis, *Papyrus in Classical Antiquity* (Oxford: Clarendon Press, 1974). Supplement, 1989.
R. Reed, *Ancient Skins, Parchments, and Leathers* (London: Seminar Press, 1972).
C. H. Roberts and T. C. Skeat, *The Birth of the Codex* (London: Oxford University Press for the British Academy, 1983).
T. C. Skeat, "Early Christian Book-Production: Papyri and Manuscripts," in *The Cambridge History of the Bible*, vol. 2 (Cambridge: Cambridge University Press, 1969), pp. 54–79.

The Interpreter's Dictionary of the Bible, ed. George Buttrick et al. (Nashville: Abingdon Press, 1962). See article on Writing and Writing Materials.
The Anchor Bible Dictionary, ed. David Noel Freedman et al. (New York: Doubleday, 1992). See articles on Codex, Literacy (Israel), Writing and Writing Materials.

Index

N.B.: Biblical passages cited in the text, items merely mentioned in passing, characters in biblical narratives, and terms like "Judaism" and "Christianity" that occur pervasively have not been indexed.

Acrostic poetry, 282–83
Acts, 205, 221, 230–37
Acts of Pilate, 205
Additions to the Book of Esther, 197
Admonitions of Ipu-Wer, 59
Adonai (Hebrew term), 313, 314
Akiva ben Joseph, 182
Aland, Kurt and Barbara, 255, 256
Alexander the Great, 157, 161, 174–75, 190, 194, 267–68
Alexandria, 175, 176, 194
Allegorical interpretation, 29–32, 182–83, 300–301, 301–2, 305
Allegory, 28–32, 160–61, 300–301
American Standard Version, 274–75, 277
Amos, 126, 131, 157
Anchor Bible, 277
Animism, 180
Annals, 55–56
Antinomianism, 249
Antioch, 176, 213, 223, 231
Antiochus III, 183
Antiochus IV Epiphanes, 71, 132, 157–67 passim, 183–84, 203, 204
Apocalypse, 155–56, 205
Apocalypse of Moses, 202
Apocalyptic
  characteristics of, 159–61
  Christian use of, 168–69

continuing appeal of, 169–70
distinguished from prophecy, 155, 157–58, 169–70
Jewish rejection of, 167–68
Apocrypha, 71, 192–200 passim, 201, 206, 264–65, 271
Apocrypha (Greek term), 198, 201
Aramaic language, 101, 173–77 passim, 234, 240, 250
Aristobulus of Paneas, 182
Ark of the Covenant, 173
Assumption of Moses, 203
Assyria, 97, 127, 172, 173
Assyriology, 44–45
Astruc, Jean, 111–12
Augustus, 190
Authorized Version, 273. *See also* King James Version
Authorship, biblical, 9–10, 138, 139, 142, 211, 298–99
Autographa, biblical, 253, 259
Azariah, 197

Babylonia, 172, 173
Babylonian Chronicle Series, 55–56
Babylonian Exile, 98–99, 134, 168, 172–74, 312
  return from, 99, 128–29, 172, 174, 178n, 305

Babylonian Theodicy, 59
Baptism, 286, 294–95, 308
Bar Kochba, revolt of, 306
Baruch, book of, 197
2 Baruch, 203
Behistun, 45
Bel and the Snake, 197
Ben Sira, Jesus, 179
Ben Zakkai, Johanan, 100
Benjamin, Don, 48
Beyerlin, Walter, 48
Beza, Theodore, 258
Bible
    as anthology, 10–11, 15, 17, 43, 93, 290
    derivation of term, 10n, 96
    historical value of, 72, 73
    as icon and talisman, 293
    as source of ecclesiastical form and
        practice, 292
    as source of eschatological concepts, 292,
        296–97
    as source of moral precepts, 291–92, 295–
        96
    as source of personal guidance, 292–93
    as source of sacred history, 291
    as source of theological doctrines, 291,
        294–95
    unity of, 10–11
Biblia Hebraica Stuttgartensia, 262
Biblical history, external witnesses to, 72
Bishops' Bible, 272
Black Obelisk of Shalmaneser III, 46
Book of the Covenant, 52, 118–19, 120
Book of the Law, 60, 97–98
Book of the Twelve, 95, 96n, 129
Botta, Paul Emile, 46
Bunyan, John, 104

Canaan, 77
Canon
    Christian, 101–4
    derivation of term, 94
    Eastern Orthodox, 192, 199–200, 220, 268
    Jewish, 95–101, 193, 201–2
    of letters, 237–38
    New Testament, 232, 237–39
    Pauline, 102
    permanence of, 105–6
    Protestant, 96, 192
    Roman Catholic, 192, 199–200, 220
Canonization, 99–104, 192–93, 204–5, 207
    definition of, 104, 192–93
Challoner, Richard, 273
Champollion, Jean-François, 44
Charles I, 257
Christ, derivation of term, 188
Christology, 245–46
Chronicler, 69–70
Church, early, 22, 205, 211–12, 232–36,
    240–41, 303–4
    social composition of, 241, 241n

Clement of Alexandria, 260–61
Code of Hammurabi, 52, 53
Codex Alexandrinus, 257–58
Codex Cantabrigiensis, 258, 261
Codex Sinaiticus, 257
Codex Vaticanus, 256–57
1 Corinthians, 248
Council of Trent, 200
Covenant, 53, 209, 210n
Coverdale, Miles, 271, 272
Creation stories, 6, 50–51, 113–14
Criticism (term), 110n
Cross, Frank, 263
Cult in ancient Israel, 140–41
Cuneiform, 45, 317
Cyrus, 128–29, 172, 174, 188–89

Daniel, 132, 156, 161, 167
Daniel, book of, 70, 101, 132, 155–61, 164–
    69 passim, 303
Darius the Great, 45
Dead Sea Scrolls, 263–65, 189, 201
Decalogue, 19, 295
Decapolis, 176
Decree of Canopus, 44
Demons, 180
De Rossi, J. B., 262
Deutero-Isaiah, 128–29, 172, 174, 188
Deuterocanical books, 200. See also
    Apocrypha
Deuteronomic historians, 67–68, 73, 141
Deuteronomic History, 21, 55–57, 67–68,
    120
Deuteronomic source, 115, 118–19
Deuteronomic writers, 99, 312
Deuteronomist, 118, 141
Deuteronomy, 98, 99, 109, 118
Diaspora, 176–78, 194, 250
Documentary hypothesis, 109
Documentary sources of the Pentateuch
    J, E, D, P, 9, 115–19
    others, 120–21
Douay-Rheims Bible, 272–73
Dualism, 159, 179–80, 218

Ecclesiastes, 138–53 passim
Ecclesiasticus, 147–48, 153, 179, 196, 197,
    264–65
Education in ancient Israel, 139
Edward VI, 271
Egyptology, 44
Eissfeldt, Otto, 119
Ekklēsia (Greek term), 37n, 225
Elizabeth I, 272, 273
Elohim, term as evidence of authorship, 112
    as title, 311
Elohist source, 115, 117–18, 119
1 Enoch, 189, 202–3, 239
2 Enoch, 203
Enuma Elish, 50–51
Epistle of Barnabas, 204, 257

Epistles of Clement, 258
Erasmus, Desiderius, 255, 257
Eschatology, 160, 228, 244–45, 292, 296–97
1 Esdras, 196, 199, 200
2 Esdras, 169, 195–200 passim
Essenes, 185, 187–88
Esther, book of, 55, 70
Estienne, Robert, 322n
Euangelion (Greek term), 210, 227
Eucharistic meal, 213, 240
Evangelists (term), 210
Ezekiel, 128, 134, 297
Ezekiel, book of, 297
Ezra, 110, 196, 206

Fertile Crescent, 86
Five Scrolls, 96n
Flood, stories of, 47–48, 49–50
Forgotten Books of Eden, 207
Former Prophets, 67–68, 95, 100
Fulfillment formula, 224

Gehenna, 179
Gematria, 301
General (Catholic) letters, 238–39
Geneva Bible, 272, 273
Genizeh, 265
Gibran, Kahlil, 105
Gilgamesh, epic of, 49–50
Goodspeed, Edgar, 207
Gospel
    as genre, 212, 218, 230, 244
    as term, 210
Gospel harmonies, 210
Gospels, 245
    authorship of, 211
    canonization of, 103–4
    context of composition of, 213
    dating and sources of, 213–15
    interpretation of Jewish scriptures in, 303–5
    literary genres in, 215–17
    purpose of, 211–13
Graf-Wellhausen hypothesis, 109
Grayson, A. K., 56
Great Bible, 271
Greek biblical manuscripts, families of, 255
Greek influence in ancient Near East, 175–77, 194. *See also* Hellenism

Haggadah, 12n
Hagiographa. *See* Writings (biblical division)
Hasmoneans, 184–85
Hebrew Bible distinguished from Jewish Bible, xi–xii
Hebrew vowels, 262, 314
Hebrews, book of, 305–6
Hell, 179
Hellenism, 174–78, 183, 194, 250–51, 303
Henry VIII, 271
Herod the Great, 184–86

Herodotus, 77
Hieroglyphics, 44
Higher Criticism, 14, 112–14, 307
Historical recital, 8, 19
History
    in Apocrypha, 71
    Bible as source of, 72–73, 291
    extrabiblical witnesses to, 72
    in Former Prophets, 67–68
    in Latter Prophets, 69
    in New Testament, 71–72
    in Writings, 69–70
Hobbes, Thomas, 110
Holiness Code, 52, 120
Hort, F. J. A., 255
Hosea, 126, 131
Hyperbole, 23–25

Immortality, 179
Infancy Gospel of Thomas, 205, 207
Inquisition, Catholic, 271
Irenaeus, 103
Irony, dramatic, 33–35
Irony of language, 35–36
Isaiah of Jerusalem, 127, 131

James, book of, 238
James I, 257, 273
Jamnia (Jabneh), 100–101, 168, 178, 187, 195
Jason (author of 2 Maccabees), 197
Jehovah (term), 314
Jeremiah, 127–28, 131, 132
Jeremiah, book of, 303
Jeroboam II, 68, 157
Jerome, 198–99, 268, 307
Jerusalem, fall of, 97, 100, 185, 186, 306
Jesus of Nazareth, 102, 103, 210–28 passim, 294–96 passim, 305, 306
    as divine son, 189, 218, 219, 222, 223, 227
    as Messiah, 188, 195, 219–24 passim
    and Paul, 211, 245
    and Sadducees, 187
    and prophetic interpretation, 216, 222, 223, 298, 303
    Stephen compared to, 7–8
Jewish Bible
    distinguished from Hebrew Bible, xi–xii
    number of books in, 96n
Jewish Christianity, 234, 239–40, 250
Jewish interpreters, medieval, 307
Jewish rebellion
    of 66–70 C.E., 100, 167–68, 186
    of Bar Kochba, 306
Job, book of, 58–59, 138–53 passim
    pious editing of, 149–50, 287–88
John, 211, 214, 225–28
John, gospel of, 71, 211, 214, 225–28, 260, 261
John the Baptist, 305
Josephus, 195
Jubilees, 202

Judaizers, 239, 248, 249
Jude, book of, 238–39
Judith, book of, 197
Julius Caesar, 190
Justification by faith, 247–48

Kennicott, Benjamin, 262
King James Version, 129, 257, 269–76
    passim, 281, 286
  limitations of, xi, 273–74
  long preeminence of, 11, 275
  translation of, 273
King, Martin Luther, Jr., 105
Koine Greek, 175, 243

L source of gospel of Luke, 215, 221
Lament as literary form, 18, 20
Langton, Stephen, 322n
Last Supper narrative, 22
Latter Prophets, 69, 95, 100
Layard, Henry, 46
Lectionaries, 254
Leningrad Codex, 262
Lepsius, Richard, 45
Letter of Aristeas, 202
Letter of Jeremiah, 197
Letter writing in ancient world, 242–43,
    242n
Letters of Paul, 102, 103, 169, 232, 239,
    242–44
  allegorical interpretation in, 303–4
  canonical order of, 244
  chronology of, 244
  disputed letters of, 237–39, 244
Lex talionis, 52–53
Life of Adam and Eve, 202
Literal vs. more-than-literal interpretation,
    298–308
Literature, definition of, 4
Literature-of-the-Bible approach, 14–16
Liturgical forms, 19
London Polyglot, 258
Lost Books of the Bible, 207
Lower Criticism, 14
Lowth, Robert, 38
Lucar, Cyril, 257
Luke, 71, 211, 213, 221–23, 224
  as author of Acts, 7–9, 230–32, 235–37
Luke, gospel of, 71, 133, 211, 213, 221–23,
    224
Luther, Martin, 199, 238, 249, 270, 307
Lyell, Charles, 46

M source of gospel of Matthew, 215, 221
Maccabees, 158, 184
1 Maccabees, 71, 196, 197
2 Maccabees, 179, 195, 196, 197
3 Maccabees, 200
4 Maccabees, 203–4
Marcion, 103, 210

Mark, 211, 214, 218–21, 221–27 passim,
    244, 305
Mark, gospel of, 71, 211, 213, 218–21, 223,
    260–61
  endings of, 220–21, 258
Martyrdom of Isaiah, 202
Mary (English queen), 271–72
Masoretic text, 262–64
Matthew, 133, 211, 214, 223–25, 299
Matthew, gospel of, 71, 133, 211, 213, 214,
    223–25
Matthew Bible, 271
Matthews, Victor, 48
Melchizedek, 305
Messiah, 102, 188–89, 190, 195. See also Jesus
    of Nazareth: as Messiah
  derivation of term, 188
Messianic secret, 220
Metaphor, 25–27
Micah, 126, 131, 298
Midrash, 228
Minuscules, 254
Miracle stories, 216
Moabite Stone, 57
Modern Apocrypha, 207
Mosaic Law, 98, 109, 181–84 passim, 187,
    206, 234. See also Pentateuch; Torah
  Paul and, 236, 239, 247–50

Name of Israel's God, 311–14
Narrative forms, 20–22
Nebuchadnezzar, 156
New American Bible, 277, 285
New American Standard Bible, 277, 285
New English Bible, 276, 282
New International Version, 277, 285, 313
New Jerusalem Bible, xi, 129, 277, 282, 285,
    313
New Revised Standard Version, 275–76, 313
New Testament, history in, 71–72
New Testament Apocrypha, 192, 204–5,
    206, 207
Nicodemus, gospel of, 205
Nineveh, 46, 47
Numbers, book of, 300

Oracle, prophetic, 19–20, 130–31
Oral tradition
  in New Testament, 22
  in Pentateuch, 116

Palestine
  agriculture in, 89
  daily life in, 88–92
  location of, 76
  as name, 77
  natural regions of, 78–84
  natural resources of, 85–87
  rainfall in, 78–81, 84–85
  religious practice in, 89–90

roads in, 83–84
seasons in, 84
size of, 77
soil and climate of, 84–85
urban life in, 87–88
Papyri, 259–60
Parable, 22, 216–17
Parables, allegorized, 29–31, 305, 308
Parallelism, 38–42, 54–55, 151–53
Parousia (Greek term), 218, 220, 245
Parvis, M. M., 260
Passion narrative, 216, 225, 227
Pastoral letters, 238
Patriarchs, date of, 64n
Patriotic poetry, 20
Paul, 10, 205, 210n, 238–44 passim, 306
    as allegorical interpreter, 303–4, 306
    key features of thinking of, 244–51
    letters of. *See* Letters of Paul
    Luke's conception of, 235–37
    as missionary, 232, 239
    as servant of Church, 235–36
    travels with Luke, 221, 232
Pentateuch, 95–96, 193. *See also* Mosaic
        Law; Torah
    Ezra as author of, 110–11
    history in, 66–67
    integrity of, 121–22
    Moses as author of, 46, 108–12
    Samaritan, 263
    sources of, 115–21
Pericope, 217–18, 222
Persia, 128
Personification, 32–33
1 Peter, 238, 239
2 Peter, 204, 239
Pharisees, 101, 185, 186–87, 194, 202
Philemon, book of, 244
Philistines, 77, 79
Philo of Alexandria, 182–83, 303, 306
Pilate, Pontius, 186
Plagiarism, 48, 61, 206–7
Poetry, Hebrew, 37–42, 129–30, 151–53
    translation of, 282–83
Polis (Greek term), 175–76, 183
Prayer of Manasseh, 197, 200
Priestly source, 115, 120
Priestly writers, 115–16, 120
Printing, invention of, 269–70, 315
Pritchard, James, 48
Pronouncement stories, 215–16
Prophecy
    distinguished from apocalyptic, 155, 157–
        58, 169–70
    poetic form of, 129–30
Prophetic interpretation, 300
    in early Christianity, 133–34, 216, 298,
        304–5
    in early Judaism, 131–33, 302–3
    in modern times, 134–36, 307, 308

Prophetic oracles, 19–20, 130–31
Prophetic paradigm, 130
Prophets (biblical division), 67–69, 95–96,
        192, 193, 302–3
    canonization of, 99–100
Prophets, divine communication to, 125n
Protestant Reformation, 199, 270–72, 307
Protests of the Eloquent Peasant, 59–60
Protoevangelium of James, 205, 207
Proverb as literary form, 151
Proverbs, book of, 138–53 passim, 295–96
Psalm as literary form, 18–19
Psalm 151, 200
Psalms of Solomon, 189, 202
Pseudepigrapha, 192, 200–204, 206
Pseudepigrapha (Greek term), 201
Pseudonymity, 161, 201, 204, 206
Ptolemies, 175, 183, 190
Ptolemy II Philadelphus, 194
Ptolemy V Epiphanes, 44
Puns, 36–37, 282

Q source of gospels, 214–15, 221–24 passim

Rawlinson, Henry, 45, 46
Redaction, 12–14
Redactors, 11–14, 119–20, 120n
Resurrection, Jewish belief in, 157, 178–79,
        187
Revelation, book of, 155, 161–64, 297
Revised English Bible, 129, 276–77, 285, 313
Revised Standard Version, 275, 276, 285
Revised Version, 274–75
Romans, book of, 244, 248
Rome, 167–68, 185–87 passim, 223, 231
Rosetta Stone, 44
Ruth, book of, 70

Sabbath, relation of, to Sunday, 240n
Sadducees, 186–87
Sanhedrin, 100, 187
Satan, the, 137, 166, 180
Sayings in gospels, 215
Scribes, 139, 181
Second Isaiah. *See* Deutero-Isaiah
Seleucids, 175, 183, 190
Seneca, 205
Senses of scripture, 298–308
Septuagint, 194n, 200–201, 263, 284, 313
    Apocryphal works in, 195, 196–98, 200–
        201
    Christian use of, 102, 181, 194–95, 196n,
        223n, 250, 268
    rejected by first-century Jews, 178, 196
    translated for Greek-speaking Jews, 176,
        194, 267–68
Sermon on the Mount, 214, 218, 219, 225
Sheldon, Charles M., 105
Sheol, 179
Shepherd of Hermas, 257

Sibylline Oracles, 189, 204
Simon, Richard, 111
Smith, George, 47, 47n, 49
Smith, Morton, 260
Solomon, 139, 142, 148
Son of God (term), 189
Son of Man (term), 8, 189, 222
Song of Songs, 300
Song of the Three, 197
Source criticism, 110
Spinoza, Benedict, 110–11
Story of Two Brothers, 52
Sumerian Job, 58
Susanna and the Elders, 197
Suzerainty treaty, 19, 53
Symbolism, 28
Synagogue, 177, 187, 240
Synoptic (term), 211
Synoptic gospels, 13, 211, 305
    John contrasted with, 225–28
    sources, 214–15

Tanakh, 96, 277
Targum, 173
Tatian, 210
Teaching of Amen-em-Opet, 57–58
Temple, 86, 312
    destruction of, 172, 173, 187, 195, 306
    discovery of scroll in, 97–98, 99
    as focus of postexilic religion, 174, 177, 180
    Herod's, 186
    Psalms as hymnbook of, 19
Testament (term), 209
Testaments of the Twelve Patriarchs, 203
Tetragrammaton, 312
Textual variants in Bible, 278–79
Textus Receptus, 256, 258
Theophilus, 71, 221
1 Thessalonians, 244
Thomas, Coptic (Gnostic) gospel of, 105, 205n. See also Infancy Gospel of Thomas
Tischendorf, Constantine, 257
Tobit, book of, 195, 197
Today's English Version, 277, 285
Torah, 95–98, 100, 184, 302. See also Mosaic Law; Pentateuch
    canonization of, 99
    in intertestamental period, 181
    legend of translation into Greek, 194, 202
Torah (Hebrew term), 95–96, 97, 181
Translation
    dynamic equivalence in, 284–86

formal correspondence in, 284–86
of idiomatic language, 281–82
of poetry, 282–83
problem of, 277–83
Translations, choosing among, 283–88
Tyndale, William, 271, 272, 275
Typological interpretation, 224, 300, 301, 302, 304n, 305–8 passim

Ugaritic tablets, 45
Uncials, 254
Uruk Prophecy, 60
Ussher, James, 47

Valley of Hinnom, 179
Verse and chapter division, 321, 322n
Virgin Birth, 133, 284
Visions in apocalyptic, 160–61
Von Soden, Hermann, 255
Vulgate, 103, 255, 268–69, 273, 313

Westcott, B. F., 255
"Wife-sister" tales, 21
Wisdom of Solomon, 147–48, 151, 153, 195–97 passim
Wise men, 138–43
Witter, H. B., 112
Wordplay, 36–37, 282
Writing, 315–16
    ancient Israelite, 319–20
    on animal skin, 318–19
    of biblical documents, 320–21
    on clay, 45, 316–17
    in codex form, 321–22
    on papyrus, 317–18
    on parchment, 322
    in roll form, 319
Writings (biblical division), 95–96
    authorship of books in, 9
    canonization of, 101, 193, 196
    history in, 69–70
Wycliffe, John, 269

Yahweh
    name as evidence of authorship, 112
    as personal name, xi, 312, 313
    translation of name, xi, 312
Yahwist source, 115, 116–17, 119

Zealots, 100, 187
Zerubbabel, 189
Zoroastrianism, 180